P 348

THE

WATCH & CLOCK MAKERS' HANDBOOK,

DICTIONARY AND GUIDE

By F. J. BRITTEN

WITH 400 ILLUSTRATIONS

———

ELEVENTH   EDITION

NEW   IMPRESSION

———

ANTIQUE COLLECTORS' CLUB

11th edition first published 1915
This edition published 1976 by Baron Publishing Ltd.
Reprinted 1988, 1993, 1995, 2001 by the Antique Collectors' Club Ltd

ISBN 1 85149 192 9

British Library Cataloguing-in-Publication data
A catalogue record for this book is available from the
British Library

For a free catalogue and details of other books on horology
please contact

ANTIQUE COLLECTORS' CLUB
5 Church Street, Woodbridge, Suffolk, IP12 1DS, UK
Tel: 01394 385501. Fax: 01394 384434
Email: sales@antique-acc.com
or
Market Street Industrial Park, Wappingers' Falls, NY 12590,
USA
Tel: 845 297 0003 Fax: 845 297 0068
Email: info@antiquecc.com

Printed in England by the Antique Collectors' Club Ltd.,
Woodbridge, Suffolk, IP12 1DS

PREFACE

In compliance with suggestions from readers of previous issues, various additions have now been made. In particular, the description of the Westminster great clock is given in greater detail. Probably the death of Lord Grimthorpe awakened renewed interest in his work, for I received several letters asking for more minute particulars.

Alterations have been made in the German equivalents on the recommendation of Mr. Heinrich Otto, who has, in addition, entirely revised the German List in the Appendix. I have also to thank Mr. Otto for corrections relating to other matters in the book. The French equivalents remain as originally furnished by Mr. Mairet.

On many occasions I have been asked to give readers some instructions respecting mechanical drawing, but really there is so little to tell. The art of watch or clock making cannot be learnt from books alone, there must be much practice if one desires to excel, and so it is with mechanical drawing. Nothing will so much assist the learner to understand the functions of a machine and of its different parts as drawing to scale, and if he has what I will venture to call the instinct of the mechanician, proficiency will come with practice.

The young draughtsman should be provided with a drawing board and tee square of half-imperial size, half-a-dozen drawing pins, drawing paper, an H. and an H.B. pencil, a 45° and a 60° set square, a good set of compasses jointed in both legs, ink and pencil bows for small circles, a ruling pen, a four-inch protractor of thin horn, a box-wood scale divided on both sides and both edges, a piece of india-rubber, two or three pallets, and either a large stick of Chinese ink or a bottle of fluid Indian ink, but not a horn centre.

For an intricate drawing the paper is best damped and fixed to the board by gluing a narrow margin all round the edge, but it is generally sufficiently secured by a pin at each corner. Care should be taken to press the pins down so that the heads clasp the paper, otherwise the least disturbance from cleaning with rubber will cause the paper to shift. The H pencil should be sharpened to a chisel-shaped edge, so as to get the full width of the lead in drawing. Keep the stock of the square close to the edge of the board with the left hand, and hold the pencil quite upright and low enough to allow the little finger of the right hand to slide along the blade of the square as you draw the line, so that if the blade is a little warped you ensure its contact with the paper. See that the central point of the compasses is sharp, and, in using the compasses, bear very lightly, turning always the same way. Keep the central leg upright, bending the joint for every different size of circle, for if the point describes a cone it will be impossible to keep the prick point in the paper small. Ink in curves and circles before straight lines, beginning with those of least radius.

Several correspondents have suggested that fuller particulars of eminent members of the craft would be of interest. But I found it would be inconvenient to include such biographical sketches, together with descriptions of early timekeepers, in a book intended primarily for workshop use, and so I was led to issue a separate volume ; and I refer those who take special interest in the history of horological instruments to " OLD CLOCKS AND WATCHES AND THEIR MAKERS," with which is incorporated a List of Former Craftsmen.

In this edition of the Handbook the remarks on Watch Springing have been extended to suit especially the needs of those who do such work occasionally. For a more comprehensive study of the subject is recommended a book entitled, " WATCH SPRINGING AND ADJUSTING."

I have received so many appreciative letters, that I venture to ask those who think well of my books to favour me by recommending them as opportunity occurs.

F. J. B.

1, Silverdale Avenue,
    Westcliff-on-Sea, Essex.
        September, 1907.

Mr. F. J. Britten died on 11th April, 1913.
The 11th edition was reprinted January, 1915, with one or two slight corrections and additions.

At my request Mr. Otto has kindly revised the whole of the German equivalents, for which I tender him my thanks.

My thanks are also due to Mr. James Savidge, Secretary of the British Horological Institute. who has given me great assistance in putting the book to press.

A. M. BRITTEN.

NOTE—French and German equivalents, printed in *italics*, refer to the first definition in cases where the title has more than one meaning.

Acceleration.—[*Accélération.*—*Die natürliche Gangbeschleunigung.*]—Gaining on its rate. It is noticed that new chronometers and watches, instead of steadily gaining or losing a certain number of seconds each day, go faster day after day. There is no certainty as to the amount or ratio of this acceleration, nor as to the period which must elapse before the rate becomes steady, but an increase of a second a month for a twelvemonth may be taken as the average extent in marine chronometers.

It is pretty generally agreed among chronometer-makers that the cause of acceleration is seated in the balance-spring, though some assert that centrifugal action slightly enlarges the balance if the arc of vibration is large, as it would be when the oil is fresh, and that as the vibration falls off, centrifugal action is lessened and acceleration ensues from the smaller diameter of the balance. Though the balances do undoubtedly increase slightly in size in the long vibrations from centrifugal action, this theory is disposed of by the fact that old chronometers do not accelerate after re-oiling. Others aver that the unnatural connection of the metals composing the compensation balance is responsible for the mischief, and that after being subjected to heat the balance hardly returns to its original position again. If true, this may be a reason for exposing new chronometers before they are rated to a somewhat higher temperature than they are likely to meet with in use, as is the practice of some makers ; but then chronometers accelerate on their rates when they are kept in a constant temperature, and also if a new spring is put to an old balance, or even if a plain uncut balance is used.

It is noticed that when the overcoil of a balance spring has been much bent or " manipulated " in timing, the acceleration is almost sure to be excessive. This is just what might be expected, for a spring unduly bent so as to be weakened but not absolutely crippled, recovers in time some of its lost elasticity. But however carefully a spring is bent the acceleration is not entirely got rid of, even though the spring is heated to redness and again hardened after its form is complete. There is little doubt that the tendency of springs is to increase slightly in strength for some time after they are subjected to continuous action, just as bells are found to alter a little in tone after use. As a proof that the acceleration is due to the bending of the overcoil, Mr. Hammersley asserts that, if the spring of an old chronometer is distorted and then restored to its original form, the chronometer will accelerate as though it were new. Helical springs of small diameter have been advocated by some as a means of lessening acceleration, on the ground that the curves are less liable to distortion in action than when the springs are larger. Springs elongate in hardening, and it has been suggested that they afterwards gradually shorten to their original length and so cause acceleration, but there does not seem to be much warrant for this assumption. Unhardened springs do not accelerate, but then they rapidly lose their strength, and are therefore not used. Flat springs do not accelerate so much as springs with overcoils. Palladium springs accelerate very much less than hardened steel springs.

Addendum.—[*Arrondi.*—*Die Wälzungshöhe.*]—The amount added to the tooth of a wheel or pinion beyond the pitch circle. (See Wheels and Pinions.)

Adjusting for Temperature.—[*Réglage aux températures.*—*Die Kompensationsberichtigung.*]—The operation of arranging the weights or screws of a compensation balance so that its vibrations will be performed as nearly as possible in the same time throughout a certain range or temperature. (See Compensation Balance.)

Adjusting Rod.—[*Levier à fusée.*—*Die Abgleichstange für die Schnecke.*]—A lever having at one extremity a hollow square with set screw for attaching it to a watch or clock fusee or barrel, and furnished with sliding weights for balancing the pull of the mainspring. Mr. Charles Frodsham balanced the weight of the rod itself by means of a small ball, and marked a scale on the stalk, so that the pull of different springs could be registered in dwts. and ozs. For the appended sketch of his adjusting rod I am indebted to Mr. H. M. Frodsham.

In cutting a fusee, the upper or small end should be left rather too large than too small. The pull of the lower and upper turns may then be equalised by setting up the mainspring, which strengthens the lower turns. But if with a fusee the lower turns are already too strong compared with the upper, it is indicative of a mainspring too weak or a fusee too small at the top, and the only choice of remedies is to change the spring or recut the fusee, unless the spring is hooked on to a pin in the barrel, or has some other kind of yielding attachment, in which case the upper turns

may be slightly strengthened by substituting a hook in the spring to form a more rigid attachment. (See Mainspring.) A more elastic mainspring is a good and easy remedy in most cases, especially where the old spring is too small in diameter when out of the barrel and opened to its greatest extent. If the spring is already sufficiently strong, and no adjustment can be obtained, a stronger one should hardly be resorted to. Fusees, in many instances, do not procure a perfect adjustment of the spring. A clinging of the coils often causes an apparent fault in the adjustment by the fusee. With the weight at the proper position on the rod, the power may be equal upon winding up the first, second, third, and fourth turns, but the spring may not let down with equal power. The fourth turn unwinding may show less power than when wound, the turns increasing in power as the spring is let down. Springs that are of too small diameter when out of the barrel, are incapable of being adjusted. For a fusee watch the mainspring should be about four times the diameter of the barrel when out of it.

Alarum.—[*Réveil.*—*Der Wecker.*]—Mechanism attached to a clock by which at any desired time a hammer strikes rapidly on a bell for several seconds. Generally, a separate weight or spring actuates an escape wheel, to the pallet staff of which a hammer is fixed to act on the inside edge of a hemispherical bell. The alarum is usually let off by a wire attached to the hour wheel lifting a detent that stops the escape wheel. (See also " Striking Work.") In the " Speedwell " alarum the same mainspring serves for the going train and the alarum as well. The great wheels of the going and alarum trains are mounted on the winding arbor, one on each side of the mainspring. The wheels are loose, and driven by ratchets and clicks to allow of the arbor being reversed to wind the spring, the outer end of which is attached to the great wheel of the going train. When the alarum is let off, the force of the spring will cause the winding arbor to turn and carry with it the great wheel of the alarum train. But the problem is to lock the spring again when the alarum has run its course. On the winding arbor is a pinion with coarse teeth taking into a segment having a strong butt which forms the alarum stop. In winding, the first turn of the arbor winds the alarum stop, if it is down, and a missing tooth in the segment allows the winding of the going train to proceed in the usual way. This will be made clear by an inspection of Figs. 1 and 2, where the pinion and segment are shown as they

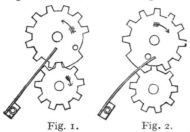

appear before and after winding. There is a light spring which presses against a pin in the segment when it is wound, thus keeping the butt away from the pinion, and ensuring the engagement of the pinion and segment teeth when the alarum is let off. The arrows in Fig. 1 indicate the course of

Fig. 1.   Fig. 2.

the wheels in winding, and the arrow in Fig. 2 the direction in which the segment will move when the alarum is in action.

All or Nothing Piece.—[*Tout-ou-rien.*—*Die Alles oder Nichts Sicherung.*]—The part of a repeating watch that keeps the quarter rack off the snail until the slide in the band of the case is pushed round. While the quarter rack is locked, the lifting piece of the hour hammer is kept free of the twelve-toothed rack, so that the hours cannot be struck till the quarter rack has fallen. The watch must therefore strike *all* or *nothing*. Invented by Julien Le Roy. Many Swiss watchmakers call this piece, as it is arranged in modern repeaters, the " ressort de crochetment," or hooking spring.

Anchor Escapement.—[*Echappement à ancre.*—*Die Anker-hemmung mit Rückfall.*]—The Recoil Escapement used in most house clocks.

(2) There is also a variety of the Lever Escapement with a very wide impulse pin, which is sometimes called an Anchor Escapement.

The Recoil Escapement, Fig. 1 (invented by Dr. Hooke about 1675), is the one most generally applied to the ordinary run of dials and house clocks. When well made it gives very fair results, but the pallets are often very improperly formed, although none of the escapements is easier to set out correctly.

There are still people who believe the Recoil to be a better escapement than the Dead Beat—mainly because the former requires a greater variation of the driving power to affect the extent of the vibration of the pendulum than the latter does. But the matter is beyond argument ; the Recoil can be cheaply made, and is a useful escapement, but is unquestionably inferior to the Dead Beat for regulators and other fine clocks with seconds pendulums.

There is no rest or locking for the pallets, but directly the pendulum in its vibration allows a tooth, after giving impulse, to escape from the impulse face of one pallet, the course of the wheel is checked by the impulse face of the other pallet receiving a tooth. The effect of this may be seen on looking at the drawing (Fig. 1), where the pendulum, travelling to the right, is allowing a tooth to fall on the left-hand pallet. The pendulum, however, still continues its swing to the right, and in consequence the pallet pushes the wheel back, thus causing the

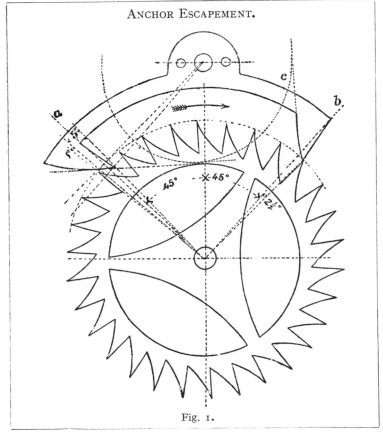

ANCHOR ESCAPEMENT.

Fig. 1.

recoil which gives the name to the escapement. It is only after the pendulum comes to rest and begins its excursion the other way that it gets any assistance from the wheel, and the difference between the forward motion of the wheel and its recoil forms the impulse.

To set out the Escapement.—Draw a circle representing the escape wheel. We will assume the wheel is to have thirty teeth, and that the anchor is to embrace seven and a half spaces.

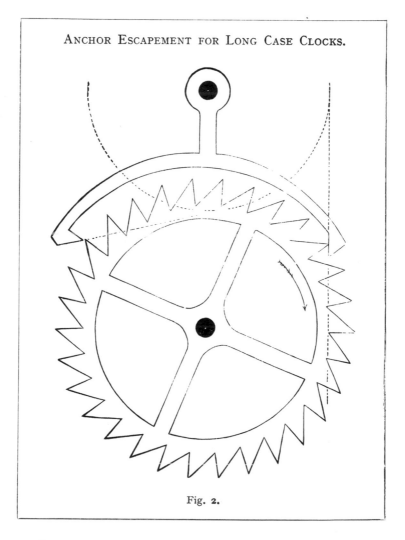

ANCHOR ESCAPEMENT FOR LONG CASE CLOCKS.

Fig. 2.

NOTE.—Space between one tooth and the next $= \frac{360}{30} = 12$; and $7\frac{1}{2}$ spaces $= 90°$. Then $\frac{92}{2} = 45°$.

Set off, with a protractor, the two radial lines *a* and *b*, each making an angle of 45° with the vertical centre line.

Disregarding recoil, the pallets in turn will be in contact with the wheel teeth for nearly the remaining half space, but not quite. The tip of each tooth may be taken to be in width equal to half a degree, and there must be freedom or drop besides. We will, therefore, take 5° to represent the acting part of the pallet.

Then 2½° on each side of *a* will serve to mark the acting part of the left-hand pallet and 2½° to the right of *b* marks the point of the right-hand pallet.

The backs of the teeth are radial and they may now be put in, starting at the line *b*.

Where the line *a* cuts the circle representing the tips of the wheel teeth draw a tangent through the vertical centre line ; this will indicate the position of the pallet staff centre.

The distance of the pallet staff centre from the centre of the escape wheel will be found to be equal to the radius of the wheel × 1·4. From the pallet staff centre describe a circle (*c*) whose radius = seven-tenths of the radius of the escape wheel, that is, half the distance between the escape wheel and pallet staff centres.

Tangents to this circle would form the faces of the pallets if they were left flat as shown by dotted lines.

Between the radial line to the left of *a* and the tip of the adjacent tooth on the left there remains a space equal to 1°, which represents the allowance for drop. If the tips of the wheel are thick and the work rough, more than 1° of drop may be necessary, and then the angle of 5° would have to be correspondingly reduced.

When a tooth has dropped off the right-hand pallet, which is the position of the escapement in the drawing, the amount of impulse is shown by the intersection of the other pallet in the wheel. The impulse, measured from the pallet staff centre, is usually from 3° to 4°.

The pallet faces are generally curved full in the middle, as shown in Fig. 1. The object of curving the pallets is to lessen the " pitting " which the wheel teeth make on the pallets. There will, however, be very little " pitting " if the wheels are made small and light, and there is not excessive drop to the escapement.

The advantage of making the backs of the escape wheel teeth radial and the foresides curved, as shown in Fig. 1, is that if the pendulum gets excessive vibration the pallets butt against the roots of the teeth and the points are uninjured.

There is another form of the Recoil Escapement often used in long-cased clocks, in which the anchor embraces ten teeth of the escape wheel, and the foresides of the teeth are radial. It is shown in Fig. 2. In other respects the construction is substantially the same as the one just described.

Annealing.—[*Recuire.*—*Erweichung durch Hitze.*]—Softening. Steel is annealed by heating it to redness and allowing it to cool gradually. Gold, silver, brass, and copper are annealed by heating to redness and plunging in cold water.

Arbor.—[*Arbre.*—*Die Welle. Der Drehstift.*]—Axis. Watch-

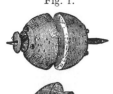

Fig. 1.

makers' arbors on which work is mounted for turning have usually either a ferrule or, since the more general adoption of hand and foot wheels, a carrier pin for communicating a rotary motion. Cork arbors such as are shown in Fig. 1 are useful for holding watch glasses that require to be reduced in size. The glass to be reduced is placed between the two corks, and the arbor may then be rotated in the turns. A broken watch glass wetted is generally used for grinding down the glass between the corks. Fig. 2 is very similar to Fig. 1, but is more adapted for holding watch dials. (See also "Eccentric Arbor" and "Screw Arbor.")

Fig. 2.

Arnold, John (born in Cornwall, in 1736; died at Well Hall, near Eltham, Kent, in 1799). He invented the helical form of balance spring (Patent No. 1113, Dec., 1775), and devised a chronometer escapement (Patent No. 1328, May, 1782) very closely resembling the one by Earnshaw, which is now in use. In Arnold's Escapement the escape wheel teeth, instead of being flat where they gave impulse, were epicycloidal curves; but they required oiling,

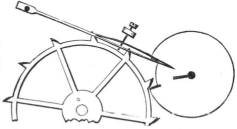

and were consequently abandoned. While Earnshaw's wheel is locked on the points of the teeth and the detent moves away from the centre of the wheel to unlock, Arnold's locked on the heel of the tooth and the detent moved toward the centre of the wheel to unlock, the sunk part of the body of the wheel allowing the locking stone to pass. His business was, after his death, carried on at 84, Strand, by his son, John R. Arnold.

Assaying.—[*Essai.*—*Die Schmelzprobe.*]—The following is the usual process of gold assay. A portion of the metal to be assayed is very carefully weighed, and with a proportion of silver and lead is placed in a cupel, that is, a small bowl made from compressed bone ash, and subjected to heat in a muffle furnace. The baser metals oxidise and pass into the pores of the cupel, leaving a button of gold and silver. This button is rolled into a thin ribbon, which is heated to redness to anneal it, and then coiled up and boiled in nitric acid to extract the silver. The pure gold remaining is then washed, again annealed, and carefully weighed; the difference between it and the original weight represents the amount of alloy. The weighing, and indeed the whole operation, requires to be conducted with exceeding exactness. The amount of silver and lead added before cupelling varies, according to the judgment of the assayer after examination of the metal to be tested; twice or three times as much silver, and five or six times as much lead as there is gold being an average proportion.

Silver.—The silver to be assayed is wrapped in lead and cupelled in the same way as in the gold assay just described, the second operation of boiling in nitric acid being omitted.

For assaying very small quantities of gold and silver a blowpipe may be used in place of the muffle furnace, and in experienced hands a very close approximation may be obtained. (See also Touchstone.)

Astronomical Clock.—[*Horloge astronomique.*—*Die astronomische Pendeluhr.*]—(1) A clock showing the comparative motion of the heavenly bodies.

(2) A clock with the hour circle divided into 24.

Ferguson's Astronomical clock is shown in Figs. 1 and 2. Fig. 1 is the dial, which is made up of four pieces. (1) The outer ring divided into the 24 hours of the day and night, and each hour sub-divided into 12, so that each sub-division represents 5 mins. of time. (2) The age-of-the-moon ring, lying in the same plane as the hour ring, is divided into 29·5 equal parts, and carries a *fleur-de-lis* to point out the time on the hour circle. A wire (A) continued from the *fleur-de-lis* supports the sun (s). (3) Within, and a little below, is a plate, on the outer edge of which are the months and days of the year; further in is a circle containing the signs and degrees of the ecliptic. Further in, on the same plate, the ecliptic, equinoctial, and tropics are laid down, as well as all the stars of the first, second, and third magnitude, according to their right ascension and declination, those of the first magnitude being distinguished by eight points, the second by 6, and the third by 5. (4) Over the middle of this plate, and a little above it, is a fixed plate (E) to represent the earth, around which the sun moves in 24 h. 50·5 m., and the stars in 23 h. 56 m. 4·1 s. The ellipse (H), which is drawn with a diamond on the glass that covers the dial, represents the horizon of the place the clock is to serve, and across this horizon is a straight line even with the XII.'s to represent the meridian. All the stars seen within the ellipse are above the horizon at that time.

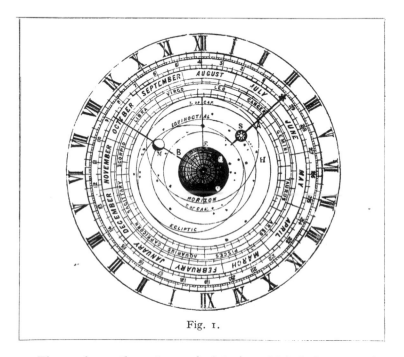

Fig. 1.

Fig. 2 shows the astronomical train, which is between the front plate of the clock and the dial. The earth (E) is stationary, and is supported by a stalk fixed to the front plate. The hollow axis of the frame that turns once in 24 hours, works on this stalk.* Fixed to the stationary stalk is a pinion of 8 (A), which gears with a wheel of 35 (B) and with a wheel of 50 (C). On the top of the axis of B is a pinion of 8 (E), which turns a wheel of 54 (F), running on the pipe of the 24-hour frame, and carrying the moon round by the wire B in Fig. 1. This wire has a support fitted to the arbor of F. On the top of the pipe

* No provision appears to be made for driving this frame in the drawing, for which I am indebted to the excellent Biography of Ferguson by Dr. Henderson. The simplest plan would be to have teeth round the edge of it gearing with a pinion driven from the train of the clock.

of the 24-hour frame is a wheel of 20, gearing with a wheel of 20, having two sets of teeth, one on the edge and the other on the face. It is pivoted to the moon wire support, and the face teeth gear with another wheel of 20 that carries the moon wire. On the axis of C is a pinion of 14 (G) that turns a wheel of 69 (H), on whose axis is a pinion 7 (I) turning a wheel of 83 (K). This wheel is pinned to the sidereal plate of the dial. The age-of-the-

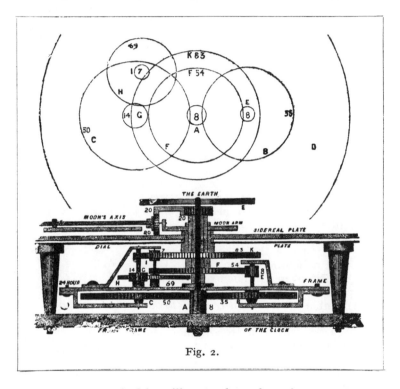

Fig. 2.

moon ring is attached by pillars to the 24-hour frame, and turns with it. The moon is represented by a round ball, half black and half white.

Auxiliary.—[*Compensation auxiliare.—Die Hilfskompensation*.]—A device attached to a compensation balance for the purpose of reducing what is called the "middle temperature error." Molyneux's and Poole's may be taken to represent the two principles on which most auxiliaries are constructed. Molyneux's (Fig. 1) is attached by a spring to each end of the central arm, and is acted on by the free ends of the rim in high temperatures only. A screw in the end of the rim and another in the auxiliary serve to adjust the action as may be required. Molyneux's patent also covered the use of a short laminated arm instead of the spring by which the auxiliary is attached to the central

Fig. 1.

arm, and many successful auxiliaries are now made in that way.

Fig. 2.

(See Mercer's Balance.) Poole's (Fig. 2) consists of a piece of brass attached to the fixed ends of the rim, and carrying a regulating screw, the point of which checks the outward movement of the rim in low temperatures.

Balance.—[*Balancier.—Die Unruh.*]—The vibrating wheel of a watch or chronometer which, in conjunction with the balance spring, regulates the progress of the hands. The time in which a balance will vibrate cannot be predicated from its dimensions alone. A pendulum of a given length always vibrates in the same time as long as it is kept at the same distance from the centre of the earth, because gravity, the force that impels it, is always the same ; but the want of constancy in the force of the balance spring, that in watches and chronometers takes the place of gravity and governs the vibrations of the balance, is one of the chief difficulties of the timer. There is another point of difference between the pendulum and the balance. The time of vibration of the former is unaffected by its mass, because every increment of mass carries with it a proportional addition to the influence of gravity ; but by adding to the mass of a balance, the

strength of a balance spring is not increased at all, and therefore the vibrations of the balance become slower.

There are three factors upon which the time of the vibration of the balance depends :—

(1) The weight, or rather the mass of the balance.*

(2) The distance of its centre of gyration from the centre of

* The mass of a body is the amount of matter contained in that body, and is the same irrespective of the distance of the body from the centre of the earth. But its weight, which is mass multiplied by gravity, varies in different latitudes.

motion, or, to speak roughly, the diameter of the balance. From these two factors the moment of inertia may be deduced.*

(3) The strength of the balance spring, or, more strictly, its power to resist change of form.

I append the usual formula for ascertaining the time of vibration of a balance, though it is difficult of application in actual practice :—

$$T = \pi \sqrt{\frac{AL}{M}}$$

A being the moment of inertia of the balance, *M* the moment of elasticity of the spring, *L* the length of the spring, and π 3·14159. (See moment of Elasticity.)

Sizes of Plain Balances generally used for full-plate watches :—

No. of watch movement—	4	6	8	10	12	14	16	18
Size of balance in inches—	·58	·61	·64	·68	·71	·74	·77	·8

Sizes of Plain balances generally used for $\frac{3}{4}$-plate watches :—

No. of watch movement—	2	4	6	8	10	12	14	16	18
Size of balance in inches—	·47	·50	·53	·56	·59	·62	·66	·69	·72

The rough rule for a full-plate watch is that the diameter of the balance should be half that of the top plate ; for a three-quarter plate watch that the balance should be the size of the inside of the barrel if a 16,200 train, and a trifle smaller if for an 18,000 train.

Though the diameter of a balance for a given movement is not absolute, it cannot be varied indefinitely even if the moment of inertia is kept the same. With a very large, light balance there is but little friction at the pivots, and the variation between hanging and lying is small. On the other hand, an unduly small and heavy balance, though it is less affected by external influence, has excessive friction at the pivots, and correspondingly large variations between hanging and lying, besides which a fall or jerk is very likely to damage the balance pivots.

Gold balances are preferable to steel. Steel has the advantage of being less affected by alterations of temperature, but, on the other hand, gold is denser than steel, and is not liable to rust nor to be magnetised. (See also Compensation Balance.)

Balance Arc.—[*Levée du balancier.—Der Schwingungsbogen.*]— The part of the vibration of a balance during which it is connected with the train. A term used only in reference to the lever and other detached escapements.

---

* The centre of gyration is that point in a rotating body in which the whole of its energy may be concentrated. A circle drawn at seven-tenths of its radius on a circular rotating plate of uniform thickness would represent its centre of gyration. The moment of inertia or the controlling power of balances varies as their mass and as the square of the distance of their centre of gyration from their centre of motion. Although not strictly accurate, it is practically quite near enough in the comparison of plain balances to take their weight and the square of their diameter measured to the middle of the rim.

B

Balance Cock.—[*Coq de balancier.*—*Der Unruhkloben.*]—The standard that in watches carries the top pivot of the balance staff. In some of the old English full-plate watches the top of the balance cock carrying a large diamond end stone is spread out to cover the balance, and often pierced and engraved in a most artistic manner.

Balance Spring.—[*Ressort spiral.*—*Die Spiralfeder.*]—Sometimes called " hair-spring " and (most improperly) " pendulum spring." A long, fine spring that determines the time of vibration of a balance. One end of the balance spring is fixed to a collet fitted friction tight on the balance staff, and the other to a stud attached to the balance cock or to the watch plate. The most ordinary form of balance spring is the volute or flat

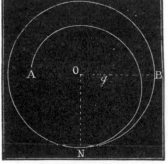

spiral, as sketched in Fig. 1. A Bréguet spring is a volute with its outer end bent up above the plane of the body of the spring, and carried in a long curve towards the centre, near which it is fixed. Fig. 2 shows an overcoil spring. If a regulator is to be used, the portion of the spring embraced by the curb pins would have to be concentric. M. Phillips,

Fig. 1.

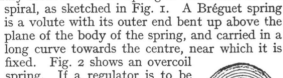

Fig. 2.

a distinguished French mathematician, laid down certain rules for the form of curve best suited for overcoils, and Messrs. George Walker and W. N. Barber devised a way of forming the curve (see Fig. 3). According to Phillips the conditions to be satisfied by a terminal curve are :—

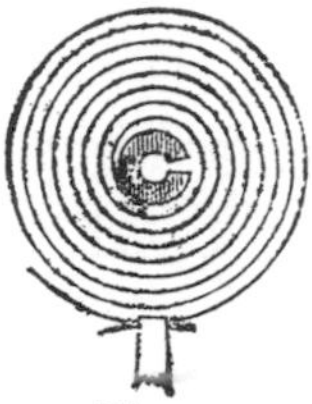

1st. That the centre of gravity of the curve fall on the line $O\ B$ making a right angle with $O\ N$ (N being the starting point of the curve).

2nd. That the distance from the centre of gravity of the curve to the centre of the spring be equal to the radius of the spring squared, divided by the length l of the curve.

Fig. 3.

$$\text{Then } O\ g = \frac{O\ N^2}{l}$$

Take the radius of the spring $O\ N$ as a constant, and l of any convenient length determined by experience ; from these data calculate the distance of the point g from the centre of the spring. From the centre O draw a circle having a radius equal to the distance of the inner side of stud hole from the axis of the balance (the same scale being used as for the rest of the construction). From centre g draw a circle of such a radius as will include the point N; cut this circle out of uniformly thick material; this circle will then balance on the point g.

Take a rod of uniform section of the pre-determined length of the curve and place one end on the point N, the other end on the circle drawn for the stud hole; this rod may then be bent to any form so that the whole system balances on the point g when the conditions for isochronism will be satisfied.

Among English watchmakers overcoil springs are generally called " Bréguet," whatever the form of curve employed. For marine chronometers helical springs, in which both ends curve inwards, are universally used. Either helical or Bréguet springs are as a rule applied to pocket chronometers, although a form of spring called " *duo in uno*," invented, I believe, by Mr. Hammersley, is sometimes preferred. The bottom of this spring is in the form of a volute, from the outer coil of which the spring is continued in the form of a helix; the upper end is curved in towards the centre as in the ordinary helical spring.

A very generally accepted rule is that the diameter of a balance spring for a watch should be half the diameter of the balance (rather under than over).

The dimensions of the spring, its form at the attachments, the position of the attachments with relation to each other, are all factors affecting its controlling power.

The length is important, especially in flat springs without overcoils. By varying the thickness of the wire two flat springs may be produced, each of half the diameter of the balance, but of very unequal lengths, either of which would yield the same number of vibrations as long as the extent of the vibration remained constant; yet if the spring is of an improper length, although it may bring the watch to time in one position, it will fail to keep the long and short vibrations isochronous. Then, again, a good length of spring for a watch with a cylinder escapement vibrating barely a full turn would clearly be insufficient for a lever vibrating a turn and a half.

The great advantage of an overcoil spring is that it distends in action on each side of the centre, and the balance pivots are thereby relieved of the side pressure given with the ordinary flat spring. The Bréguet spring, in common with the helical and all other forms in which the outer coil returns towards the centre, offers opportunities of obtaining isochronism by slightly varying the character of the curve described by the outer coil and thereby altering its power of resistance.

The position of the points of attachment of the inner and outer turns of a balance spring in relation to each other has an effect on the long and short vibrations quite apart from its length. For instance, a very different performance may be obtained with two springs of precisely the same length and character in other respects, but pinned in so that one has exactly complete turns, and the other a little under or a little over complete turns. This property, which is more marked in short than in long springs, is depended upon by many for obtaining isochronism. A short spring as a rule requires to be pinned in short of complete turns, and a long one beyond the complete turns. In duplex and other watches with frictional escapements, small arcs of vibration and short springs, it will be found that the spring requires to be pinned in nearly half a turn short of complete turns. Marine chronometer springs are found to isochronise better and act truer when pinned in at about a quarter of a turn short of complete turns.

There is no doubt that the less a spring is "manipulated" the better. Mr. Glasgow contends that the whole question of isochronism resolves itself into the adoption of a spring of the correct length, and recommends for a lever watch fourteen turns if a flat, and twenty turns if a Bréguet spring is used. He argues that if a spring is too short, the short vibrations will be fast and the long vibrations slow, and that all bending and manipulation of the spring with a view to obtain isochronism are really only attempts to alter the effective length of the spring.

Many who have been in the habit of obtaining isochronism with a lesser number of turns demur to the length given on the ground that alterations in the form and position of the attachments are more effective in short springs. This is true ; but then it is asserted that short springs are more likely to become distorted in use, and it may be taken as a very good rule that a balance spring should be half the diameter of the balance, and have twelve turns if it is a flat spring or eighteen turns if a Bréguet for a lever watch of ordinary size. For small lever watches a turn or two less will suffice. These lengths, it will be understood, only apply where the work is good ; with coarse work a shorter spring is required in order to get the short arcs fast enough. Springs for cylinder watches should have from eight to twelve turns.

Watch springs of thick and narrow wire are apt to cockle with large vibrations, while springs of wide and thin wire keep their shape and are more rigid. It is of even greater importance that the springs of marine chronometers subjected to the tremor of steamships should be of wide and thin wire.

Helical springs for ordinary two-day marine chronometers are made from ·4 to ·54 of an inch in diameter, and about a quarter of a turn short of either eleven, twelve, or thirteen turns. The smaller diameter of spring is advocated on the ground that the acceleration or gaining on their rates noticed in new chronometers may be lessened with a short overcoil. If the overcoil in a watch or chronometer is subjected to any considerable amount of bending, the piece is sure to gain on its rate. Some manufacturers expose new chronometers to a high temperature

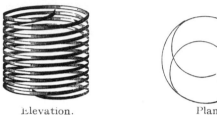

Elevation.   Plan.

Fig. 4—Helical spring.

in the oven for a time, so as to get them to settle down quickly to a steady rate. Other eminent makers declare there is no benefit in doing so. Mr. T. Hewitt tells me that after a chronometer has been brought to time he has removed the balance spring, collet and stud, and subjected them all together to sufficient heat to lower the colour of the spring without materially altering its form or at all affecting the acceleration of the chronometer. Still there is no doubt it is desirable to submit all overcoil springs, whether for watches or chronometers, to a bluing heat after the curves are formed, for the purpose of " setting " the curves. Timers of experience say the time-keeping is much more steady and regular when the springs have been so treated.

It is remarkable that while in watches the difficulty is generally to get the short arcs sufficiently fast, precisely the reverse is the case with the marine chronometer, in which the trouble is usually to get the short arcs slow enough. The escapement is not entirely responsible for the difference, because pocket chronometers follow the same rule as watches with lever escapements. The size of the pivots in proportion to the size of the balance is partially the cause, for in very small watches, where of course the pivots are relatively large, the slowness of the hanging position is proverbial, and a shorter spring by a turn or two has often to be substituted. Very quick trains should be avoided on this account. Watches have occasionally been made with 19,800 vibrations, carrying of course correspondingly light balances, and the great trouble has always been to get them fast enough in the short arcs. A balance staff pivot slightly too large for the hole is occasionally the cause of undue slowness in the hanging position of a watch ; a very small reduction in the diameter will in such cases reduce friction and quicken the short arcs.

The relative slowness of the long arcs in marine chronometers is much greater if the rim of the balance is narrow and thin, and the weights large, than if a wider and stouter rim with proportionately smaller weights is used ; for in the former case the greater enlargement of the rim in the long arcs from centrifugal tendency will be more marked. Mr. Kullberg, by substituting a chronometer balance of ordinary proportions for an uncut one, has demonstrated that the effect of centrifugal tendency by increasing the arc of vibration from three-quarters of a turn to a turn and a quarter amounts to twelve or fourteen seconds in twenty-four hours. Mr. Dennison some years ago advocated a method of making the long arcs slower in watches having compensation balances by drilling a small hole in each half of the balance rim, close to the arm, and broaching these holes out as much as should be found necessary to produce isochronism. The larger the holes, the more the balance would expand from centrifugal tendency, which, of course, would have more effect in the long than in the short arcs.

Mr. Robert Gardner insists that the relative proportional inertia of the fourth wheels in marine chronometers and watches accounts for the difference observed in the long and short arcs, and there is no doubt that watches with small fourth and escape wheels are comparatively easy to time. J. F. Cole suggested a stronger mainspring to quicken the short arcs in going-barrel watches.

With single beat escapements, such as the chronometer and duplex, the short arcs are quickened if the drop of the escape wheel tooth on the pallet is decreased ; if the drop is increased the long arcs are quickened. The theory of this appears to be that with more drop in the long arcs (when the balance is travelling faster), the pallet gets farther away before being overtaken by the wheel than it does in the short one, and therefore the amount of impulse given before the line of centres is proportionately less ; and Mr. Robert Gardner asserts that the same effect may be produced by putting the piece out of beat, e.g., if it were desired to quicken the short arcs, the balance spring collet would be shifted so as to bring the pallet more away from the unlocking when the balance spring was quiescent ; but it must not be forgotten that if the piece is out of beat it is more likely to set.

Fig. 5 is a careful reproduction of the curves of a marine chronometer spring. a is the lower coil. The two short lines crossing the spring denote the face of the collet and the face of the stud respectively, and the dotted lines the direction of the ends of the spring, which form nearly a right angle. Occasionally, if the short arcs are fast, the upper turn is slightly bent just as it enters the stud so as to throw the end outward. The pin is then placed on the inside.

Fig. 6 is the lower coil, and Fig. 7 the upper coil of a pocket chronometer spring. It will be observed that the sweep is longer than in the marine chronometer, and that the lower curve is carried rather farther back into the spring than the upper. The spring makes just complete turns, but the upper turn just as it enters the stud *has a slight sharp bend throwing the end inwards.* This little bend is of the utmost importance, for it has the

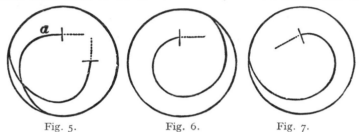

Fig. 5.   Fig. 6.   Fig. 7.

effect of quickening the short arcs. Pocket chronometer springs are made half the diameter of the balance, and from seven to ten turns.

Bréguet springs are now often used for pocket chronometers, instead of the helical form. Just as good a result can be got with the Bréguet as with the helical, and the latter takes up height, and in consequence is often made too short. There is, though, one advantage in using helical springs for pocket chronometers. The escapement may be banked through the spring, and this is done more readily in the helical form. Mr. Kullberg's method is to place two nearly upright pins on the balance arm, so close to the spring as to prevent it expanding more than is required for a sufficient vibration. These pins should be slightly inclined to the centre so as to touch the top of the spring first, and thereby stop the balance more gradually.

Two pairs of pliers with curved noses lined with brass are generally used for forming the overcoil of marine chronometer springs. The coil of the spring where the overcoil is to start is grasped by one pair curved exactly to correspond with the spring, and the other pair used to bend the overcoil. The operation looks easy enough, but it really requires great skill to get at once an overcoil of the desired shape.

Palladium Springs.—M. Paillard has introduced balance springs of palladium and some alloy which possess the two great advantages that they do not rust, and are not liable to be magnetised. These springs, therefore, are often applied to watches and chronometers. They do not accelerate so much as hardened steel springs, and trials seem to show that the middle temperature error is with them considerably reduced. The fifth chronometer on the 1883 Greenwich trial was fitted with a palladium spring and an ordinary compensation balance without auxiliary. No chronometer maker would expect such a result with a steel spring. Whether palladium springs are as lasting as hardened steel remains to be seen. Being heavier than steel they droop more, and in Bréguet springs the over-coil has to be lifted farther above the plane of the spring, so as to avoid the possibility of contact if the watch is roughly turned on its face. Their limit of elasticity is less than that of tempered steel springs, so that they will not bear so much bending and they more readily show marks of the pliers.

Steel-Nickel Alloys.—The researches of Dr. C. E. Guillaume, at the Sèvres Office of Weights and Measures, have revealed the fact that steel liberally alloyed with nickel produces a compound with a very small co-efficient of expansion. Dr. Guillaume found the co-efficient of expansion of steel alloyed with 36·2 per cent. of nickel (named Sèvres alloy) to be but 8, in comparison with brass 189 ; steel or iron, 108 to 122 ; glass, 86 ; wood (*bois de sapin*), 44.

Professor M. Thury conducted some experiments with a view of determining the variations in the elasticity of Sèvres alloy when subjected to changes of temperature ; and he discovered that in a range of temperature of 22° Centigrade (+ 15° to + 37° Cent.), the elasticity actually increased with a rise in temperature, thus reversing the behaviour of steel under similar circumstances. Such a startling result leads at once to conjectures of a revolution in the methods of compensating watches. Accepting the Sèvres alloy with its low co-efficient of expansion as a suitable material for the balance of a watch, it is easy to suppose that by some slight variation in the proportion of nickel employed a substance may be obtained for the balance spring with a ratio of expansion calculated to entirely neutralise the temperature error. Recent tests have shown that watches furnished with plain uncompensated balances and balance springs of nickel-steel alloy have scarcely one-fifth of the temperature error which is observed when steel springs are used. So that for watches of the lower and medium qualities the temperature error is practically overcome. For fine watches and chronometers a compensation balance with the arm and inner part of the rim of steel-nickel alloy and the outer part of the rim of brass is recommended. Though the steel-nickel alloys cannot be fire-hardened, they are readily compressed and stiffened by hammering or rolling. They are less affected by magnetism than steel, but can hardly be classed under the head of non-magnetic.

Applying a Flat Spring.—The number of vibrations required depends, of course, on the train. For a modern watch with a seconds train in which the fourth wheel turns once in a minute, divide the number of the fourth wheel teeth by the number of leaves in the escape pinion ; multiply the quotient by 30 (double the number of escape wheel teeth), the product will be the number of impulses the balance will receive in a minute. If it is an 18,000 train the number will be 300, that is 5 beats a second. A 16,200 train gives 270 a minute ; 4·5 a second.

Should the watch not have a seconds train we must go back to the centre wheel which rotates once in an hour. In this case multiply together the numbers of centre wheel, third wheel, fourth wheel teeth and 30. Also multiply together the number of leaves of third pinion, fourth pinion, escape pinion and 60. Divide the first sum by the last and the quotient will be the vibrations per minute.

Having ascertained the number of vibrations required proceed to select what appears to be a suitable spring. Lay the spring with its centre coinciding with the cock jewel and mark on the coil that is to be the outer one exactly where it would enter the stud hole, bearing in mind that the spring should be rather small in diameter than large ; FOR A SPRING TOO LARGE IN RELATION TO THE INDEX PINS AND STUD IS PRETTY WELL SURE TO BE SLOW IN THE SHORT ARCS.

The next thing is to count the vibrations of the balance when connected with this spring.

Put the eye of the spring over the balance staff down on to the balance, and press the collet on to the spring so as to confine the inner portion that will be broken away for the eye. If it should happen, as in rare cases it may, that the collet when pushed on the staff its proper way fails to hold the spring, the collet may be reversed and pushed on with the smaller side of the hole first.*

Place in a convenient position on the bench a watch that is known to be keeping correct time, then take hold of the spring firmly with tweezers at about the distance of the curb pins short of the spot marked for the coil to enter the stud ; then lift the spring so that the balance hangs horizontally and the lower

---

* In the absence of the collet a bit of beeswax or putty powder the size of a small pin's head may be placed upon the end of the lower balance staff pivot to catch the eye of the spring under, in order to count the vibrations ; but the use of wax or powder, a portion of which may be afterwards transferred to the watch, is not to be recommended. Besides, clasping the spring between the balance and the collet places that part of the eye which will be broken off out of action, and the result will be more exacting than catching up the spring close to its centre.

balance pivot is just above the watch glass. Give the balance about half a turn so that it will vibrate for over a minute. Owing to the contraction and expansion of the spring the balance will also acquire an up and down motion.

With your eyes on the watch dial count every alternate vibration registered by the contact of the pivot with the glass till a spring is obtained that gives 75 double vibrations, if it is for an 18,000, and from 67 to 68 during half-a-minute, if for a 16,200 train. The observation may extend over a lesser or a greater period, always remembering that for an 18,000 train there are to be 5 DOUBLE vibrations every TWO SECONDS, and for a 16,200 train $4\frac{1}{2}$ in the same time. It is better for the spring to give, say, half a vibration less in a minute rather than more, for it is sure to be a little faster when the eye of the spring is broken out to suit the collet. Extend the observation and counting to one minute if any doubt exists respecting the watch which is taken as a standard. It should not only be a good timekeeper but also have the seconds dial correctly spaced, and the seconds arbor concentric therewith, if exact results are to be obtained for shorter periods than a minute. Of course, if a chronometer is available it may be used instead of a watch. Many prefer to listen to the taps of the pivot on a watch glass and to watch the seconds hand of a regulator. Others will dispense altogether with the audible report on a watch glass and catching up the spring will give a slight twist of the hand holding the tweezers, and watch the number of vibrations given during 4 or 5 seconds while they listen to the beats of a regulator.

Yet another plan is to hold to the ear a watch having the same number of vibrations as it is desired the balance under trial should have, and note the discrepancy, if any. The first method given is, I think, best suited to a tyro, and will, I am sure, result satisfactorily if carefully conducted.

Although very accurate results may be obtained by " vibrating " on the watch when continued for, say, two minutes, the use of a " Vibrator " has the advantage that the attention may be withdrawn from the balance under trial for a few seconds without the observation being lost.

The " Vibrator," Fig. 8, consists of a vertically pivoted box, with glass top, containing a balance and spring accurately adjusted to give a standard number of vibrations. If adjusted to, say, 18,000 vibrations, another small box is provided to give 16,200, and sometimes a third giving 14,400. Any of the two latter may be slipped over the first when these numbers are required. Attached to the side of the pivoted box there is a vertical pillar from which projects a horizontal arm carrying tweezers to hold the outer end of the spring. The construction of the tweezers is such that the spring may be drawn out or in by simply turning a small milled nut.

The balance and spring to be tested are adjusted over the standard balance, the outer end of the spring being held by the tweezers so that the bottom pivot remains in contact with the glass throughout the entire vibration. The two balances are then set in motion, exactly together, by a push from a lever pivoted to the stand. With a little practice it is easy to quickly adjust the length of spring so that the balances are running accurately together.

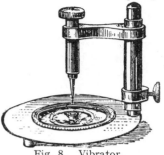

Fig. 8. Vibrator.

When a spring of the right strength is found it may be pinned to the collet. To break away sufficient of the eye for it to pass over the collet place the spring on the board paper ; grasp it with tweezers at the point where it is to be severed ; with a screwdriver, or other suitable implement, bend the eye close to the tweezers, and it will break away.

Assuming the position of the stud to be fixed, and that the spring should start from the collet on a level with the centre of the cock jewel if the watch were on edge with the pendant uppermost, place the spring on the cock, with the mark already made for the outer attachment coincident with the face of the stud, and mark on the eye of the spring the point where it is to spring from the collet. The spring should start away from the collet with an easy curve, and must not hug the collet, or isochronism will be out of the question. With tweezers having a rounded grip rather than a sharp edge, hold the eye a little farther up the spring than the mark just made, and with a peg gently bend round to the requisite angle the piece which is to enter the collet. Do not press with the peg close to the tweezers, or the spring may break. A sketch of the collets showing the hole for the spring may be made on the board paper as a guide for the angle. The piece bent round, of course, is curved. Hold it in the tweezers, and, with the assistance of the peg, straighten it.

Occasionally it may be that the only spring of the required strength available is too small with relation to the stud. If unhardened a spring may be expanded by placing it on a flat plate previously warmed, and heating it to a bluing temperature. It must not be heated beyond a bluing temperature, but the operation may be repeated more than once, and the spring will expand at each heating. To ensure the flatness of the spring remaining undisturbed it may be covered by a small flat piece of brass or glass, which must not be too heavy, or it may prevent distention of the spring.

The Collet.—The depth of the collet should not be less than twice the diameter of the central hole, if room allows. The taper of the hole and the taper of the balance staff where it fits should exactly correspond, so that the collet does not rock, and that part of the staff should be left grey to allow a tighter grip for the collet. The bottom of the hole should be rounded, so that it readily enters on the staff ; the bottom outer edge of the collet should also be rounded, to facilitate the entrance of the tool used for raising it. The hole for the reception of the spring must be exactly at right angles to the central hole, for if the spring has to be forced or bent to suit the hole it is hopeless to expect good results. Beginners are often in such a hurry to pin the spring on that they proceed when the hole is obviously not in the proper direction. The end of the spring should be passed into the hole, just kept in position with a temporary pin and examined. If any alteration is needed, the spring is to be taken out and the hole broached in the direction required till it is right. A round pin of hard brass is then fitted to the hole. When this is done, take a waste piece of the spring, insert it in the hole, and flatten away the pin till it and the spring together fill the hole properly. Then cut off the pin, leaving but little to project, so that there is no danger of its touching the spring in action.

The usual plan of pinning in is to put the collet on a broach held between the thumb and finger of the left hand while the pin is fitted and the spring pinned on, taking the precaution to push a piece of paper on the broach before the collet or the spring may touch the fingers, and in the case of a damp hand the spring would be likely to be spoilt by rusting. By previously fitting the pin as directed, there is now very little fear of injuring the spring. The pin can be pressed home with a small joint pusher.

Another mode of pinning on is to place the collet on the board paper, adjust the spring to the collet, and with a short piece of boxwood sloped away at the end press the collet on the board ; the pin can then be fitted with comfort, and without danger of shifting the collet.

Yet another way of proceeding to pin the spring to the collet is to get a plate of brass about an inch square, and tap a hole in the middle of it less in size than the hole in the collet. A screw passing through the hole in the collet fixes it to the plate. The advantage of this plan is that the surface of the plate serves to show at once if the spring lies in the proper plane. Care must be taken with the underside of the head of the screw, so that it does not scratch the collet.

The collet may now be put on an arbor, and the arbor rotated in the turns to ascertain if the spring is true. In setting the spring it must only be touched close to the eye. Steady time-keeping will be out of the question if the spring is bent to and fro in reckless attempts to get it true. The eye should be brought gradually round to get it in circle, taking care not to overdo it. When this is right, and the spring is also true on the face, many good timers heat the spring and collet to a blue to set the eye. The outer coil may now be pinned into the stud.

It will be prudent to first pin the spring in temporarily and notice if the eye is true with the cock jewel, or very slightly towards the stud. If it is not, and the stud cannot be shifted, the stud hole had better be broached in the required direction. Or, of course, the spring can be slightly bent close to the stud hole to bring the eye right with the cock jewel. The largest part of the pin should be towards the body of the spring, and therefore the hole should be broached from that direction. First fit a round pin, and then flatten it to suit with a waste piece of the spring as directed when pinning to the collet. The pin at the stud may be left fairly long. The waste outer part of the spring should not be broken off close to the stud, but sufficient should be allowed to project through the stud to allow for any letting out of spring that may be required.

The curb pins should be of brass, only just free of the spring, parallel inside and tapered back from the point on the outside, without burr or roughness that would allow a lodgment of the spring. Mr. Kullberg considered the curb pins should not reach below the bottom of the spring, so that there is no room left for the second coil to jump in. The curb pins should not extend far beyond the stud, or the outer coil may jump out. But, apart from the jumping of the spring, it is objectionable to have the curb pin or pins far from the stud. It is noticed that watches in which this distance is great are very difficult to time.

The watch should be tried, lying and hanging, for twenty-four hours in each position. If it gains in the short vibrations the spring may be taken up, and if it loses in short vibrations let out a little. But if the alteration has the opposite effect to that desired, as in some instances it may, proceed in the contrary direction.

If the short arcs are slow, some recommend bending the outer curb pin away from the spring and pressing the inner one hard against it, so that the outer coil is really dragged out of its position. This seems to be a most objectionable nostrum, which doubtless originated through a spring too large having been applied.

If the short arcs are fast the curb pins are sometimes slightly opened as a remedy, but this is also open to objection, for the pins then obviously cease to be parallel with each other, and with undue play between the curb pins the spring will get worn where they touch it, and the going of the watch will not be satisfactory.

Unless there is some grave constructional fault the variation will not be so excessive but that the watch may be brought to time with the index, dividing the error between the long and short vibrations.*

Where there is no seconds hand, the usual plan is to make a mark on the fourth wheel ; but Mr. Bickley recommends placing the watch to the ear and counting every alternate beat for a minute or two, and comparing with a regulator or watch going to time. However, a small dot on the fourth wheel is an almost imperceptible disfigurement, and will generally be found useful as the watch is getting pretty close to time.

The method of hardening a flat spring without distorting it after it had been applied was at one time kept a secret by the few who practised it. It is, however, exceedingly simple. The spring is placed on a round, PERFECTLY FLAT plate of German silver ; the space between the coils being filled in with black lead, the spring is covered with a plate similar to the first. These plates, with a spring between them, are then laid on a small press of German silver very much resembling in form the presses used for copying letters. (See Fig. 9.) The plates are kept together by means of a screw. The spring, press and all, are then heated to a cherry red and plunged into water. When cool a thin slip of bright steel is laid on the press, which is heated till the slip of steel is brought to a blue colour, when

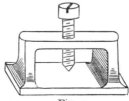

Fig. 9.

it is plunged into oil. It is not imperative that the press should be of German silver, but it must not be of steel or wrought iron, or the springs will be spoiled. A press of this sort is also useful for restoring to shape a spring distorted from flatness. This may be placed between the flat plates and heated to a blue, as already described.

Macartney's Collet Adjuster.—For twisting a balance spring

* All watch balances should, I think, be furnished with screws for ready adjustment of weight. The extra cost, even for cheap watches, is not worth consideration. There is no doubt that the proper way of bringing the piece to time is by varying the balance instead of lengthening or shortening the spring at the stud, which upsets any prearranged position for the inner point of attachment.

collet to position, the special tweezers shown in the sketch (Fig. 10) are admirable. One jaw is concave, and the other in the form of a knife edge.

Fig. 10.

Plose's Collet Lifter.—This is an excellent form of pliers for raising a balance spring collet. The jaws are curved and have knife edges. The leverage is much more even and effectual than

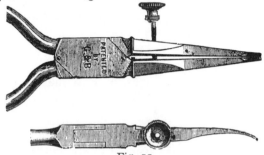

Fig. 11.

with the old style by one-sided wedging. There is a set screw to prevent the edges closing too far and so injuring the staff (Fig. 11).

For pressing a collet home, a table with holes or slits, such as is depicted under the head of " Stud," will be found useful.

Balance Spring Holder.—This tool (Fig. 12) forms a kind

Fig. 12.

of self-acting tweezers, and is useful for holding the outer coil of the spring, especially when it is desired to get the length accurately with the balance in its place in the watch before pinning to the stud, as in the case of Bréguet spring prior to turning the overcoil.

Breguet Spring, or Flat Spring with Overcoil.—The selection of a spring, counting its vibrations with the balance and pinning to the collet, may be conducted in the same way as for a spring without overcoil, except that if it is decided to keep the point of attachment to the collet in a horizontal line with the cock jewel when the watch is placed pendant up, allowance must be made for the bending in of the spring to reach the stud, which, of course, is nearer to the centre than a stud for a flat spring would be. Experienced springers after selecting a suitable spring, by vibrating, break away the eye very accurately, to ensure the desired result. The overcoil will not materially affect the number of vibrations in a given time. It is, however, difficult to frame a rule of any practical use to the beginner, because there are so many factors to be considered : the diameter of the spring, the length of the overcoil, the distance of the stud from the centre, the thickness of the wire from which the spring is made, all affect the amount to be broken away at the eye. To be quite certain the spring selected will bring the watch to time after the overcoil is formed, it may be desirable to try it in the watch, and for this purpose the holder (Fig. 12) will be useful for gripping the outer coil.

The overcoils of watch springs are usually turned with curved nosed steel tweezers like Fig. 13. These curves should be of

Fig. 13.

suitable shape for the particular spring, and be well polished. Beginners are not sufficiently careful in forming the curved noses. When a bit of spring wire is gripped between them they must show an exactly circular outline. If defective they would twist the spring, even in the hands of an experienced operator. In forming the Phillips' curves, some watchmakers use hot pliers of the requisite shape to set the curve to the required form, but this is exceptional treatment, and does not appear to be necessary.

Fig. 14 is a useful little clamp of brass by Mr. T. D. Wright, for holding a spring while turning the overcoil, the jaws of which are curved to the outer coil of a spring of average size. The upper screw passes through a clearing hole in the loose jaw and is tapped into the body of the tool : the pin below is a steady pin ; the bottom screw tapped into the loose jaw serves to keep the opening

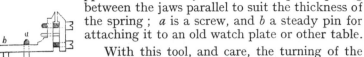

between the jaws parallel to suit the thickness of the spring ; *a* is a screw, and *b* a steady pin for attaching it to an old watch plate or other table.

With this tool, and care, the turning of the overcoil will be comparatively easy. Having

Fig. 14.

decided on the length and form the overcoil is to take, and of which you will have either a pattern or a photograph, or a drawing on paper or metal, break off a piece of the waste outer coil and bend it to the shape required ; then pull it out again to coincide with the outer coil, and from the point where the spring enters the stud, or if there is to be an index, the middle position of the curb pins, mark off the length of the curve. Then fix the outer coil in the clamp just short of the point where the overcoil should start, and place underneath it a sketch of the overcoil for guidance. There are now two distinct operations—first, to raise the overcoil above the plane of the spring, and then to form the curve. If the two are attempted at once, it is most likely that in unskilled hands the spring will be twisted and practically spoiled. Therefore with strong tweezers raise the spring sufficiently, and then with the curved-nosed tweezers bend the curve very gradually without twisting the wire, taking care not to overdo it in any part, for bending to and fro is sure to deteriorate the spring.

Presuming that the copy be a theoretically isochronal curve, let the arch of the overcoil, if it deviate at all, be rather nearer the centre than the other way, so as to be, if anything, quicker in the short arcs. When the curve is correct it may be set by heating the spring to a bluing temperature. Grandjean suggests gently heating the spring on a white enamel dial till a drop of oil placed alongside the spring begins to smoke.

If the spring is to be free, without an index, pinning to the stud is a simple affair requiring no instruction beyond what has already been given when dealing with a flat spring ; but with an index there are two methods of procedure, one of which is to be recommended. The plan formerly most generally followed was to carefully shape that part of the outer end of the curve where the index pins would operate into a circular form of a radius equal to the distance of the curve from the centre of the index ; that is, from the centre of the cock, and then proceed to pin to the stud. But in almost every case this last operation distorted ·the circular part on which pains had been bestowed, and the shaping of this had accordingly to be done over again. The best plan is to pin to the stud first,

C

and then shape the circular part to suit the path of the index pins. This way is more expeditious, and the spring is subjected to less bending.

If the balance spring has been carefully selected the watch will not show a considerable variation from mean time. Leaving the final adjustment to be made by screwing the mean time screws in or out, it may now be brought to within about two minutes a day by, if necessary, adding to or taking weight from the balance. Usually two opposite screws are removed and heavier or lighter ones substituted, as may be required. Platinum screws are used when extra heavy ones are needed. The overcoil of the spring should not be touched for mean time adjustment.

Possible tests and corrections now to be made may be classed under the heads of :—

> Isochronism.
> Compensation for temperature.
> Position errors.
> Mean time adjustment.

There is no absolute rule as to the sequence of these operations. Some adjusters proceed in one way while others, equally successful, pursue a somewhat different course. Isochronism is, as a rule, dealt with first, because it is possible more easily to carry it to completion in presence of errors coming under the other heads enumerated. Much, however, depends on the character of the watch, and the performance required from it. Between the straightforward procedure adopted with a medium class piece and a fine watch which has to be subjected to the exacting ordeal of an Observatory test there is a wide difference. In the latter case the margin of error allowed is so narrow that perplexing constructional inaccuracies, quite distinct from the balance and spring, have to be located and rectified.

We will first take, as an example, a going barrel watch not required to be adjusted closely for positions.

Timing Box.—A brass box should be provided for the reception of an uncased watch movement while it is being timed. The edge of the pillar plate rests in a rebate, and a cover with a glass let into the top screws on to the box, and keeps the movement in position.

Rack or Holder for Watches on Edge.—The primitive and most usual rest for watches placed on edge for positional adjustment consists of a rack or rest of wood having V-shaped notches to suit various sized watches.

A convenient movement holder sometimes used is shown in the annexed cut (Fig. 15). It is adjustable, and capable of being turned in any desired way for position timing.

ISOCHRONISM ADJUSTMENT.—Tests for isochronism may extend over different periods, but variations are always reduced to 24 hours for comparison. Although approximations may be first obtained by observations at short intervals, 24 hour trials should, if possible, be allowed as the rates become close. Some adjusters check the watches just when they happen to be at leisure, and record the duration of each trial in minutes, reducing the variation to 24 hour periods, by the aid of logarithms, or a slide rule, to save extended calculation. Aliquot parts of 24 hours can, however, generally be managed by systematic procedure, and there is then less liability to error from miscalculation. Trials of less than four hours' duration can scarcely be regarded as a reliable guide.

Fig. 15.

Dial up is generally taken to represent the long vibrations, and the mean of two opposite vertical positions the short vibrations. With short trials of going barrel watches the same turns of the mainspring should be used for each test of the same kind. If we begin with four hour trials it will be convenient to start in the morning, to wind the mainspring but one turn so as to get the shortest vibration the balance will have in ordinary use, and to set the watch to mean time or in agreement with a reliable regulator. After four hours pendant up the watch is found to have lost three seconds, and its variation from mean time — 3 is noted. The mainspring should then be wound about half-a-turn, and the watch placed pendant down. After four hours' running it is found to be — 8, having lost 5 secs., and its variation from mean time — 8 is noted. The mainspring may then be fully wound to get the longest vibration, the watch placed dial up and allowed to run for four hours ; it is then found to be — 4, having gained 4 secs. The mean variation of the two positions representing the short vibrations is represented by — 4, and the variation in the long vibrations by + 4 ; as 4 hours, the duration of each trial, are one-sixth of 24 hours, each of these results s multiplied by 6 to obtain the variation in 24 hours, giving respectively — 24 and + 24, which added together = 48 ; the short arcs are said to be 48 secs. slow.

If the variations had been plus in each case or minus in each case, one total would have been subtracted from the other, and the remainder would have represented the daily rate.

The most usual way of making the long and short vibrations isochronous is by altering the form of the overcoil. When the short arcs are slow, as in this case, the arc of the overcoil is closed slightly, and a little more of the body of the spring added to the overcoil. When the short arcs are fast, the arc of the overcoil is made a trifle flater, and a little of the overcoil taken back into the body of the spring. Fig. 16 shows clearly what is meant. If the full line represents the original form of curve, it would be altered in the direction of *b*

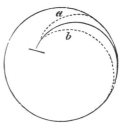

Fig. 16.

to quicken the short, and in the direction of *a* to quicken the long arcs. To alter the shape of the overcoil the spring is held with a pair of tweezers with brass faces curved exactly to suit it, while the alteration is made with a similar pair. Alterations must be made very gradually, for if bent too much the spring is likely to be spoilt in bending it back.

After the requisite alteration is made, the watch may be again set to mean time in the morning with the mainspring fully wound, and after 24 hours' running, dial up, it is found to have gained 35·5 secs., this variation + 35·5 secs. having been noted, the watch, with about two turns of the mainspring wound, may be tried 12 hours pendant up ; it is then found to be + 50·5 secs. fast, having gained 15 secs. ; this is noted : the mainspring wound about a turn and a half, and the watch placed pendant down for 12 hours ; it is then found to be 59·5 secs. fast, having gained 9 secs. The short arcs are now slow, at the rate of 11·5 secs. in 24 hours ; the balance spring, therefore, requires a slight alteration in the same direction as before.

After this has been done the watch is fully wound, again set to mean time, and after 24 hours' trial, dial up, is found to be, say, 36 secs. fast. The watch is noted + 36 s. With the pendant up for 12 hours, it gains 21 s., making it + 57 s. And with the pendant down 12 hours, it gains 15 s., and is then + 72 s. So that the variation in the long vibrations is + 36, and the sum of the variations in the short vibrations is + 36 also. The watch is now said to be isochronous, but there is still a position error between pendant up and pendant down of 6 seconds in 12 hours.

Twelve hours' trial with pendant to the right shows a gain of 27 s., and 12 hours' trial with pendant to the left a gain of 13 s., making a position error of 14 seconds in 12 hours in the quarter positions.

The mean time screws may be drawn out sufficient to give a loss of 38 seconds in 24 hours, thus dividing the error. The piece would then show a variation from mean time in 24 hours running in each position :—Dial up — 2 s. ; pendant up + 4 s. ; pendant down — 8 s. ; pendant right + 56 s. ; pendant left — 12 s.

This is not at all a bad result, and the watch being a little faster pendant up than dial up is likely to approach still nearer in these positions as the oil thickens. We may, therefore, leave it as sufficiently near.

Position Errors, which are due to escapement faults, bad jewelling and other constructional inaccuracies, are often confounded with a want of isochronism ; but a watch may be perfectly isochronous, and yet have very large position errors.

If the short vibrations are not more than a turn in extent, the vertical positions could be equalised by drawing out the quarter screws of the balance nearest the top in the fast position, and this is the remedy generally adopted. For instance, to take an open-faced watch vibrating rather under a turn when placed on edge, which, when at rest, has a quarter screw pointing to the pendant. This watch, when pendant right, is faster 10 secs. in 12 hours than when pendant left ; the quarter screw nearest the III., which, looking at the back of the watch, is nearest the top in the fast position, would be drawn out, say, half a turn, and the opposite screw turned in a similar amount. The larger the short vibrations the more the screws would have to be operated upon to give the same effect, until a limit is reached, when the remedy is inoperative. If the short vibrations are a turn and a quarter, altering the screws will be useless. Should the short vibrations be a turn and a half, which is not very likely, the opposite method would have to be resorted to, viz., the screws furthest from the fast positions must be set in, and those furthest from the slow position drawn out. This is called " timing in reverse." The best authorities strenuously object to tampering with the balance, and prefer to leave the position errors alone. No doubt it is a grave fault to set the balance out of poise, and as the extent of the vibration falls off from thickening of oil and dirt the correction is destroyed and an additional error introduced. Of course, with a first-rate watch, where everything is of the best, the position errors will be very small.

There is another method employed by some to lessen the position errors, but it is not less objectionable, and is certainly not sufficiently efficacious to be worth consideration if the position errors are large. It consists of bending the balance spring (not the eye), so that the eye, instead of being true with the cock jewel, presses against the staff in the fast positions ; in the preceding example, the spring would be moved from the pendant, and to the right of it. The effect of this is to cause more friction on the balance staff pivots at those points.

A watch out in positions may often be brought nearer by turning the roller of the escapement half-way round on the balance staff, slight defects in the pivots and holes being thus neutralised.

There should not be much difference whether the short vibrations are caused by decrease of motive power or by change of position from lying to the rack, yet the results will not be absolutely identical. If the discrepancy is very great, it would probably indicate defective jewelling, holes too long or too large, or not sufficiently polished. The length of the straight part of the jewel hole should be about equal to its diameter.

In adjusting marine chronometers and chronometer clocks for isochronism, the long and short arcs are not obtained by changing the position of the instrument, but the arc of vibration is reduced about one-quarter by letting down the mainspring, or reducing the power in any other convenient manner. For instance, a stiffish spring may be brought to bear on the fourth wheel pinion with sufficient pressure to absorb as much of the force as may be desired.

The method of recording observations varies with different adjusters. The subjoined arrangement appears to be convenient. Symbols are used to indicate the positions, as follows :—

Dial up = **O** ; dial down = **O** ; Pendant up = **T** ; pendant right = **⊣** ; pendant left = **⊢** ; pendant down = **⊥**. The time is set down in hours and quarter hours.

Date.	Mainspring, how wound.	Time of day trial begins.	Watch time.	Position,	Time of day trial ends.	Duration of trial.	Watch time.	Daily rate.	Result. What alterations.
Oct 2	1 turn	8a	R	T	12	4	—3	—24	Shorts 48 secs. slow. Spring set.
,,	,,	12	—3	⊥	4P	4	—8		
,,	Full	4	—8	O	8p	4	—4	←24	
3	Full	7¾a	R	O	7¾a	24	←35·5	←35·5	
4	2 turns	7¾a	←35·5	T	7¾p	12	←50·5	←24	Shorts 11·5 secs. slow. Spring set.
,,	,,	7¾p	←50·5	⊥	7¾a	12	←59·5		Set to mean time.
5	Full	8a	R	O	8a	24	←36	←36	
6	3 turns	8a	←36	T	8p	12	←57	←36	
,,	,,	8p	←57	⊥	8a	12	←72		

Adjustment of a Fine Watch.—Where the best possible performance is to be obtained from a watch in every position and in varying temperatures, an exact standard of time for comparison is essential. Without this it would be foolish to undertake the task. Even with an Observatory time signal at frequent intervals there should be two good regulators to check a doubtful signal before making any alteration to the watch as it is approaching completion.

However, assuming all the necessary apparatus to be at hand the whole of the mechanism of the watch should undergo a rigorous scrutiny prior to proceeding with the timing and adjustment. It is often the next to nothing, the very slight errors of construction, which, if overlooked, give trouble at the end. Taking, as before, a going barrel watch with a lever escapement, we will suppose that an over-coil balance spring has been applied as already directed. The escapement should then be examined in accordance with the procedure given under the head of the particular escapement in use.

The escapement having been passed as correct, and the mainspring fully wound, the balance should have a vibration of not less than a turn and a half nor more than a turn and three-quarters with the dial up, and not less than a turn and one-eighth with the pendant up.

If with the pendant up the arcs are only a turn in extent a stronger mainspring should be tried, and this remedy will probably prove effective ; of course, if a balance too large or heavy has been selected, a fresh balance must be obtained and the watch resprung.

Assuming the vibration to be satisfactory, the watch may be brought somewhere near to mean time—say, to within two minutes a day—by, if necessary, adding to or taking weight from the balance.

Bonniksen recommends springing the watch so that there is at this stage a loss of four minutes a day to allow for weight to be taken from the balance in poising it, but this is not the usual practice.

Then the compensation should be proceeded with and almost perfected. If tests for isochronism were attempted first variations due to changes of temperature might easily be mistaken for want of isochronism. The compensation cannot be quite concluded at this stage, because subsequently poising the balance may slightly derange it. On the other hand, it is useless to think of finally poising the balance now, because re-arrangement of the screws and a possible set of the balance during the compensation adjustment will probably affect the poise. But the fact of the balance being untested for poise renders it necessary to provide *perfectly level* surfaces in the refrigerator and oven on which to lay the watch so that want of poise will not affect its rate.

With the compensation correct, or nearly so, try the balance for truth in the callipers or on the poising tool as directed in the chapter devoted to the subject ; get it as perfect as possible, and then proceed finally to test the poise of the balance and spring in the watch. The mainspring should be wound but little, so as to give only a full turn, or rather less, when pendant up. It is very rarely that the vibration cannot be sufficiently reduced if the stop work is removed. In an extreme case it may be necessary to substitute a weaker mainspring.

Suppose the watch, when horizontal, gains at the rate of 30 seconds a day. Try it pendant up ; its rate is then + 36. Pendant down it is + 16.

It may be assumed that the balance is too heavy at the bottom when the watch is pendant up, and as a ready improvement for a watch which is not to be adjusted in positions, the timing screws nearest this position are proportionately screwed in, and the opposite ones similarly withdrawn.

But clearly the heaviest part of the balance may not be exactly opposite the pendant, therefore let us try the quarters :—

We find with pendant right the daily rate is +32.

    ,,   ,, pendant left   ,,   ,,   + 21.

Probably the quarters would not be so close on account of constructional errors, and the possible presence of these renders it unsafe to accept the deviation as due wholly to want of poise in the balance.

Therefore wind the watch fully, so that the vibration is quite a turn and a half, and try round the vertical positions again.

    Pendant up, its daily rate is + 40.

    ,,   down,   ,,   ,,   + 20.

    ,,   right,   ,,   ,,   + 35.

    ,,   left,   ,,   ,,   + 25.

It is, again, hardly likely there would be no other disturbing elements, but accepting these figures as an example, we may ascertain, if it is an open face watch, that the balance when at rest is too heavy between the IV. and V. This may be adjusted by screwing in a timing screw, or nut, at this point, or by proportionally screwing in the adjacent ones.

Absence of Isochronism.—A spring with theoretically perfect terminal curves would be correct if applied to a balance also perfect, and not subjected to external influences. But such conditions do not, of course, exist in watch work.

An external force applied to the balance at the dead point, that is, when the spring is quiescent, has no effect

ON THE TIME OF OSCILLATION, AND THE EFFECT OF A FORCE IS MUCH ACCENTUATED AS THE DISTANCE BETWEEN THE DEAD POINT AND THE POINT OF APPLICATION INCREASES.

AS A RULE, THE DISTURBING EFFECT OF A GIVEN EXTERNAL FORCE WILL VARY ACCORDING TO THE AMPLITUDE OF THE OSCILLATIONS, AND WILL BE MUCH MORE MARKED AS THE AMPLITUDE DIMINISHES.

AN IMPULSE DELIVERED BEFORE THE BALANCE REACHES ITS DEAD POINT QUICKENS THE TIME OF VIBRATION.

AN IMPULSE DELIVERED AFTER THE BALANCE HAS PASSED ITS DEAD POINT EXTENDS THE TIME OF VIBRATION.

SIMILARLY, ANY ABSTRACTION OF FORCE, IF TAKING PLACE BEFORE THE DEAD POINT IS REACHED, EXTENDS THE TIME OF VIBRATION, WHILE ABSTRACTION OF FORCE AFTER THE DEAD POINT IS PASSED QUICKENS THE VIBRATION.

IN THE CYLINDER ESCAPEMENT, AND IN THE LEVER ESCAPEMENT ALSO, THE IMPULSE IS NOT EQUALLY DIVIDED ON BOTH SIDES OF THE DEAD POINT, THEREFORE, SO FAR AS THE INFLUENCE OF THE ESCAPEMENT GOES, ALL CYLINDER AND LEVER WATCHES MUST LOSE IN THE SHORT ARCS.

Variations between Hanging and Lying. Summary of Causes and Remedies.—Various escapement faults ; excess of shake in pivot holes, or insufficient freedom in pivot holes ; escape wheel too heavy ; bad train depths ; escapement out of beat.

Variations between Dial up and Dial down.—Balance spring badly pinned to collet. Any twist in the spring here is almost sure to cause variation in these positions. Probably the hole in the collet is not exactly in the proper direction, and the best plan is to have a new collet.

Want of uniformity in escapement end shakes, causing the end of the impulse pin to graze the safety finger of the lever ; or the balance to graze the escape cock.

If the impulse pin is not perfectly upright, it may in one of these positions be free as it passes in and out of the horns, and in the other positions may graze.

A short balance-staff pivot may in one position just rest on its conical shoulder.

A defective or unevenly set end stone.

Excessive or careless oiling may cause a drop of oil on some part of the escapement to spread and create sufficient adhesion in passing an adjacent surface to retard the balance and diminish its vibration. See especially that contact between the lever and banking pins is free and not sticky from oil.

Variations in the Quarters. Positional Errors.—Want of poise of the lever and pallets.

Too little draw, causing the safety pin or finger to rub on the roller ; this may occur only in one position, owing to want of poise in the lever and pallets.

Extra Long Arcs Defective.—If the extra long arcs are fast, the fault may be that the impulse pin knocks the outside of the lever. To correct this a weaker mainspring may be substituted for the existing one.

If the balance spring is pressed out of its proper position by one curb pin, though there is play between the two, it may be that the spring leaves the curb pin and becomes free in the extra long arcs, causing a loss.

The balance, when expanded to its fullest extent through natural action in low temperature and by centrifugal force, may just touch some part of the mechanism.

Cylinder watches and carriage clocks with this escapement, after cleaning, will often so increase in vibration that a banking error is introduced. A weaker mainspring is one remedy, but the matter is more fully dealt with under the head of Cylinder Escapement.

In a cylinder watch which has been cleaned a sticky banking will sometimes give trouble ; brushing is often not sufficient to remove the holding tendency. Both pin and block should be scraped.

Isochronism.—The short arcs may be quickened by rectifying faults in the escapement.

By increasing the angle of draw on the pallets.

By altering the overcoil of the balance spring.

By altering the point of attachment of the spring to the collet.

By centrifugal force, *e.g.*—

A small hole is drilled through each of the half rims of the balance close to their fixed ends, and, after trial, broached out till the desired equality is obtained. This remedy was, I know, successfully adopted by the late E. B. Denison. The greatest care must be taken to broach the holes alike, for if one half of the rim is weakened more than the other the balance will be out of poise when moving quickly. This method is open to the objection that the balance is rendered less rigid and more likely to be bent.

By using a shorter balance spring.

A shorter balance spring to give the same number of vibrations will, of course, be *thinner*, and to reach the hole in the stud will have more open coils.

If there is play between the curb pins, close them till the spring is but just free ; there should be no apparent motion of the spring when its action is viewed through a glass.

By using a stronger mainspring.

As already explained (p. 37), the short arcs may be quickened by putting the balance out of poise, but this must be regarded as an illegitimate method, because the remedy is but ephemeral.

Making and Polishing Balance Springs.—In making balance springs the first requirement is good wire. After being drawn to suitable sizes, the round wire is flattened between rollers and then drawn through jewelled clams, or rollers, to ensure uniformity of substance. The most general fault is that, during the frequent annealing necessary in drawing it so small, the steel has lost too much of its carbon, and is difficult to harden except at an excessive temperature. Before making balance springs it would be prudent to test a piece of the wire for hardening and tempering. If the wire is suitable it should harden well after heating to what is called a cherry red, that is, before it attains a brilliant red hue. Flat spiral springs are coiled up in a circular box, like a small watch barrel, and cover (see Fig. 17), made preferably of aluminium-bronze, or of platinum. Copper and German silver are sometimes used, but platinum and aluminium-bronze are found to best retain their form after heating. Obviously, the top and the bottom of the box must be perfectly flat, if flat springs are to be produced. Though rigid, the box and cover must be as thin as possible, because, if a considerable body of metal surrounds the spring it has to be raised to a greater heat to harden. Either two, three, or four springs are coiled at a time, generally three ; if coiled up four at a time the coils would be more open than usual ; springs for overcoils are generally closer, and would be coiled two at once.

The following is substantially the process described in the *Horological Journal* by Mr. Glasgow :—" The winder (Fig. 18) is

Fig. 17.   Fig. 18.

of steel, with a pivot formed on the end to pass through a hole in the bottom of the box, and project into the box just the width of the wire of which the springs are to be made. Across the pivot of the winder are slits to receive the ends of the wire. For springs, coiled up two together, there is one straight slit across the centre of the pivot. If three springs are to be coiled up together, three equidistant radial slits are cut (as at *a*, Fig. 18), and for open springs coiled up four together, two slits at right angles across the centre are made. The pivot should be snailed from the slits according to the wire to be used, so as to ensure the truth of the eyes of the springs. A hole is drilled and tapped in the centre of the pivot to receive a small screw, the head of which passes through a hole in the cover. Holes not radial, but almost tangential, are drilled through the side of the box to thread the wire through. The ends of the wire passed into the slits are secured by the screw, the cover is gently pressed on, and the winder rotated till the box is full. The ends of the wire are cut off, the screw and winder removed, and the cover bound tightly with wire. The holes are then stopped, usually with a mixture of soap and animal charcoal, to exclude the air, and the box put in an open iron vessel containing charcoal, which has been placed in the fire of an ordinary grate." Animal charcoal is usually prepared from burnt bones or burnt leather, reduced to powder. The object of using it in this connection is to further enrich the steel, but some discretion is necessary, for if the steel is already sufficiently rich in carbon it may be rendered too brittle. Difficulty would then be experienced, especially in bending overcoils, if such were needed. If the steel is found to harden readily at a proper temperature, vegetable charcoal may be used to exclude the air. If soap is used it should not be wetted, but softened by heating. Either beeswax or oil appears to be a more suitable vehicle than soap. When the spring box is cherry red, it is dashed into a vessel containing plenty of cold water. The spring box is then placed in a boiling out pan or an old metal spoon filled with oil, and held over a flame till the flame just flickers over the oil.

After cooling, the springs may be turned out of the box. If they do not come apart quite readily they should not be forced, as they would be likely to be permanently distorted thereby. If thrown into a pill box and rattled for a minute, the slight concussion against the sides of the box is usually enough to induce separation unless the springs have been burnt. Water or oil is the medium generally selected for plunging steel in to cool it when it has been heated to the requisite redness for hardening. Either mercury or petroleum may be recommended if extra hardness is desired. Salt water will give great hardness, but the steel is rendered brittle. Oil and mercury are considered the two best media for hardening steel if toughness is desired. If the steel has been protected from the air, and not overheated ' during the process of hardening, its surface will not be scaled nor materially injured, and its brightness may be restored by polishing, but if the surface is intricate, as is the case with balance springs, a quicker method of cleaning may be employed. The steel may be washed in a strong solution of hydrochloric acid (about one-third of pure acid to two-thirds water is recommended), and immediately afterwards rinsed in a saturated solution of cyanide of potassium. In one respect springs finished with a high polish are inferior to others left grey, for they are more liable to rust.

To avoid oxidation of the surface during hardening and tempering the springs, melted crystals of potassium cyanide may, with advantage, be employed as a heating medium. The cyanide having been placed in a wrought-iron pot on a stove, and raised to a red heat, the spring box is immersed, and allowed to remain till it becomes of the temperature of the bath, when it is removed and plunged into cold water. A film of cyanide clings to the exposed surface, when the box is removed from the bath, and so the air is entirely excluded from contact with any part of the springs. This protecting film leaves the surface when the box is plunged in the water.

To temper the springs, the box on being removed from the water may be again momentarily plunged in the cyanide bath, and then subjected to an even heat in a small stove or iron box. When the film of cyanide peels off the box it is considered the desired temper is obtained.

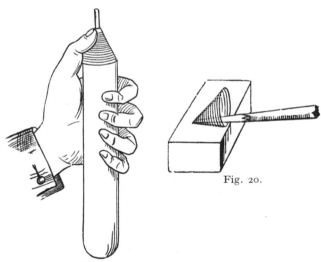

Fig. 20.

Fig. 19.

This is, I believe, the process adopted by some of the Swiss spring manufacturers, and in America. Of course, wire of uniform quality must be selected to be sure of always attaining the desired temper. The difficulty generally is to ensure uniformity of quality, and for this reason no exact tempering temperature for balance springs can be given, because the steel varies in quality. Continual care is necessary to ensure success. Mr. Arthur Webb, on taking a fresh bobbin of wire, hardens a piece, and then finds the temperature at which that particular wire yields the desired temper, by placing it in the oil bath with a thermometer. When satisfied with the temper he notes the temperature which he assumes will be the best for tempering so long as that wire lasts.

Steel is less oxidised by tempering in an alloy than if tempered in the air, and the required temperature is obtained with much greater certainty. Olive oil, which boils at 600°, is now most generally used as a bath for tempering balance springs. But great care must be taken to ensure the purity of the oil. Olive oil adulterated with cotton seed oil, when used as a tempering bath, has been found to rust springs, possibly from the presence of some chemical employed in the preparation of the oil. Pure olive oil may be used with confidence. It may be heated in a wrought iron pot, which must not be tinned. The cover may have a hole in it for the insertion of a thermometer, and hooks on the underside from which the springs may be hung.

Springs are more liable to distortion when perfectly hard than when tempered. No explanation appears to have been given of this curious fact, but mention of it may prevent a beginner attempting to polish watch springs before tempering.

The most usual method of mechanical polishing is as follows :—

The flat faces are polished by gently pressing the tip of the finger on the spring, and moving it in a circular direction on a piece of writing paper on which red stuff has been rubbed.

The spring is drawn down over a piece of wood with a conical end like Fig. 19, with a pin at the apex for the eye to go over, and the outsides are polished with a well-worn brush charged with red stuff.

The inner sides are much more difficult to polish ; but if the wire is good, and has not been raised to an excessive heat in hardening, and if the air has been properly excluded during that operation, but little polish will be needed. The spring is placed on a flat piece of cork, and the inner sides rubbed to and fro from side to side with a finely-pointed peg charged with red stuff, the spring taking the form of a cone the while, as shown in Fig. 20.

After polishing, the spring is washed in PERFECTLY CLEAN benzine, and blued on a flat plate. The plate should be heated before the spring is brought into contact with it and care taken that no dust is allowed to settle on the spring. A glass tube may be placed over it during the operation.

An alternative method of polishing flat springs on the surface
and between the coils at the same time is to place the spring on
an india-rubber pad under two plates with radial sides terminating
in a point, as in the subjoined sketch, Fig. 21. Over the spring
is a holder on a universal joint. Through a hole in the holder
passes freely the polisher spindle, carrying at its lower end a
thin strip of willow wood cut almost to a knife edge where it
bears upon the spring. A little pressure on the knob at the top
of the polished holder will cause the coils of the spring to make
serrations in the edge of the willow. Charged with a little fine
emery the polisher is now pressed on the spring and rotated as
far as the radial edges of the plates will allow for, say, six times
backward and forward; the position of the spring is then
altered so as to bring the part at first under the plates into
contact with the polisher and treated in like manner. The

Fig. 21.

spring is then turned over and the other side operated on. The
india-rubber pad is on a circular metal table, and underneath the
tool is a bridge; one end of a helical spring around the stalk of
the table presses against the underneath surface of the table
and the other end against the bridge, and so keeps the spring on
the rubber pad sufficiently tight against the plates. Care must
be taken not to rub the spring more than is necessary and to
subject each part of the surface to the same number of rubs.
Excessive and indiscriminate rubbing will spoil the spring.

Fig. 22 shows a different arrangement for winding springs,
for which I am indebted to Mr. R. B. North. Here the winding
machine stands on the bench, and the centre arbor is actuated by

turning a handle at the side with which it is connected by a pair of mitre wheels. The spring box without the cover is placed on the table of the machine and prevented from turning by a notch at the end which engages with a pin in the table. At the end of a hinged arm is pivoted a little bar which may be pressed on the wire while it is being wound in. In this way the winding is under perfect control, and can be watched as it proceeds. When the box is full the cover, with a hole to pass over the centre arbor, is placed in position and fastened with three screws. There is a special feature in the box which is worthy of mention. It has a loose disc or false bottom which can be readily trued or a fresh one substituted. The cover is flat, and does not enter the box, therefore narrow wire would not fill the space between the disc and the cover, but the central hole in the bottom of the

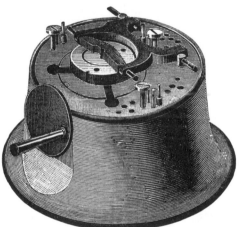

Fig. 22.

box is larger than the hole in the disc and tapped, so that when the cover is screwed on and the centre arbor removed a screw can be inserted to press the disc till the whole surface of the wire is in contact with both the disc and the cover.

J. F. Cole used instead of a spring box a square plate, rising from which were four half-round pins instead of a rim, keeping the wire within the required limit.

Helical Springs are formed by winding the wire on a thin solid drawn tube of brass or aluminium-bronze. If the same tube is to be used for tempering it has shallow grooves cut to the pitch the spring is to be, and in depth about half the thickness of the wire. The ends of the spring are fastened to the tube under the heads of brass screws which must be made left-handed, so as to draw the wire tighter on the tube as they are screwed up. The tube is mounted on an arbor in the turns ; one end of the wire is fixed by one of the screws, the remainder of the wire hangs freely in the air, a weight of about 12 ounces being fastened to the lower end to keep the wire taut. A piece of thin sheet platinum or sheet copper is wrapped round the spring and tube and kept close with binding wire prior to hardening. It is hardened by being heated to redness and plunged into cold water. Immediately it is withdrawn from the water one of the screws is slackened, the spring drawn tight on the tube and held again in position by the screw. It is then tempered by boiling oil.

A grooved tube is not essential for coiling the spring. Mr. Mercer uses a plain thin tube of solid drawn brass or aluminium-bronze. The wire is coiled closely, covered with a thin sheet copper wrapper on the other side of which has been smeared either beeswax or a little lard ; the ends of the copper are tucked over the ends of the tube, which is bound round with wire. A light wire holder, sufficiently long to protect the operator from the heat, is attached, and the tube is subjected to a charcoal fire heated by gas. The operator turns the tube, watching till it attains a cherry red, when it is dashed into mercury, which Mr. Mercer prefers as a hardening medium. During hot weather the pot containing mercury should be surrounded with cold water or the steel may not harden satisfactorily. After hardening, the spring is tempered on a grooved tube. The lard should be melted fresh pork. Lard, as purchased, is liable to contain salt or other impurity which might engender rust.

The outsides and edges are polished on a block a little larger than the tube used for coiling the springs. For polishing the inside, the spring is mounted on a piece of wood charged with red-stuff and rotated in the turns, the spring being held between the thumb and the finger the while. Before bluing, the spring is thoroughly cleaned with soap and water, and afterwards with benzine. Unless the spring is perfectly clean it will not blue evenly. For bluing, the springs are secured by screws to a block with very shallow grooves very similar to the tube used for tempering, except that it is solid and a shade larger in diameter. The block is placed end upwards on a bluing pan, which is heated over a spirit lamp. The spring may be encircled with an open glass tube to keep the temperature uniform, and keep off dust which might make the spring specky.

Mr. T. Hewitt's procedure in coiling and hardening is a little different. He turns down his piece of solid-drawn brass tubing, leaving the surface plain without grooves. On this he winds the wire, the coils following each other close together. He smears

D

a mixture of oil and vegetable charcoal on the coiled wire, and over all slips another thin brass tube, stopping the ends of the two with a luting of the oil and charcoal. It is then hardened. After hardening, the spring is wound on to another brass tube having shallow grooves of the desired pitch and of exactly the diameter at the bottom of the grooves as the hardening tube was, fixed with left-handed screws and tempered. Tempering sets the spring satisfactorily to the pitch decided on.

The late F. Knudsen, a very successful chronometer springer, wound his springs on a plain tube, and heated them for hardening by hanging the tube on a pin in a piece of charcoal and directing a flame through the tube with a blow-pipe. After plunging in water he heated the tube to nearly the ultimate tempering heat to avoid breaking the spring in loosening the screw. He set the springs by tempering them on a grooved tube, and after polishing finally blued them on a plain block, the spring being secured at one end by a screw.

After helical springs are tempered the greatest care is taken to avoid rust and stain. I have heard of springers conducting subsequent operations necessitating handling, with stalls of oiled silk on their fingers to ensure that when the terminals of a spring are formed, and it is pinned on, the even colour remains without blemish. This, though, is not the usual practice, and is, I should say, not necessary unless the operator has very damp hands. Indeed, rust is generally the result of dirt or impurity met with in course of manufacture.

Balance Spring Tester.—This gauge, designed by Mr. T. Hewitt, is invaluable to the springer of high-class timekeepers. Dr. Hooke, the inventor of the balance spring, in explanation of its isochronous property, gave utterance to the now well-worn maxim, *ut tensio sic vis*. (As the tension is, so is the force.) And in theory this is true of springs perfect in construction, though it sometimes happens that a spring apparently faultless contains some latent defect that renders its action not regularly progressive. With the tool shown in the engraving, the behaviour of a spring under different degrees of tension may be tested before it is applied to the timekeeper, and if it prove to be bad it may be at once rejected, and much labour saved that would be otherwise thrown away. The eye of the spring is held in a spring collet, and its outer end caught in a split nose at the end of the arm on the left. This arm is held in a split bearing, and is capable of adjustment to suit springs of various kinds. The circular scale is divided into degrees and into radians, and turns friction-tight on its axis so that it may be readily set to zero. By means of the knurled knob at the back of the tool the spring may be wound up, weights being gradually added to the scale pan to balance it. A fair average of the extremes of winding a balance spring would be subjected to in the long and short vibrations of a watch, would be from three to four radians, and if for every tenth of a radian between these points the same increment of weight were required, the progression of the force of the spring under trial would be regular. The bending moment of any spring may be ascertained, and in the case of substituting one balance for another in order to obtain increased vibration, the relative moments of inertia of the old and the new balance may be readily found. The central staff holding the eye

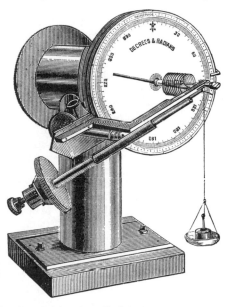

of the spring is very nicely fitted into jewelled holes. It carries the lever to which the scale pan is hnng at exactly 1 inch from the centre of motion, and then the bending moments, &c., are very readily calculated in inch-grains.

Balance Spring Buckle or " Guard."—[*Clef de spiral.—Der Spiralschlüssel.*]—A small stud with a projecting tongue attached to the index arm so as to bridge the curb pins, and prevent them engaging two coils of the balance spring. It is used chiefly in Swiss watches.

Balance Staff.—[*Axe de balancier.—Die Unruhwelle.*]—The axis of the balance. In the best lever watches, the balance collet is solid with the staff. For commoner work, the collet is of brass, and fitted tightly on.

Replacing Balance Staff.—In running in a balance staff it will be prudent to note if the balance seat and escapement roller are correctly placed in the old staff, or in what direction alterations would be desirable. Remove the spring-collet and roller, and in the case of a colleted staff drive the staff out of the collet. With a solid staff, if the balance rivet is much spread, turn it away before attempting to punch out the staff. For a colleted staff centres may be turned or filed on a piece of steel, which may then be finished in the turns. For a solid staff, those who have not a lathe will avail themselves of the rough staffs to be obtained at the tool-shops. One of these may be hardened and tempered and finished in the turns. A workman who has a lathe will prefer to mount a piece of plain steel in a split chuck or in some other convenient way, and turn it all over nearly to size, with conical centres on the ends. It may then be hardened and tempered and finished in the turns. Mark the position for the balance seat, reduce it carefully till the balance fits well, and be sure the shoulder of the seat is flat and square ; if left hollow the balance will probably be bent or will work loose. Turn the part that fits the balance to height and form the undercut for the rivet. Let on the spring collet and the roller, leaving the staff a little taper so that each piece passes freely to its place and fits there. Rest the balance collet on a staking tool and rivet on the balance with a hollow punch having a rounded face. Now measure from the balance seat to the end of the top pivot with the self-registering callipers or other gauge. Polish the angular face of the flange, and finish off back slopes and pivots as advised under the head of Polishing, allowing a sufficiency of end shake from the measurement shown in the staff gauge.

Balance Wheel.—[*Roue de rencontre.—Das Gangrad einer Spindeluhr.*]—The escape wheel of the verge escapement. This term is often applied by amateurs to the " balance " proper.

Banking.—[*Rebattement.—Prellen.*]—In a lever watch, the striking of the outside of the lever by the impulse pin owing to excessive vibration of the balance. In a cylinder or verge, the striking of the pin in the balance against the fixed banking stud or pin.

Banking Pin.—[*Goupille de renversement.—Der Ausschwung-stift.*]—A pin for circumscribing the motion of the balance in verge and cylinder watches.

Banking Pins.—[*Goupilles de renversement.—Die Prellstifte.*] —(1.) In a lever escapement, two pins that limit the motion of the lever.

(2.) In a pocket chronometer, two upright pins in the balance arm which limit the motion of the balance spring.

(3.) The curb pins that embrace the balance spring of a watch are occasionally spoken of as banking pins.

Banking Screw.—[*Vis pour arret de detente.—Die Prell-schraube.*]—An adjustable screw in the chronometer escapement, the head of which regulates the amount of locking by forming a stop for the pipe of the detent.

Barleys.—[*Grains d'orge.—Das Korn im Guilloehe.*]—The little projections formed by the operation of engine-turning.

Barlow, Edward.—A clergyman who invented the rack

striking work for clocks in 1676. With this mechanism clocks could be made to repeat the hour at will, and its popularity on this account led to the introduction of repeating watches a few years later. Barlow and Quare both applied for a patent for repeating watches, and the Government decided in favour of the latter in 1687.

Bar Movement.—[*Mouvement à ponts.—Das Klobenwerk.*]— A watch movement in which the top plate is omitted and the upper pivots of the movement are carried in bars. A bar movement is sometimes called a " skeleton " movement.

Barometric Error.—[*Erreur barométrique.—Der Luftdruck fehler*.]—The alteration in the timekeeping of a clock due to changes in the density of the atmosphere through which the pendulum has to move. The barometric error is very small. It has been stated to be about a third of a second a day for a change of one inch in the

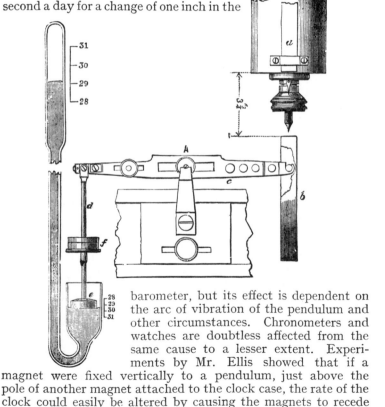

barometer, but its effect is dependent on the arc of vibration of the pendulum and other circumstances. Chronometers and watches are doubtless affected from the same cause to a lesser extent. Experiments by Mr. Ellis showed that if a magnet were fixed vertically to a pendulum, just above the pole of another magnet attached to the clock case, the rate of the clock could easily be altered by causing the magnets to recede from or approach each other, as the attractive power of dissimilar poles caused it to gain. Taking advantage of this fact he devised a barometric compensation (as shown in the engraving) for the standard sidereal clock at the Greenwich Observatory, where, I believe, it answers admirably. Two bar magnets, each about six inches long, are attached to the pendulum bob, one behind and one (*a*) in front, with their dissimilar poles towards a horseshoe magnet (*b*) carried by a lever resting at A on knife edges, so that the horseshoe is *always* attracting the pendulum and increasing the acceleration due to gravity, having of course its least effect when most distant. The poles of the horseshoe are exactly under the bar magnets and about 3.75 inches below them. At the other extremity of the lever is a rod (*d*) carrying a float (*e*) which rests on the mercury in the short leg of a barometer, as shown. The area of the cistern part of the short leg is four times the area of the upper part of the barometer tube, so that a variation of one inch in the barometric pressure would affect the height of the mercury in the cistern but ·25 of an inch. As the clock lost with a rising barometer, the bar magnet over the south pole of the horseshoe magnet is placed with its north pole downwards, and the bar magnet over the north pole of the horseshoe magnet with its south pole downwards, so that there should be attraction between the bar magnets and the horseshoe magnet. The bracket supporting the knife edges can be shifted to increase or diminish the action of the magnet, and the lever is balanced by placing weights in the pan *f*.

Barrel.—[*Barillet.—Das Federhaus.*]—A circular box for the reception of the mainspring of a watch or clock. (See Mainspring.)

Barrel Arbor.—[*Arbre de barillet.—Der Federstift.*]—The axis of the barrel, round which the mainspring coils.

In the absence of a suitable tap or screw plate when turning in a Geneva barrel arbor, if the collet is good it may be used as a plate. Soften the collet, and file two slight passages across the threads with a fine three-sided file ; screw a piece of brass wire through the collet so as to clear the threads of burr ; then re-harden the collet, and cut the screw on the arbor with it. A pair of pliers with faces curved to suit the collet are used to hold it. In an emergency the old arbor may be prepared for use as a tap if the old collet is not available.

Barrel Arbor Turnscrew.—[*Bonde a vis.—Der Federkern.*

schlüssel.—The drawing sufficiently explains the nature of a handy adjustable turnscrew for barrel centres of Geneva watches by Mr. John McKay.

Barrel Contractor.—[*Outil à resserrer les serges de barillet.—Das Werkzeng zum Verengern des Federhauses.*]—Boley's tool for restoring distorted mainspring barrels to shape, consists of a die with a series of tapered holes, as shown in Fig. 1, and corresponding punches. The defective barrel is placed in a hole of the proper

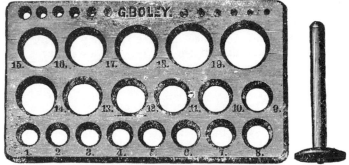

Fig. 1. (Barrel Contractor.)

size, as in Fig. 2, and a few light taps from a hammer on the punch quickly will bring the barrel to the desired form, and of a size to suit its cover. It may be desirable to interpose paper between the barrel and the tool, but if the holes have been well polished this is unnecessary. In some instances it will be better to turn out the groove in the barrel afresh, and to shape the edge of the cover to snap in properly. The groove should be slightly undercut, and the edge of the cover formed to suit, but the bevel must not be too sharp, or the cover after snapping in may be loose. Besides its primary use, this tool will be found useful for contracting rings and watch bows. The smaller holes also answer for closing the sockets of hour hands, the collets of carriage clock hammers and such like pieces.

Barrel Cover.—[*Couercle de barillet.—Der Federhausdeckel.*]—A lid that snaps into a rebate in the barrel.

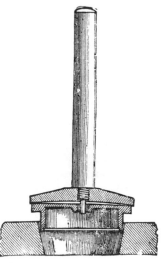

Fig. 2. (Barrel Contractor.)

Barrel Hole Closer.—See Riveting·Stake.

Barrel Hollow.—[*Noyure du barillet.—Die Federhausaus-drehung.*]—The sink, cut either in the top plate or the pillar plate of a watch to give freedom to the barrel.

Barrel Hook.—[*Crochet de barillet.—Der Federhaushaken.*]— A pin in the barrel to which the mainspring is attached. Sometimes the attachment is made by a hook on the mainspring fitting into a hole in the barrel. (See Mainspring.)

Barrel Ratchet.—[*Rochet de barillet.—Das Sperrad (Feder-haus).*]—A wheel on the barrel arbor kept from turning back, when the mainspring is wound, by a click pivoted to the watch plate or bar, which thus becomes the resisting base for driving the train. By means of the rachet the lower turns of the mainspring may be set up as required for adjustment.

In going-barrel watches there should be some provision to allow a slight recoil of the ratchet after winding, and where the stopwork is omitted this is essential to avoid banking error after the spring has been tightly wound. Sometimes a little play is given to the screw in the click hole for this purpose, but many

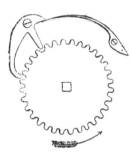

special forms of clicks have been devised to permit a slight recoil of the mainspring. Among the best is the one by Mr. Walsh, which is here shown. When the rachet wheel is turned in the direction of the arrow, as in winding, the click proper is lifted out of the wheel by the straight arm, and when the wheel is released, it goes back before the click is engaged. The amount of this recoil can easily be regulated by the amount of the intersection of the straight arm with the wheel.

Bascine Case.—[*Boîte bassine.—Abgerundete Form des Gehäuses.*]—A form of watch case in which there is no bead or projection on the outside of the cover.

Bastard Cut.—[*Lime bâtarde.—Mittelfeiner Feilhieb.*]—A file between rough and smooth.

Bearer.—[*Contre-charnière.—Der Scharnierträger.*]—A piece of metal soldered to the " middle" of a watch case as a support for the joint.

Beat.—[*Coup.—Schlag.*]—The blow of the escape wheel as it falls on the pallet or locking surface.

Beat Pins.—[*Die Abfall. Einstellschrauben (Pendel).*]—Small screws to adjust the position of the crutch with relation to the pendulum. In a gravity escapement the pins at the end of the gravity arms that give impulse to the pendulum.

Beckett.—(See Denison.)

Bell Crank Lever.—[*Levier à équerre.—Der Glockenzug-hebel.*]—A lever whose two arms form a right angle.

Bell Punch.—[*Poinçon à centrer forme entonnoir.—Der Mittelpunkttreffer.*]—A centring punch set in the centre of a conical shield. If the shield is placed over the end of a rod, and

the punch struck, the centre of the rod is accurately marked. This is especially useful for centring pieces that have to be turned in a lathe with male centres. The Bell Punch shown in the figure has a spring to return the punch.

Bench.—[*Etabli.—Der Werktisch.*]—An excellent arrangement of a watchmaker's bench and lathe by Boley is shown in the cut below. His stool, capable of adjustment to height, is very con-

venient. Mahogany is the material most often used for a bench ; but birch, beech, and many other kinds of wood are suitable. Cedar should never be used either for bench or drawers, for it exudes a sort of gum, which forms a sticky deposit on work and tools, and rapidly spoils any lubricating oil that may be exposed to it. Some watchmakers sit and some stand to their work. Others just rest the body on the edge of a high stool and assume an attitude between sitting and standing. There is no doubt that continuous sitting is bad, and a good medium height for a bench which will allow of either sitting or standing, and be right for working at the lathe is 3 feet 7 inches, and if the bench is about 4 feet in length, and 22 inches in width, it will afford ample room. The stack of drawers on the right hand should be 13 or 14 inches from front and back, and graded in depth, the deepest at bottom as shown. The top three may with advantage have divisions running diagonally from the front right-hand to the back left-hand corner. These will be found handier for picking up a tool than the more common rectangular partitions.

Benzine.—[*Benzine.—Das Benzin.*]—A hydrocarbon obtained from coal tar. It is an absorbent of grease, and is used for cleansing watch work. It should not be used close to a gas or lamp flame. A glass pot like the sketch is used to contain the benzine for immersing the article in. The cover should be well fitted, or the benzine will evaporate. Dirty benzine may be again rendered fit for use by filtering it through animal charcoal. Mr. Arthur Webb stops the end of an ordinary pint glass funnel with paper, and presses therein calcined bone dust till it is about three-parts full. On the top of this is poured the benzine to be filtered.

Bezel.—[*Lunette.—Der Glasreif.*]—The internally grooved ring of a watch or clock case that contains the glass.

Bezel Clearer. —This is for clearing the groove of a bezel to facilitate the fitting of a glass. There are two guides and one file-like cutter. When the guides are adjusted to the groove the cutter is drawn in by means of the screw and the tool rotated. It is recommended that  the bezel be shut when the clearer is used.

Berthoud Ferdinand (born 1727, died 1807), an eminent French watchmaker, author of *Essai sur l'Horlogerie, Traité des Horloges Marines, Histoire de la Mesure du Temps*, and other works containing a mass of useful information concerning the history, theory, and practice of the horological art, dealing with Harrison's, Sully's, and Le Roy's inventions, and, indeed, everything known in Berthoud's time.

Blind Man's Watch.—[*Montre pour les Aveugles.—Uhr für Blinde.*]—A watch in which the progress of the hands may be ascertained by touch.—One of the most simple plans is to put a short hour hand between the inner and outer bottoms of the case, and twelve knobs on the inner bottom arranged in a circle just outside the path of the hand. The subdivisions of an hour can be very closely estimated after practice from the position of this hand. A more usual device is to place the knobs round the band of the case, and to have a pointer on the outside of the outer bottom. This pointer, pivoted at the centre of the outer bottom, does not travel with the going of the watch, but may be moved easily till its position coincides with the hour hand of the watch, when its further progress is arrested. There is a large motion wheel turning once in twelve hours above the top plate, carrying a finger arranged like the passing spring of a chronometer detent, and it is this finger that catches a stop on a prolongation of the arbor of the pointer when the latter is turned the way the watch is going. But when this finger *follows* the stop on the pointer, as it does in the going of the watch, it passes easily without obstruction. The objection to this form of touch watch is that if the pointer is pressed hard against the finger, it is apt to advance the hands of the watch.

Another plan is illustrated below. There are knobs on the band of the case for the hours, with the spaces between divided into intervals representing five minutes each. The dial of the watch is somewhat smaller than usual, and outside of it is a movable ring having a little pointer fixed to it. The edge of the dial has fine teeth cut round it, and pivoted to the under  side of the movable ring is a click to take into these teeth, but kept away from them by a very weak spring. The hour hand extends a little beyond the dial, and if the ring is turned round till the click meets the end of the hour hand, the hand sends the click into the teeth, which then take any pressure which may be put upon the ring.

Blowpipe.—[*Chalumeau.—Das Lötrohr.*]—In its simplest form a tapered tube, from the smaller end of which a stream of air is projected into a flame, serving the double purpose of supplying it with oxygen and directing it to some particular

point. The larger end of the blowpipe is held in the mouth, and the intensity of the current

Fig. 1.

may thus be regulated with great nicety, which is of much importance in soldering. The blowpipe shown in Fig. 1 has a hollow bulb introduced to arrest any moisture from the mouth. In another arrange-ment air is alternately admit-ted into and ejected from an indiarubber ball squeezed by the hand. Fig. 2 shows Bar-thel's soldering lamp, which is

Fig. 2.

automatic, with but one spirit reservoir, and most effective. For melting metals a bellows worked by the foot is often used to supply the blast to the blowpipe.

Bluestone.—[*Pierre à eau.—Der Wasserstein.*]—A fine grit soft stone used for giving a surface to brass before polishing.

Bluing.—[*Bleuir.—Das Anlassen.*]—The process of changing the colour of steel by heat. The articles to be blued are placed in a bluing pan, which is held over the flame of a lamp and protected from draughts or currents of air. The bluing pan should be well warmed to disperse all moisture before the work is placed on it. Many pieces to be blued are not sufficiently flat on the under side to become equally coloured on a flat pan. If fine brass filings are laid thickly in the pan, and the part to be blued is pressed into them, the colour will then be more even. (For bluing temperatures, see Tempering.)

Steel for bluing should be finished with medium red stuff, and the last few rubs given not with a metal polisher, but with a piece of boxwood or horn and plenty of red stuff and oil. Pieces that have been cleaned in dirty benzine (*i.e.*, benzine charged with oil) will become specky in bluing.

Difficulty is often experienced in bluing soft screws, but Mr. Kullberg tells me a good even colour may always be ensured if they are finished with a slightly soapy burnisher.

Bluing Pan.—[*Revenoir.—Das Anlassblech. Die Anlass-pfanne.*]—A piece of thin copper, shaped something like a frying-pan, in which steel articles are placed to be blued.

For bluing screws, an old barrel with holes drilled through the bottom is used. This is placed bottom upwards in the bluing pan, and forms a holder for the screws, the holes being of such a size as to allow the thread to pass freely through.

Bob.—[*Lentille.—Die Pendellinse.*]—The mass of metal at the bottom of a pendulum.

Boiling-out Pan.—[*Casserole.—Der Auskochtiegel.*]—A copper vessel shaped like a stewpan. It is used principally for tempering steel articles by boiling them in oil, and for removing cement from pieces of watchwork by the same means.

Bolt and Joint.—[*Charnière de mouvement.—Scharnier und Schlussfeder.*]—The contrivance for attaching a watch movement to a double bottom case. The joint is a hinge formed between the pillar plate and the middle of the case, and the bolt is a spring catch fixed to the pillar plate so as to snap into the middle of the case opposite to the joint. To get at the movement, the extremity of the bolt which projects above the dial is pressed back by the thumb-nail.

Bolt and Shutter.—[*Entretien à volet.—Das Aufzugsvorgelege (Turmuhren).*]—An obsolete contrivance for keeping clocks going while winding. During the going of the clock the shutter —a plate of metal—stood in front of the winding square, so that the winding handle could not be applied. In moving this aside to wind, a click supporting a weight engaged with the great wheel. A modification of the old bolt and shutter introduced by Sir E. Beckett, in which a toothed segment was substituted for the click, is still used in some turret clocks, but is inferior to the " Sun and Planet " and other maintainers.

Bouchon.—[*Bouchon.—Das Futter.*]—Hard brass tubing inserted in watch and clock plates to form the pivot holes.

To Fit Bouchons.—After repairing the pivot a bouchon is selected as small as the pivot will admit, for the smaller the bouchon is, the neater will be the job. Open the hole of the plate or cock so that the bouchon, which should be previously lightly draw-filed at the end, will stand with a slight pressure upright in the opened hole of the plate or cock. Then with a knife cut it across at the part where it is to be broken off, so that it may break very readily when required to do so. Press it in the plate on the side the pivot works, break off, and then drive it home with a small centre-punch. In every repair of this nature, notice should be taken of the amount of end shake of the pinion, and allowance made by leaving the bouchon so that any excess may be corrected. To finish off the shoulder end a small chamfering tool like

the sketch should be used. It has a hole smaller than the pivot one to receive a fine brass wire serving as a centre to prevent the tool changing its position while being used ; or the wire may be put through the bouchon hole, and then the hole of the tool may be left open. The above is a far more expeditious way than using the mandril.

Bow.—[*Archet.*—*Der Drehbogen.*]—(1) A thin slip of whalebone or cane used to rotate the mandril of a lathe, a drill, or an arbor, by coiling a line attached to its extremities once round a ferrule or pulley on the object rotated ; this line, according to dimensions of the work, is of gut, horsehair, or human hair. (2) The ring of a watch case by which it is attached to the watch guard is also called a bow.

Bow Closer.—[*Die Bügelspann Vorrichtung.*]—For closing a keyless watch bow on the pendant, or putting a distorted bow into shape, the hollowed block and

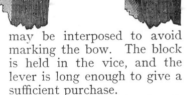

round lever shown answer admirably, being more under control than pliers shaped for the purpose. A bit of leather may be interposed to avoid marking the bow. The block is held in the vice, and the lever is long enough to give a sufficient purchase.

Bow Pivoter.—[*Die Bügelzapfen-Fräsmaschine.*] —When fitting a new bow to the watch this tool is of assistance. There are several rose cutters of different sizes, and a suitable one may be at once selected and brought into position. The ends of the bow are in turn subjected to the cutter, and good pivots of uniform size formed instantly. A spring holder keeps the bow against the cutter.

Broach for Keyless Buttons.—[*Der Kronenfräser.*]— This will be found useful when fitting a new button or crown to a keyless watch, and when from wear the inside of the crown fouls the pendant. The cutting nose is slit after the manner of a split chuck, and the screw at the right hand will, when turned in the right direction, expand the cutter as desired.

Holding the body of the tool in one hand and the button to be opened in the other, a few twirls do all that is needed.

Turner's and Platnauer's Watch Bows.—Mr. Turner and Mr. Louis Platnauer have each patented a better method of attaching the bow to keyless watches than springing the ends of it into the pendant. Necks are formed at the ends of the bow, and in Turner's hollow plug dovetails into the pendant and

is secured by a screw placed where marked with a dot in the left hand figure. In Platnauer's a thin metal cylinder (Fig. 2) with slots terminating in a bayonet joint is slipped inside the pendant so as to clasp the necks. If desired, the ends of the pendant may be passed through holes in a jacket on the outside of the pendant so as to form a turning or thief-proof pendant. In that case, a groove is turned in the pendant to receive the collars at the ends of the bow, and the cylinder is then slipped

Fig. 1.   Fig. 2.

down outside of the pendant between it and the jacket.

Lange's Guard Attachment.—Here the old style of bow is discarded. The winding pinion is fixed direct to the button, which is pierced to admit a loose shank secured by a nut at the lower end and carrying a small ring at the upper end for the attachment of the guard.

Box Chronometer.—[*Chronométre de marine.* —*Das See Chronometer.*]—Marine Chronometer.

Boxing-in.—[*Embôiter.*—*Das Einpassen des Taschenuhrwerkes in das Gehäuse.*]—Fitting the watch movement to its case. A term generally used in connection with keyless watches, where it implies, in addition to the actual fixing of the movement, the fitting and connecting the winding stem or pinion, and the push-work for setting the hands.

Brass.—[*Laiton.*—*Das Messing.*]—An alloy of about 64 parts of copper to 36 parts of zinc. A little lead is sometimes added for smoothness of working in the lathe. Hard brass called gunmetal is composed of 85 parts of copper to 15 of tin.

Blackening Brass.—Dissolve copper wire in nitric acid weakened by adding, say, three or four parts of water to one of acid. The article to be blackened is made hot and dipped into the solution ; it is then taken out and heated over a Bunsen burner or spirit lamp. When the article is heated, the green colour of the copper first appears, and as the heat is increased the article becomes of a fine dead black. If a polished surface is desired, finish with a coat of lacquer. The process is the very best for fine work, though articles soft-soldered cannot be safely subjected to it. For such, and rough work generally, the following, which is equally applicable for zinc and other metals, may be substituted :—Mix lampblack on a stone with gold size ; if a dull black is desired, make it to a very stiff paste ; if a more polished surface, then use more gold size. Add turpentine to thin it, and apply with a camel's hair brush.

Brass Edge.—A brass rim fitted round the pillar plate of watch movements. It projects above the level of the plate so as to form a recess for the motion work, and beyond the plate so as to rest on the seat of the case. It also serves as an attachment for the dial. Rarely used in modern watches.

Bréguet, Abraham Louis (born 1747, died 1823), a French watchmaker of rare attainments and inventive power. Berthoud, who was Bréguet's senior by two years, ends a brief notice of his brilliant contemporary thus, " Il n'a rien publié." Bréguet lived sixteen years longer than Berthoud, but unfortunately for us, it must still be recorded " he published nothing."

Bréguet Key, Tipsy Key.—[*Clef Bréguet.—Der Bréguet-schlüssel.*]—A watch in which the upper and lower portions are connected by means of a ratchet clutch kept in gear by a spring, so that the upper part will turn the lower part in the proper direction for winding, but if the upper part is turned in the opposite direction, the ratchet clutch slips without moving or straining the lower part of the key. This form of key was patented in England in 1789 by S. B. Harlow.

Bréguet Spring.—[*Spiral Bréguet.—Die Bréguet Spiralfeder.*] —A balance spring in the form of a volute, the outer coil of which is raised above and carried towards the centre of the rest of the spring. (See Balance Spring.)

Broach.—[*Eccarrissoir.—Die Reibahle.*]—A long tapered steel cutter used for enlarging holes already drilled. Watchmakers incline to the five-sided shape, but a half-round broach really has a better cut and leaves a rounder hole. A fluted broach also makes a good hole. A very taper half-round is a good enlarger. The handle of a broach should be perfectly straight with it, and moderately stout ; if thin and springy, the hole will not be round. Paper wrapped round a broach is useful for ensuring roundness, especially with the five-sided variety. Bottoming broaches, used chiefly for verge work, have chisel-shaped ends. If a hole has to be opened very much, the work should not be held in the same position all the time, but turned round at regular intervals, and the broach withdrawn each time. A round broach is used for burnishing brass holes.

Brocot Suspension.—[*Suspension Brocot.—Brocot's Pendelaufhängung.*]—A method of suspending a pendulum so that it can be regulated from the front of the dial. Two chops, free to slide between guides, embrace the pendulum spring. A vertical screw running through these chops, and pivoted top and bottom in the casting that supports the whole, carries a bevel wheel at its upper end. The pinion taking into this wheel is fixed to an arbor running to the front of the dial and terminating in a square. By means of a key placed on the square, the chops may be drawn further up or let further down the pendulum spring, thus practically lengthening or shortening the pendulum.

Buff.—[*Cabron.—Die Lederfeile.*]—A polisher of leather. Soldiers' old belts make very good buffs. French buffs are a soft kind of chamois. Sticks coated with emery paper are also called buffs.

Bull's Foot or Box Bottoming File.—[*Lime à sabot.—Die Bodenfeile.*]—A circular file used for filing sinks and other depressed surfaces. The handle rises from the top and is bent over to a right angle.

Burnisher.—[*Brunissoir.—Der Polierstahl.*]—A hard polished surface, generally steel, which is rubbed against metal to smooth it. The surface to be burnished must be free from scratches, which the burnisher would not remove, but render more distinct by contrast, and the burnisher must be kept highly polished, for the surface burnished can never be smoother than the burnisher. Burnishing polished pivots with the glossing burnisher preserves them from wearing. Very little, if any, of the metal is removed by burnishing in the ordinary way, although watchmakers sometimes use what are called cutting burnishers to form pivots. The cross-section of these burnishers matches the outline of the pivot it is desired to form, and they are roughened by rubbing on a lead block charged with coarse emery. The pivot is finished with a smooth burnisher of the same form as the cutting one. Silversmiths use burnishers of agate.

Butting.—[*Arc-boutement.—Das Aufsetzen in einem Eingriffe.*]—Two wheels touching on the points of the teeth when entering into action with each other. The tendency of pinion leaves to butt the wheel teeth when coming into contact is caused either

E

by the bad shape of the teeth or the leaves, or by using a pinion of an improper size, or by the wheel and pinion being placed at an incorrect distance from each other. (See Depthing Tool.)

Button.—[*Couronne.—Die Aufzugkrone.*]—On pendant-winding keyless watches, the serrated knob by which the watch is wound.

Button Turn, which is used chiefly in Swiss watches, is a brass block pivoted in the index arm and covering the curb pin. (See Balance Spring Buckle.)

Calcinas.—A compressed block of calcined bone and starch used for cleaning watch brushes.

Calendar Clock, Calendar Watch.—[*Horloge à calendrier. —Die Kalenderuhr.*]—A clock or watch that denotes the progress of the calendar. In a simple calendar, the mechanism has to be adjusted at the end of all months having less than 31 days. A perpetual calendar performs correctly at the end of the short months, and also during leap year

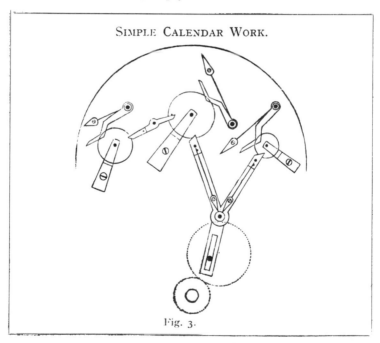

SIMPLE CALENDAR WORK.

Fig. 3.

Simple Calendar Work.—A simple calendar work is shown in Fig. 1. Gearing with the hour wheel is a wheel having twice its number of teeth, and turning therefore once in 24 hours.

A three-armed lever is planted just above this wheel ; the lower arm is slotted, and the wheel carries a pin which works in this slot, so that the lever vibrates to and fro once every 24 hours. The three upper circles in the drawing represent three star wheels. The one to the right has seven teeth corresponding to the days of the week ; the centre one has 31 teeth for the days of the month ; and the left-hand one has 12 teeth for the months of the year. Every time the upper arms of the lever vibrate to the left, they move forward the day of the week and day of the month wheels each one tooth. The extremities of the levers are jointed so as to yield on the return vibration, and are brought into position again by a weak spring as shown. There is a pin in the day of the month wheel which, by pressing on a lever once every revolution, actuates the month of the year wheel. This last lever is also jointed, and is pressed on by a spring so as to return to its original position. Each of the star wheels has a click or jumper kept in contact by means of a spring.

For months with less than 31 days, the day of the month hand has to be shifted forward.

Perpetual Calendar Work.—Figs. 2, 3, 4 and 5 illustrate M. Brocot's arrangement of calendar, lunation, equation work. The various parts are planted on a circular plate, of which the inner side containing the pieces for indicating the days of the week and the days of the month, is shown in Fig. 3. Fig. 5 shows the outer side of the plate, which is devoted to the mechanism for producing the phases of the moon and the equation of time. To this side of the plate the dial (Fig. 4) is attached.

The calendar is actuated by means of a pin *e* fixed to a wheel of the movement which turns once in 24 hours. Two clicks G and H are pivoted to the lever M *m*, G is kept in contact with a ratchet wheel of 31 teeth, and H with a ratchet wheel of 7 teeth. As a part of these clicks and wheels is concealed in Fig. 3, they are shown separately in Fig. 2.

When the lever M *m* is moved by the pin *e*, the clicks G and H slip under the teeth, their beaks pass on to the following tooth, and the lever then, not leaning on the pin *e*, falls quickly by its own weight, and makes each click leap a tooth of the respective wheels of 7 and 31 teeth. The arbors of these wheels pass through the dial (Fig. 4), and have each an index which, at every leap of its own wheel, indicates on its special dial the day of the week and the day of the month. A roll or jumper, kept in position by a sufficient spring, keeps each wheel in its place during the interval of time which separates two consecutive leaps.

This motion clearly provides for the indication of the day of the week, and would be also sufficient for the days of the month if the index were shifted by hand at the end of the short months.

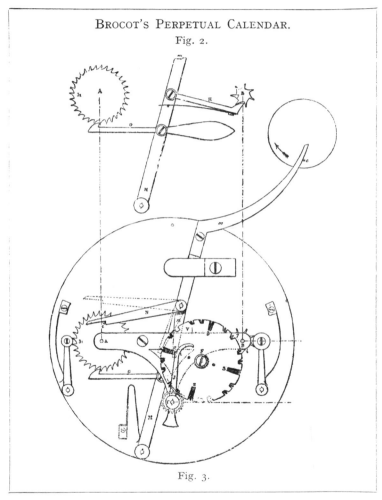

BROCOT'S PERPETUAL CALENDAR.
Fig. 2.

Fig. 3.

To secure the proper registration of the months of 30 days, for February of 28 during three years, and of 29 in leap year, M. Brocot makes the following provision. The arbor A of the month wheel goes through the circular plate, and on the other

side is fixed (see Fig. 5) a pinion of 10. This pinion, by means
of an intermediate wheel (D), works another wheel (centred at *e*)
of 120 teeth, and consequently turning once in a year. The
arbor of this last wheel bears an index indicating the name
of the month. The arbor C goes through the plate, and at the
other end (see Fig. 3) is fixed a little wheel gearing with a wheel
having four times as many teeth, and which is centred at F.
This wheel is partly concealed in Fig. 3 by a disc (V), which is
fixed to it, and with the wheel makes a turn in four years. On
this disc are made 20 notches, of which the 16 shallowest corre-

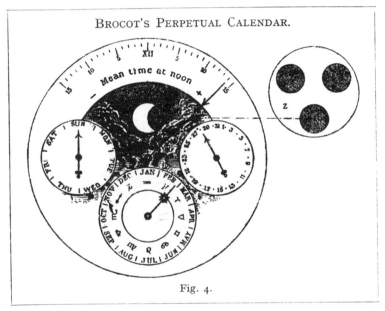

BROCOT'S PERPETUAL CALENDAR.

Fig. 4.

spond to the months of 30 days, a deeper notch corresponds
to the month of February of leap year, and the last three deepest
to the month of February of common years in each quaternary
period. The uncut portions of the disc correspond to the months
of 31 days in the same period. The wheel of 31 has a pin (*i*)
placed before the tooth which corresponds to the 28th of the
month. On the lever M *m* is pivoted freely a bell-crank lever
(N *n*), having at the extremity of the arm *n* a pin (*o*) which leans
its own weight upon the contour of the disc *v*, or upon the bottom
of one of the notches, according to the position of the month, and
the arm N is therefore higher or lower according to the position of
the pin *o* upon the disc.

It will be easy to see that when the pin o rests on the contour of the disc the arm N of the bell-crank lever is as high as possible, as it is dotted in the figure, and then the 31 teeth of the month wheel will each leap successively one division by the action of the click G, till the 31st day. But when the pin o is in one of the shallow notches corresponding to the months of 30 days, the arm N of the bell-crank lever will take a lower position, and the inclination that it will have by the forward movement of the lever M m will bring on the 30th the pin i in contact with the bottom of the notch, just as the lever has accomplished two-thirds of its movement, so the last third will be employed to make the wheel 31 advance one tooth, and the hand of the dial

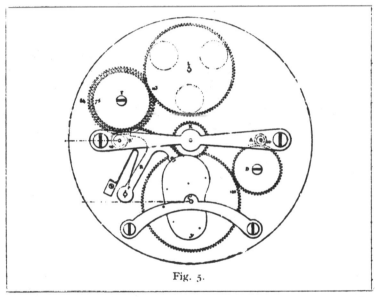

Fig. 5.

by consequence marks 31st, the quick return of the lever putting this hand to the 1st. If we suppose the pin o in the shallowest of the four deep notches, that one for February of leap year, the end of the arm N will take a position lower still, and on the 29th the pin i will be met by the bottom of the notch, just as the lever has made one-third of its course, so the other two-thirds will serve to make two teeth of the wheel of 31 jump. Then the hand of the dial will indicate 31, the ordinary quick returns of the detent putting it to the 1st. Lastly if, as it is represented in the figure, the pin o is in one of the three deepest notches, corresponding to the months of February in ordinary years, the pin will be in the bottom of the notch on the 28th just at the moment the lever begins its movement, and three teeth will pass before the return of the lever makes the hand leap from the 31st to the 1st.

The pin o easily gets out of the shallow notches, which, as will be seen, are sloped away to facilitate it doing so. To help it out of the deeper notches there is a weighted finger (j) on the arbor of the annual wheel. This finger having an angular movement much larger than the one of the disc G, puts the pin o out of the notch before the notch has sensibly changed its position.

Phases of the Moon.—The phases of the moon are obtained by a pinion of 10 on the arbor B, which gears with a wheel of 84 teeth, fixed on another of 75, which last gears with a wheel of 113, making one revolution in three lunations. By this means there is an error only of ·0008 day per lunation. On the wheel of 113 is fixed a plate (z), on which are three discs coloured blue having between them a distance equal to their diameter, as shown in Fig. 4 ; these discs slipping under a circular aperture made in the dial, produce the successive appearance of the phases of the moon.

Equation of Time.—On the arbor of the annual wheel is fixed a brass cam or " kidney piece " (y), on the contour of which leans the pin s, fixed to a circular rack (R). This rack gears with the central wheel (K), which carries the hand for the equation. That hand faces XII. the 15th April, 14th June, 1st September, 25th December. At those days the pin s is in the position of the four dots marked on the kidney piece. The shape of the kidney piece must be such as will lead the hand to indicate the difference between solar and mean time, as given in the table under the head of time.

Manner of Adjusting the Calendar.—Firstly, the return of the lever M m must be made at the moment of midnight. To adjust the hand of the days of the week, look at an almanac and see what day before the actual date there was a full or new moon. If it was new moon on Thursday it would be necessary, by means of a small button fixed at the back, on the axis of the hand of the week, to make as many returns as requisite to obtain a new moon, this hand pointing to a Thursday, and after to bring back the hand to the actual date, passing the number of divisions corresponding to the days elapsed since the new moon. To adjust the hand of the month, see if the pin o is in the proper notch. If for the leap year, it is in the month of February in the shallowest of the four deep notches (o) ; if for the same month of the first year after leap year, then the pin should be, of course, in the notch 1, and so on.

Perpetual Calendar Watch.—The wheel H (Fig. 6), driven by the minute wheel, makes one turn in 24 hours and carries a movable finger, *a*, which by contact with a pin moves the armed lever D by its extremity *p*. This lever, which has its centre of motion at *i*, acts through its different arms. Firstly at *c* it moves the day of the week star wheel (7 teeth). Secondly,

Fig. 6.

at *b* the star wheel for the day of the month (31 teeth). The finger *a* makes engagement and passes one tooth each day of the star wheel E (59 teeth), for showing the phases of the moon.

The part of the mechanism which renders the calendar perpetual is composed of a wheel of 31 teeth F, engaging with

NOTE.—For this arrangement, which is, I believe, the invention of Mr. C. H. Audemars, I am indebted to the *Journal Suisse d'Horlogerie.*

the star wheel c. This wheel, which makes one turn per month, passes at each turn, by means of the movable finger n, one tooth of the star wheel G (48 teeth), which latter by this means makes a revolution in four years. The circumference of account disc fixed to this star wheel corresponds to the months of 31 days, the shallowest notches to those of 30 days, and the four quarter notches to the month of February. At e, which is for February in leap year, the notch is hardly so deep as the other three quarter notches.

Each day, after moving the day of the week and the day of the month, the lever D, solicited by the spring h, returns its arm r to rest on the circumference of the count disc or in one of its notches according to the position of the disc.

The point of the piece u pressed by its spring rests on the snail k. Before the last day of the month it falls on to the small part of the snail, and then its action is substituted for that of the arm b ; the point of the piece u presses against the notch of the snail, and advances the star wheel the number of teeth necessary for the hand to indicate the 1st of the following month. It will be understood that the distance the point of the piece u falls is regulated by the position of the arm r on the disc or in one of its notches.

In the engraving the mechanism is set to the 1st December of the second year after leap year. The two pieces m and t are at the disposition of the watch wearer ; the first for adjusting the day, and the second for the age of the moon. The finger a is movable, to permit of putting the hands back without fear of deranging the mechanism. When the wheel H is turned back, the finger is arrested by the arm p, and, as it is sloped at the back, the pin carried by the wheel is able to pass easily, because the flexibility of the piece s permits it to give a little. The wheel should be the same diameter as the star wheel c.

Caliper.—[*Calibre.*—*Das Kaliber.*]—The disposition of the parts of a watch or clock. The arrangement of the train.

Callipers.—[*Calibre.*—*Der Federzirkel.*]—Two curved steel or brass fingers pivoted together to form a tool for measuring diameters. (Fig. 1.)

Callipers are used by watchmakers for readily observing if wheels are true. The wheel mounted on an arbor is held between the points of the callipers. A toucher made of brass or German silver with a passage for the wheel or balance to pass between is held against or attached to the side of the callipers. The part out of flat is detected by *very gently* turning the balance or wheel round, and noticing the spot nearest to the passage.

To ascertain if a balance is correctly poised, the balance staff is centred between the points of the callipers and one limb of the callipers gently tapped, when if the balance is not in poise the heaviest part is drawn to the lowest position. The sinks in the limbs must not be too large or worn. In Fig. 2 is shown an adjustable link connecting the upper and lower portions. By means of this, after the limbs have been opened to suit any particular balance staff, the link can be set so that if the callipers are opened to remove the balance they can afterwards be closed to clasp the staff again without fear of damaging the pivots, because the link stops them from going beyond their former position. As a poising tool for fine work, jewelled callipers

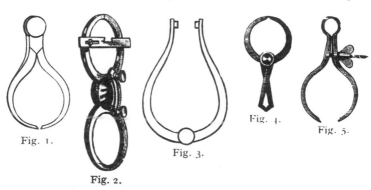

Fig. 1.

Fig. 2.

Fig. 3.

Fig. 4.

Fig. 5.

are preferable. The ends of the limbs of an ordinary pair of brass callipers may be straightened as in Fig. 3, and a hole drilled and tapped in each ; to these holes brass screws without slits in the heads are fitted ; the screw heads jewelled with good endstones and holes large enough to take any ordinary pivot. Some prefer V-shaped jewels instead of holes and endstones. One outside edge of the callipers is roughened, and when the tool is in use it is slightly rubbed with a screw-driver or other tool that may be handy, to cause the heavy part of the balance to travel to the lowest point. Plose's pattern is a very superior kind. It has ruby pins, with polished sinks and an adjusting screw to prevent the callipers closing too far and causing injury to the pivot ends from careless handling. (See also Poising Tool.) The " Excelsior " callipers, having the limbs crossed as shown in Fig. 6, embody a distinct improvement over the ordinary kind. When held in the left hand, with the thumb and forefinger placed at opposite sides of the upper part, a slight pressure will close the callipers, while a pressure with the ball of the hand below the crossing of the limbs will open them as easily. That the old-

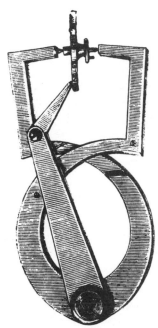

75

fashioned callipers can only be opened or closed by the use of two hands is a disagreeable experience to every watchmaker, especially when the joint of the callipers works stiffly. Therefore this advantage in the "Excelsior."

Fig. 7 are very useful inside and outside registering callipers. For instance, in selecting a mainspring for a clock barrel, the inside of the barrel may be gauged and the size noted on the right hand scale. The callipers are then opened for gauging the outside of springs and by observing the index on the left hand scale, a corresponding spring can be quickly chosen.

The plyer form of callipers shown in Fig. 8 are excellent for poising. They are furnished with male and female cen-

Fig. 6. "Excelsior" Callipers.

Fig. 7.

tres and a spring for keeping the object to be trued in position without undue pressure.

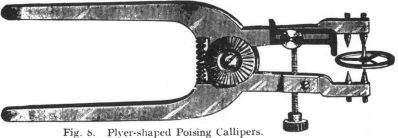

Fig. 8. Plyer-shaped Poising Callipers.

Spring callipers, as shown in Fig. 5, are useful when it is desired to retain a measurement.

Double-ended callipers (Fig. 4) give on the lower limbs an inside measurement corresponding to the outside measurement shown in the upper limbs.

The registering callipers by Mr. R. E. Chambers, shown in

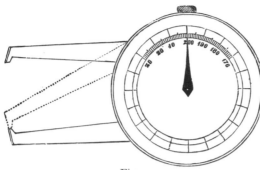

Fig. 9.

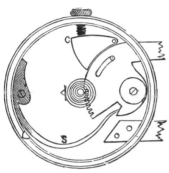

Fig. 10.

Figs. 9 and 10, are most handy. The back of the dial is recessed and arranged as in Fig. 10. One limb is fixed : the other is pivoted, and has a few rack teeth taking into a centre pinion. The pinion carries the hand, which should make a revolution in closing the callipers. The spiral spring attached to the pinion is to keep it and the hand banked in one direction for shake. The spring *s* is to keep the jaws open. The milled headed screw and the clamp *c* are to fix the jaws in case it is required to do so. A cover is snapped into the recess, and takes the back pivot of the pinion.

Cam.—[*Doigt de levée.—Die Hebescheibe.*]—A piece of mechanism, moving from a centre, whose contour is formed to give irregular motion to a lever or other device with which it is in contact.

Cannon Pinion.—[*Chaussée.—Das Viertelrohr.*]—The pinion with a long boss or pipe to which the minute hand is fixed. (See Motion Work.)

Cannon Pinion, snapping on (full-plate). Broach the hole of cannon smooth. Turn the arbor and polish it with oilstone dust to fit the cannon just easily. Square sink the bottom of cannon, to allow of the free entry of projecting centre pivot and centre hole. The sink must be wider than the hole. Cut this sink with the cannon shellaced on a plate in the mandril, or on an arbor in the turns with the ferrule previously knocked off, and a brass screw ferrule fixed on the cannon. For the first method use a cutter in the slide rest, for the second a graver or right-angled sinking cutter with flat cutting face. Top of the cannon leaves must be slightly above the top of the minute wheel teeth, and bottom just free of the plate when both are in position ; the hour and minute wheels will then run freely. If the cannon leaves are much higher than the minute teeth, it will be best to reduce their height from below, leaving a little over to polish away at the finish. Should the cannon be very taper, turn and polish it until it is of almost an imperceptible taper, otherwise the hour wheel will not fit it properly. File and polish a hollow in one side of the cannon, or turn one in it, between the top of the leaves and the bottom of the square. Take care it is not too deep : three-quarters of the thickness removed is mostly sufficient. Polish any burr off that may have been made by hollowing. When the cannon is snapped on to the centre arbor, it should be as far down as necessary, and should not rise up when the hands are set. For this to be so, care is necessary in gauging the position of the hollow to be turned in the centre arbor, and the width of the hollow. Make the centre arbor a frosty grey, not bright, by using the polisher charged with very liquid oilstone dust ; then rub the polisher with but little pressure from left to right a few times as the arbor slowly rotates. Bread off clean. Peg the pipe of the cannon perfectly clean, and put it on to the centre arbor. Rest the cannon on a bed of a brass stake fixed in the vice, and give a smart, but not hard, tap on a prick punch that is held exactly in the centre of the hollow. When the hollow is a turned one, a chisel-shaped punch (not quite sharp) is used. The sketch shows a useful form of stake by Mr. Plose, with three hollows or beds for various-sized cannons. There is a back plate to hold the cannon firm. The bolt from this back plate passes through the block and terminates in a knob or button. Between the face of the block and the collar of the button is a spiral spring strong enough to give a sufficient grip to the pinion head. The button is pressed in to insert the

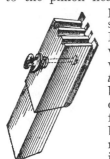

pinion in position or to release it. The sinks should be continued through the back plate. Press the centre pinion tightly into the cannon while the blow is given, then a decided mark will be made. Remove the centre and cannon *together* from the stake, and if the blow has been sufficiently hard the cannon will not fall off ; twist it a little way round backward and forward, then turn half-way and pull off. A bright mark will be seen, of the exact width required, where the cannon has been knocked in on to the centre arbor by the punch. Turn a groove in the arbor of very shallow depth, the width of the bright mark, and a little deeper in the middle than the sides. If the groove is too deep the cannon will go on tightly, but when it snaps into the groove, will turn too easily. Should the fit be too easy, everything else being correct, an additional tap with the punch, without the centre being in the cannon, will prove sufficient. But this time a lighter tap than the first, as there is nothing to resist the blow and keep the punch from driving the indentation in too far.

When the pinion is properly snapped on it should turn with sufficient tightness all round, and not jump.

Cannon and Set Hands Arbor (¾-plate, hollow-centre pinion to fit.—Clear a possible burr out of centre pinion hole with a broach. See that the cannon and set hands arbor run true; if the cannon does not, proceed as in " full-plate " advice. Should the set arbor be out of true in places, or " lumpy," stretch the hollow side by hammering it on a stake with light blows from the pane of the hammer. The blows must not follow in one place, but at regular intervals close on one another and a few at a time before observing in the turns if they have altered the shape from hollow to level. Any attempt to straighten a *hard* set arbor, or any kind of arbor, by hammering on the rounding, lumpy, or long part is mostly useless, and often ends in a breakage. When trued in this manner, turn quite smooth, and very little taper, until the arbor just enters the centre pinion ; reduce it with oilstone dust nearly to a fit. Presuming the leaves of the cannon to be the right height, sink it square underneath as in " full-plate." Rest the cannon on the plate in position and put the dial on, gauge the quantity of superfluous pipe and turn it away. This is necessary, as some cannons, if fitted first and shortened afterwards, are then too easy ; when the cannon has to be snapped on to the arbor, this shortening first is not so necessary. Broach the cannon until it is near the fit wanted. The centre pivots, if too long, must be shortened in the first place, the top pivot just to be above the sink in the top plate, and the lower one to project double as much. Polish the arbor with red stuff, diamantine or rubytine, to fit the centre just friction tight, at the same time polishing the other part to fit the cannon, which must be a tighter fit than the centre, or it will slide off by hand setting should the key fit tightly. The cannon should be knocked on to the arbor up to the centre pivot with a hollow punch, and if tried out of the plates it must not allow the arbor to work back out of the pinion. Each time the fit of the pieces is tried, bread and clean very nicely, otherwise a proper fit will be impossible. When the cannon is being opened, hold it in brass faced pliers with hollows filed to fit the pipe, or it will become bruised or misshapen, and at the time of removal from the arbor use these too.

With a solid centre pinion the cannon must be snapped on in a ¾-plate in a similar manner to the full-plate. In keyless watches the set hands arbor requires very exact fitting ; if too tight the wheels will jump in their depths, and sometimes break or spoil teeth, and it is awkward to set the hands to time. All hand works should be fitted so that the train is not driven back when the hands are set back. When the watch is wound the train should not be perceptibly interfered with, if the hands are set. In keyless minute repeaters the springs to be overcome by the cannon must not be too strong, or the hand may fail to advance.

Set hands push pieces should never be oiled, or they stick in the hole in time.

Cap.—[*Calotte.—Die Kapsel.*]—A cover. The full cap to full-plate watches covers the top-plate, to which it is attached by studs and a sliding bar or spring, and also encloses the circumference of the movement. The rim cap encloses the circumference only. With three-quarter-plate movements the cap covers the escapement and balance, and is sometimes made to open with a spring.

Capillarity. — [*Capillarité. — Die Haarröhrchenkraft. Die Kapillarität.*]—The attraction which exists between some fluids and solid bodies. Oil sinks are formed in watch and clock plates so that by capillary attraction the oil is kept close to the pivot instead of spreading over the plate. These sinks should be rather deep than wide ; some are much too large and shallow. Back slopes are formed on arbors having conical pivots, so that the oil may not be drawn all up the body of the arbor ; though they hardly seem to be necessary, for the amount of oil applied must be very excessive for it to be drawn to the root of the pivot. If anything of the kind is needed, a small hollow between the pivot proper and the largest part of the cone would seem to be best ; similarly, a slight groove in the shoulder of a straight pivot near its circumference would form a good receptacle for oil. The " attraction " is sometimes negative and becomes a repulsion, as in the case of mercury in a glass tube. It is still called capillarity, whether the fluid is raised above its natural level or depressed below it.

Capped Jewel.—[*Trou en pierre à contre pivot.—Steinloch mit Deckstein.*]—A jewel with endstone, see "Jewelled."

Carborundum.—An exceedingly hard abrading and polishing agent, composed of carbon and silicon, which has recently come into use.

Carriage Clock.—[*Horloge de voiture.—Die Reiseuhr.*]—A small portable clock, usually in a brass rectangular case with glass panels, and an ornamental hinged handle at the top for carrying.

Carrier. — [*Taquet. — Der Mitnehmer.*]—Watchmakers use

this word indifferently to describe a pin projecting from a ferrule to drive a wheel or other work to be turned, and a carrier proper such as is screwed to the work to engage with the driver. (See Ferrule.)

Case Springs.—[*Ressorts de carrure.—Die Gehäusefedern.*]—The springs in a watch case that keep the outer bottom closed

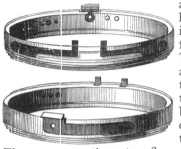

and cause it to fly open. The lock spring fits in a groove formed in the band of the case, and is fastened at one end by a screw. The other end, which projects above the band of the case near the pendant, is sloped off and by the closing of the cover is pressed back, but when the closing is completed springs into a nick in the under side of the outer cover.

There are sometimes two fly springs. They are fixed at one end in the same way as the lock spring, and their free ends press the joint of the outer cover. August Guye fixed the case springs to a thin brass ring between the movement and the case, which allowed of the use of a lighter case, and also preserved the movement from dust. Sketches of this are herewith given.

Case Stake.—[*Tas à boîte.—Der Gehäuse Amboss.*]—Stakes with heads flat, as in the engraving, and of different degrees of rounding, made of steel or of boxwood, are used for restoring

bruised watch cases to shape. The case-bottom is laid over a stake of suitable form and carefully hammered till the bruises disappear. For sharply-defined dents a hammer must be used, but a boxwood mallet will be sufficient to get rid of more superficial defects. It is almost impossible to restore battered engine-turning except by engine-turning it again. After a bottom is staked it may be, if necessary, unpinned and mounted on the lathe in a wooden chuck and " spun," that is, brought to a smooth circular form by. moving a burnisher over its surface while it is rapidly rotated.

Cement.—[*Siment.—Der Kitt.*]—The cement generally used by engravers and engine turners to fix their work is composed of four parts of pitch, two of plaster of Paris, and one of resin. These proportions are given to me by Mr. Warman. This cement is very strong, but has an offensive odour, and after its use the work requires to be cleaned in spirits of wine. Some watch engravers, therefore, now use gutta percha, which holds sufficiently firm and leaves the work quite clean, or with only slight marks which a little oil will remove. The gutta percha is slightly warmed in a flame and the work put on cold ; if the work is heated, it will not afterwards leave the gutta percha so clean.

Jewellers' Cement.—Dissolve gum mastic in spirits of wine. Soften an equal weight of isinglass in water. Dry it, and dissolve it separately in spirits of wine. Rub in with it, till dissolved, half the weight of gum ammoniacum. Add the dissolved gum mastic, and heat till thoroughly mixed. Keep tightly corked, and set the bottle in boiling water before use.

Cementing Metal to Glass.—(1) Mix some of the best plaster of Paris thoroughly dried with best gum Arabic previously dissolved in water. Apply quickly. This soon sets. (2) A very tenacious cement, which takes longer to dry than the preceding, may be prepared as follows :—Mix two parts of finely powdered litharge and one part of fine white lead ; mix three parts of boiled linseed oil with one part of copal varnish ; stir the powder into the liquid until it has the consistency of a stiff dough. Spread the cement on the metal, press it against the glass, and scrape off the surplus.

Cement made of quicklime mixed to a thick cream with white of egg is useful for mending amber and the like.

Paper on Metal.—Paper labels gummed or pasted on to metal will adhere much more securely if the metal is first treated with a strong and hot solution of soda and then wiped dry.

Centre of Gravity, Centre of Mass.—[*Centre de gravité.— Der Schwerpunkt.*]—That point in a body around which the mass is evenly balanced.

Centre of Gyration.—[*Centre de rotation.—Der Rotationsschwerpunkt.*]—That point in a rotating body in which the whole of its energy may be concentrated. A circle drawn at seventenths of its radius on a circular rotating plate of uniform thickness would represent its centre of gyration. The moment of inertia or the controlling power of balances varies as their mass and as the square of the distance of their centre of gyration from their centre of motion. Although not strictly accurate, it is practically quite near enough in the comparison of balances to take their weight and the square of their diameter.

Centre of Oscillation.—[*Centre d'oscillation.—Der Schwingungsmittelpunkt.*]—That point in a vibrating body in which, if all the matter composing the body were collected into it, the time of the vibrations would not be affected. In a straight bar

F

suspended at one extremity the centre of oscillation is at two-thirds of its length, and in a long cone suspended at the apex at four-fifths of its length from the apex. From the irregular form of most pendulums the position of the centre of oscillation is not easy to calculate, but it is always situated below the centre of gravity or centre of mass of the pendulum. In constructing a pendulum it will be sufficiently near to assume the centre of oscillation to be coincident with the middle of the bob.

Centre of Percussion.—Centre of impact. (See footnote under the head of Pendulum.)

Centre Seconds.—[*Seconde au centre.*—*Sekunde aus der Mitte.*] —A long seconds hand moving from the centre of a watch dial. An ordinary centre-seconds watch has the train arranged so that the fourth wheel pinion which carries the seconds hand is planted in the centre of the movement. This necessitates an unusual arrangement of the motion work, the minute wheel being driven by an extra wheel (of the same number as the cannon) attached to the second wheel pinion. The cannon wheel works on a pipe screwed to the plate. If this kind of centre seconds is to be a stop watch, a slide in the band of the case, when pushed round, presses a thin wire brake against the roller of the escapement. Owing to the extra shake from the additional wheel in the motion work, the minute hand is not so exact in its movement, unless the minute wheel is composed of two thin ones kept one in advance of the other by a spring, or there is some other provision to avoid backlash. But altogether this is a most unsatisfactory way of obtaining a centre seconds. (See " Independent Centre Seconds," " Chronograph," and " Split Seconds.")

Centre Wheel.—[*Roue de centre.*—*Das Minutenrad.*]—The wheel in a watch or clock the axis of which in ordinary movements carries the minute hand ; it is so called because it is planted in the centre of the frame.

Centre Bottom Hole.—To put in new centre bottom hole in a watch plate. First repolish the pivot, taking care to burnish it well. Put the frame in the mandril, holding by the top plate, and centre from the top hole. Peg the bottom hole quite true, and turn it to take a small stopping. Turn a stopping to fit the hole tightly, and take care not to have it longer than enough to rivet properly. Care will be required in riveting to avoid hammer marks and bending the plate. It may be done best with a steel punch slightly hollow on the face, and with a pivot in the centre to enter the hole of stopping. Replace pillar plate in mandril (the top plate having remained there meanwhile) and turn out the hole quite true, testing it by the peg, and to nearly right size for the pivot. Broach it in carefully, finishing with a clean and well-polished round broach, and turn off and finish stopping with POLISHED cutters. Fit the stopping from the inside of plate, tap it slightly to drive it in, resting the plate on a piece of peg-wood held in the vice, and having a hole drilled up it of a size to fit the stopping.—H. B.

Centrifugal Tendency.—[*Force centrifuge.—Die Zentrifugal-kraft.*]—The endeavour of every particle in a revolving or rotating body to fly from the centre of motion. The centrifugal tendency varies as the mass and as the square of the velocity of the body, and inversely as the radius of the circle described.

$$\text{Force} = \frac{WV^2}{g\,r}.$$

Compensation balances, if made thin in the rim, alter in diameter in the long and short vibrations from variation in the amount of centrifugal action.

Chain, Fusee Chain.—[*Chaîne.—Die Kette.*]—A small chain of steel that connects the barrel and fusee of a chronometer, watch, or clock. By the act of winding the fusee is rotated and the chain coiled thereon ; when the timekeeper is going the chain is drawn off the fusee on to the barrel. Gruet, a Swiss, is said to have introduced the chain in place of catgut about 1664.

Fusee Chain, to mend.—First find out from which side the pin or rivet is put in. In a new chain the side that lies to the largest part of the fusee should be the entry side of the pin. If there is much rivet, remove it with an oilstone slip. Rest the chain on a smooth lead block or stake, so that the thin end of the rivet is uppermost. Start the rivet with a point of a graver, by gently tapping on the handle. When the rivet shows through on the other side, lay the chain on a riveting stake, and gently punch it through the link. Ludwig's Fusee Chain Stake shown in the

sketch is the best for this purpose. It consists of a stake with a tapered slit formed to receive the link of any ordinary watch fusee chain, and hold it securely without distortion while the rivet is being punched out. A few thin steel punches driven into brass handles about three times as thick as the punches, projecting half an inch, and as thick as the centre arbor of a lever watch, filed tapering to pass easily through the chain link, are handy for this and other jobs. Clear the useless portion of link away without straining the other rivet. If a hook is required, stone it until it fits the link in every respect. The hole not being exactly in a line with the link holes causes a bung-ling rivet. Form the new rivet from a good needle let down to a light blue. File to fit the hole and stone smooth. Particular attention must be given to the shape of the rivet : it must be as straight as possible ; if taper, the hole will not contain it in each of the three joints, and no hammering will keep the chain together.

Push the pin into the chain from the larger side of the hole, while the chain lies on the board, pressing the chain down with thumb and forefinger of the left hand at the same time. Twist while pushing the pin, and when the link moves with the pin, the pin is sufficiently far in. Nip off the thin end close to the link, and also the thick end. Stone the ends flush with the link. Hold the longest part of the chain while riveting it, the short end is less likely to jump. Lay the part to be riveted on a nice flat stake with the thin end of the pin to the top, and with a succession of light blows from a flat-faced hammer the rivet will spread quite sufficiently, and the link will close to its normal state. Beginners mostly leave too much of the pin to be riveted, or hammer it too much, and very often use a pin too taper, too soft, or too hard, and the parts are in many cases not fitted to each other.

An uneven chain, or one unevenly strained, will often show a tendency to fall over on its side in working. This can sometimes be cured by reversing the hooks and turning the chain end for end.

Chain Hook.—[*Crochet de chaîne.—Der Kettenhaken.*]—The hook fixed at each end of the chain to attach it to the fusee and barrel.

Chamfer.—[*Biseau.—Die Schräge.*]—The small surface caused by the removal of a sharp aris. The chamfering tool with which the aris is removed is often spoken of as a " chamfer."

Chamois Leather.—[*Peau de chamois.—Das Waschleder.*]— To clean a chamois leather, make a solution of weak soda and warm water, rub plenty of soft soap into the leather and allow it to remain in soak for two hours, then rub it well until it is quite clean. Afterwards rinse it well in a weak solution composed of warm water, soda, and yellow soap. It must not be rinsed in water only, for then it would be so hard when dry as to be unfit for use. It is the small quantity of soap left that allows the finer particles of the leather to separate and become soft like silk. After rinsing, wring it well in a rough towel, and dry quickly, then pull it out and brush it well. In using a rouge leather to touch up highly polished surfaces it is frequently observed to scratch the work ; this is caused by particles of dust, and even hard rouge, that are left in the leather, which may be removed by a clean rougy brush.

Chapters.—[*Poinçon à chiffers.—Römische Zahlen.*]—The Roman characters used generally to mark the hours in watch and clock dials.

Chariot.—[*Chariot.—Der Schlitten.*]—A brass bar screwed to the pillar plate of a cylinder watch to carry the lower pivot of the cylinder and to afford a seat for the balance cock. Slight alterations in the intersection of the cylinder and escape wheel are made by shifting the chariot.

Chimes.—[*Carillon.*—*Das Glockenspiel.*]—The following gives the Cambridge Chimes, which are used in the Westminster Great Clock. They are founded on a phrase in the opening symphony of Handel's air, " I know that my Redeemer liveth," and were arranged by Dr. Crotch for the clock of Gt. St. Mary's, Cambridge, in 1793.

Chiming Barrel.—[*Rouleau de carillon.*—*Die Stiftenswalze.*]—A brass or wooden cylinder studded with pins for lifting the hammers in a chiming train. (See Quarter Clock.)

Chiming Clock.—[*Horloge à carillon.*—*Uhr mit Glockenspiel.*]—A clock that plays tunes or runs through musical notes periodically. (See Quarter Clock.)

Chops.—[*Tenaille pour ressort de suspension.*—*Federbacken* (*Pendel*).]—Two plates of metal which clasp the end of a pendulum spring.

Chronograph. — [*Chronographe.* — *Der Chronograph. Das Chronoscop.*]—Strictly a timekeeper that leaves a record of its going, though the term chronograph is now generally applied to those watches that have a centre-seconds hand driven from the fourth wheel, which may be started, stopped, and caused to fly back to zero by pressing either the pendant or a knob at the side of it. The chronograph hand generally beats fifths of seconds, and to permit of this an 18,000 train is necessary.

(2) A revolving barrel driven by clockwork with a conical pendulum. It is used by astronomical observers. The barrel is covered with paper, on which a prick is made at the instant

of the transit of a star. The position of the prick on the barrel indicates the time of the passage.

Centre-Seconds Chronograph.—Fig. 1 shows one of the most common forms in which the chronograph mechanism is arranged on the pillar plate under the dial. The chronograph hand is fixed to the pipe of a brass wheel which runs freely on

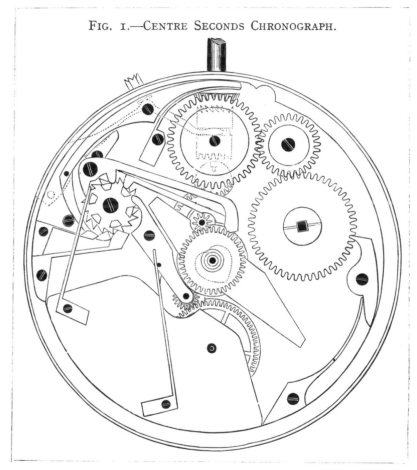

FIG. 1.—CENTRE SECONDS CHRONOGRAPH.

the centre arbor under the cannon pinion. This wheel has a finely serrated edge, and is usually driven by a smaller wheel having its edge serrated in the same manner. This latter is attached to the pinion which gears with the fourth wheel. The two serrated wheels bear the same proportion to each other as the fourth wheel does to the pinion already mentioned, so that the chronograph hand travels round the dial in a minute, which is the time that the fourth wheel takes to make a rotation. The smaller serrated wheel and the pinion to which it is attached are mounted on a pivoted carriage with a projecting tail. On the left of the engraving is a " castle ratchet," which has eighteen ratchet-shaped teeth round its edge and six projections or castle teeth rising from its upper face. In the figure the two serrated wheels are in contact, and the chronograph hand is consequently travelling. If now the button in the pendant is pressed, the shorter end of the bent lever which is lying around inside the case is depressed, and the hooked end of the lever draws the ratchet round so that the tail of the carriage on which the small serrated wheel is mounted is moved far enough to take it from contact with the larger wheel, and the chronograph hand consequently stops. At the same time a castle tooth, which has been keeping a circular brake off the larger serrated wheel, is moved out of the way so that the brake drops, thus keeping the chronograph hand from being shifted by accidental motion of the watch. When the button is again depressed, the castle ratchet is shifted round still further, and the returning lever with the pointed end is allowed to drop on to the heart shaped cam, which is fixed to the larger of the serrated wheels. As the returning lever drops, its tail lifts the brake off the serrated wheel, and the lever impelled by a spring, as shown in the drawing, turns the cam from whatever position it may happen to be in till the lever rests on that part of the edge of the cam which is nearest to its centre of motion. The chronograph hand is then at zero. Each time that the bent lever is pressed it draws the ratchet round one tooth, and as there are three ratchet teeth to one castle tooth, it is evident that all the pieces in contact with the castle return to their original position after every three movements of the lever.

The chronograph, on account of its ability to measure fractions of a second, has almost displaced the independent centre seconds watch, but in this form it is by no means a perfect construction. The serrated wheels are not calculated to withstand continuous wear, and it is evident that, however fine the serrations, they would cause the chronograph hand to jump backward or forward when brought into contact, unless a projection and groove happen to exactly coincide. This is often aggravated by minute portions of broken glass or other grit getting in the serrations.

If the small serrated wheel is removed, and the hand brought to zero, it should not move, however often the button is pushed in, but it often occurs that if the spring pressing on the returning lever which falls on the heart-shaped cam, or on the brake lever, is too strong, and the lever itself rather weak, the lever gives a little, and causes the hand to shift. This is one of the most common faults met with, and may be rectified by weakening the spring or bringing it to bear more on the end of the lever farthest from its centre of motion. The same jerking of the hand is sometimes observed when the pivots of the levers do not accurately fit the holes. (See also Split Seconds, and Repeater, for the examination of Repeaters contains useful hints equally applicable to Chronographs.)

Minute Chronograph.—The mechanism just described provides only for registering seconds. An addition for actuating

Fig. 2.—BAUME'S MINUTE CHRONOGRAPH.

a minute hand in connection with the chronograph will be understood from Fig. 2. Here the seconds chronograph work, instead of being under the dial, is all arranged on the top of the movement.

Above the centre of the figure may be seen a double ended
lever following nearly the curve of the centre cock. Each
time the point of the heart-piece passes, it lifts one end of this

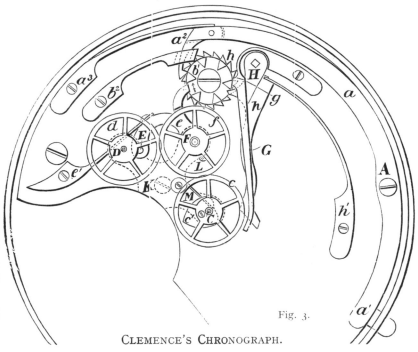

Fig. 3.

CLEMENCE'S CHRONOGRAPH.

lever, causing the other end to move forward one tooth of a star
wheel of sixty teeth, which is underneath the dial. This star
wheel is kept steady and helped forward by a jumper, and to its
arbor is attached the chronograph minute hand, which travels
round a small circle on the dial opposite to, and matching the
circle for the usual seconds hand on the fourth wheel arbor. The
end of the lever pressing on the point of the heart-piece serves
also to keep it perfectly steady when the seconds hand is jumped
back to zero, and so takes the place of a spring which is often
applied for this purpose to seconds chronographs. There is a
heart-piece on the pipe of the chronograph minute hand for
returning the hand to zero. When the chronograph is put out of
action, a lever presses on this heart-piece in the same way, and
with the same effect as in the case of the heart-piece for the
seconds hand.

I notice that in this arrangement for driving the chronograph

hands there is a serrated wheel on the fourth wheel arbor, and
that this and the other two serrated wheels are large, and all
three are of the same size. The serrations on the driving and
intermediate wheel are three times as coarse as those on the
centre wheel, and the teeth are distinctly and accurately formed.
By this means a safe working depth is secured for the driving
and intermediate wheels, which always run together ; yet when
the chronograph is started, contact is made with the same certainty and deadness as if all the serrations were fine, the teeth of
the intermediate wheel taking into every third space of the centre
wheel. By this improvement the uncertainty and trouble arising
with the more usual small serrated wheel, to which I have already
referred, are avoided.

Clemence's Chronograph.—On the previous page is shown
another form of chronograph. The distinctive feature of this
invention is the manner of driving the minute recorder from the
third wheel arbor, which will be clearly understood from the drawings and references.

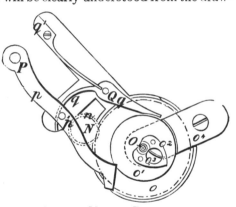

Fig. 4.—MINUTE RECORDER.

A, axis of lever *a*, which
is actuated by the push
piece at *a¹*, and carries a
click at *a²*. *B*, axis of castle-
ratchet b, *b¹*, *b²* jumper. *C*,
axis of serrated wheel *c* and
heart-piece *c¹*. This axis
passes through the centre
pinion and carries on the
other end the seconds chronograph hand. *D*, axis of
fourth wheel carrying serrated wheel *d*. *E*, axis of
carriage *e* carrying serrated
wheel *f*, and kept in contact with castle at *e* by
spring *e¹*. *F*, axis of wheel
f. *G*, brake. *H*, axis of
returning lever *h* held
against castle by spring *h¹*. *K*, bridge over chronograph work, removed
and indicated by dotted lines. *L*, banking screw for *e*. *M*, thin spring
bearing against arbor of chronograph hand to ensure steadiness. In Fig. 4,
which is a view of the minute recorder work under the dial, *N* is the axis of
n, a pinion on the third wheel arbor gearing with the wheel *O*, on the pipe
of which are *o¹* brake disc and heart-piece *o²* screwed to it ; *o³* is a flexible
blade fixed to the heart-piece and pressing sufficiently against a groove
formed in the pipe of the wheel *O* to carry the heart-piece round with the
wheel . On the pipe of the heart-piece the minute recording hand is carried.
P, axis of returning lever *p*. This axis is really a prolongation of the axis *H*
in Fig. 3, consequently the seconds and minute hands are returned to zero
concurrently. *Q*, axis of brake *o¹*, which is kept from the disc by the
spring *q¹* and brought to bear by a pin *p¹*.

Chronograph for Recording Twentieths of a Second.—
For timing the flight of projectiles, races, or any observations

requiring greater accuracy than can be obtained by the use of the ordinary chronograph beating one-fifth seconds is the one shown in the sketch below.

The long hand rotates once in a minute, the short one once in a second. It is claimed that by fitting the heart pieces friction tight upon the centre and escape wheel arbors a perfect start is ensured, as errors due to the loss of time in making contact and shake in the depths of ordinary chronograph mechanism are eliminated, and observations can be taken with certainty to the twentieth part of a second. A cylinder escapement is used, giving 20 vibrations per second, thereby corresponding with the divisions on the small dial, and as the balance can be stopped at any part of a vibration, or during the impulse, it is even possible to get the one-fortieth of a second.

It might be thought that with 72,000 vibrations per hour, the escapement would soon wear out, but it is said that experience has shown this is not to be the case. In all other respects the mechanism is the same as that of a minute recording chronograph, with the start, stop, and return to zero actions.

Chronometer. —[*Chronomètre.— Das Chronometer.*] —Any exact time-keeper. More generally under-stood to mean a timekeeper with the spring detent escapement.

Chronometer Escapement. —[*Echappement à détente.—DieChro-nometerhemmung.*] —(Invented by Le Roy about 1765. Perfected by Earnshaw and Ar-nold about 1780.) An escapement in which the escape wheel is locked on a stone carried in a detent, and impulse is given by the teeth of the escape wheel to a pallet on the balance staff once in every alternate vibration. This escapement, which is unexcelled for timekeeping, is represented on page 93.

Action of the Escapement.—A tooth of the escape wheel is at rest on the locking pallet. The office of the discharging pallet is to bend the detent so as to allow this tooth to escape. The discharging pallet does not press directly on the detent, but on the free end of the gold spring, which presses on the tip of the detent. The balance, fixed to the same staff as the rollers, travels in the direction of the arrow, with sufficient energy to unlock the tooth of the wheel which is held by the locking pallet. Directly the detent is released by the discharging pallet, it springs back to its original position, ready to receive the next tooth of the wheel. There is a set screw to regulate the amount of the locking on which the pipe of the detent butts. This prevents the locking pallet being drawn further into the wheel. It is omitted in the drawing to allow the locking to be clearly seen. It will be observed that the impulse roller is planted so as to intersect the path of the escape wheel teeth as much as possible, and by the time the unlocking is completed, the impulse pallet will have passed far enough in front of the escape wheel tooth to afford it a safe hold. The escape wheel, impelled by the mainspring in the direction of the arrow, overtakes the impulse pallet, and drives it on until the contact between them ceases by the divergence of their paths. The wheel is at once brought to rest by the locking pallet, and the balance continues its excursion, winding up the balance spring as it goes, until its energy is exhausted. The balance is immediately started in its return vibration by the effort of the balance spring to return to its state of rest. The nose of the detent does not reach to the end of the gold spring, so that the discharging pallet in this return vibration merely bends the gold spring without affecting the locking pallet at all. When the discharging pallet reaches the gold spring, the balance spring is at rest ; but the balance does not stop, it continues to uncoil the balance spring until its momentum is exhausted, and then the effort of the balance spring to revert to its normal state induces another vibration ; the wheel is again unlocked and gives the impulse pallet another blow.

Although the balance only gets impulse in one direction, the escape wheel makes a rotation in just the same time as with a lever escapement, because in the chronometer the whole space between two teeth passes every time the wheel is unlocked.

By receiving impulse and having to unlock at every other

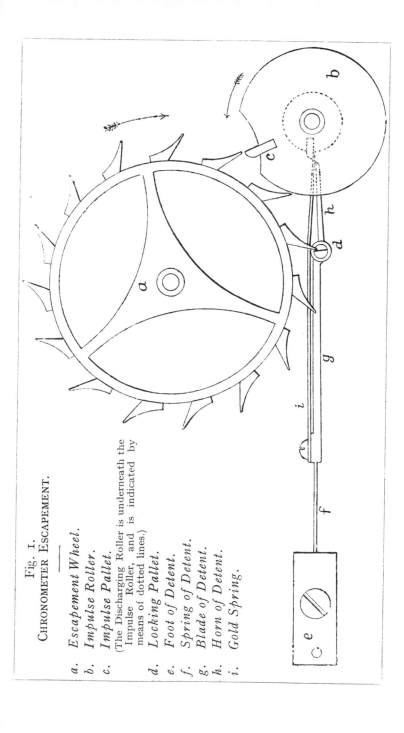

Fig. 1.
CHRONOMETER ESCAPEMENT.

a. *Escapement Wheel.*
b. *Impulse Roller.*
c. *Impulse Pallet.*
(The Discharging Roller is underneath the Impulse Roller, and is indicated by means of dotted lines.)
d. *Locking Pallet.*
e. *Foot of Detent.*
f. *Spring of Detent.*
g. *Blade of Detent.*
h. *Horn of Detent.*
i. *Gold Spring.*

vibration only, the balance is more highly detached in the chronometer than in most escapements, which is a distinct advantage. No oil is required to the pallets and another disturbing influence is thus got rid of. If properly proportioned and well made its performance will be quite satisfactory as long as it is not subjected to sudden external motions or jerks. For marine chronometers it thus leaves but little to be desired, and even for pocket watches it does well with a careful wearer ; but with rough usage it is liable to set, and many watchmakers hesitate to recommend it on this account. It is much more costly than the lever, and would only be applied to the very high-priced watches, and in these the buyer naturally resents any failure of action. Its use in pocket pieces is therefore nearly confined to such as are used for scientific purposes, or by people who understand the nature of the escapement, and are prepared to exercise care in wearing the watch. There is another reason why watchmakers, as a rule, do not take kindly to the chronometer escapement for pocket work. After the escapement is taken apart, the watch does not so surely yield as good a performance as before. In fact, it is more delicate than the lever.

Construction and Proportion of the Escapement.—For the ordinary 3-inch two-day marine chronometer movements, three sizes of escape wheels are used—viz. : ·54, ·56, and ·58 of an inch in diameter ; for eight-day marine chronometers the sizes are—·48, ·50, or ·52 of an inch. The escape wheel has fifteen teeth and the diameter of the impulse roller is half that of the escape wheel. The roller is planted as close between two teeth of the escape wheel as possible, so that theoretically the roller intersects the path of the teeth for 24° of the circumference of the wheel. This gives theoretically a balance arc of 45°.* Practically it is less ; there must be clearance between the roller and wheel teeth, an allowance must also be made for the side shake of the pivots. In Fig. 1 the impulse pallet is just opposite a tooth of the escape wheel when the discharging pallet is resting on the end of the gold spring. The balance moves through about 5° to accomplish the unlocking, and by the time that is done the impulse pallet will be 5° in advance of the tooth, and the tooth will drop through this space and more before it reaches the pallet, because after the wheel is unlocked it takes some time to get into motion at all, and at first its motion is slower than the motion of the pallet, which had not ceased to travel. The drop must be enough to allow the pallet to safely intersect the path of the tooth, and is arranged generally as shown, so that the pallet is 5° in advance of the tooth when the unlocking is completed. But many authorities insist on even more drop, so as to give the

* The balance arc is the amount that the edge of the impulse roller inter-

95

impulse more nearly on the line of centres. It is argued that the drop is not all mischievous loss of power, as it is in the lever escapement, for with a greater amount of drop the wheel attains a greater velocity when it does strike the pallet. However, most makers adhere to the 5°, although it may in some instances be advisable to vary it. If there is fear of over-banking, the arc of vibration may be reduced by giving more drop ; and if the vibration is sluggish and the drop can be safely reduced, the vibration will be increased thereby.

The rim of the escape wheel may extend to five-sixths of the total diameter ; the rim and arms are for lightness reduced to about one-half the thickness of the teeth. The fronts of the teeth diverge about 30° from a radial line so that the tips, being more forward, draw the locking stone safely in. The locking face of the stone is also set at a sufficient angle to ensure perceptible draw. The edge of the impulse roller acts as a guard to prevent the wheel teeth passing in the event of accidental unlocking at the wrong time. There is a crescent-shaped piece cut out of the roller to clear the teeth of the wheel. It should be very little behind the pallet, and less than the distance between two teeth of the escape wheel in front of it, to avoid the danger of running through or passing two teeth when such accidental unlocking occurs. It is important to see that there is enough cut out in front of the pallet to clear the wheel tooth at all times. When the balance is travelling very quickly—i.e. with an unusually large vibration—the pallet gets a long way in front of the tooth

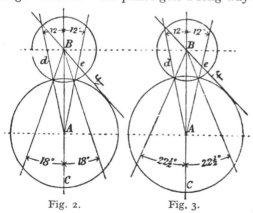

Fig. 2.   Fig. 3.

sects the path of the wheel teeth, measured from the centre of the balance staff. Fig. 2 (36° of balance arc) is a usual proportion for pocket, add Fig. 3 (45° of balance arc) a usual proportion for marine chronometers. Let A B be the distance of es-cape wheel and balance centres. From A, the centre of the escape wheel, set off 12° (half the distance between two teeth) on each side of the centre line, and draw d e. From B set off on each side of the centre line half the amount of the given balance arc and draw two other lines, as shown. The circles representing the tips of the escape wheel teeth and the impulse roller are drawn to cut the intersections of these four lines. The line f, being a tangent to the wheel circle through the centre B, will indicate the direction for the detent to be planted.

before the tooth starts, and then if the crescent is not cut far enough beyond the face of the pallet, the tooth would butt on the roller.

The radius of the discharging pallet is a trifle less than one-half that of the impulse pallet. If made too small the locking stone cannot return quick enough to catch the tooth.

The detent is made very light, and of about the proportion shown in the drawing. The spring of the detent is thinned down so that when the foot is fixed and it stands out horizontally, one penny-weight hung from the pipe deflects it about a quarter of an inch. If the spring is made too thin, it will cockle and give trouble, the detent may very easily be made too long from the point where it bends to the locking pallet, and would then be too sluggish and allow the wheel to trip by not returning quick enough after the unlocking to receive the next tooth of the wheel. The distance from the shoulder of foot to pipe to be equal to the diameter of the wheel is recommended by Mr. T. Hewitt as a very good rule.

The escape wheel is of hard hammered brass, the rollers of steel, the detent of steel, carefully tempered, with the point of the horn left softer to allow of bending. The pallets are all of sapphire or ruby, fastened in with shellac. A brass plug is fitted in to occupy the space in the pipe of detent not filled by the locking pallet. The gold spring is hammer hardened.

Pocket Chronometer.—The escape wheel for pocket chronometers varies from ·29 to ·35 in. in diameter. The impulse roller is made larger in proportion to the escape wheel than in the marine chronometer, so as to lessen the tendency of the escapement to set. If the chronometer escapement is brought to rest by external motion just as the unlocking is taking place *it must set*, for the balance spring is then quiescent. In the lever escapement the tooth of the escape wheel is in the middle of the impulse plane of the pallet when the balance spring is quiescent, and in this respect the lever has the advantage. If the velocity of the balance in a chronometer is much reduced when the unlocking is completed, then a large impulse roller is of great assistance to the wheel in overcoming the inertia of the balance.

As the diameter of the roller is increased, the balance arc, and also the intersection of the path of the wheel teeth by the impulse pallet, are decreased. The velocity of the edge of the roller, too, more nearly approaches the velocity of the wheel tooth, so that less of the power is utilised. It is, therefore, not prudent to adopt a much less balance arc than 28° or 30°.

The tendency of pocket chronometers to set is also lessened by adopting a quick train. 18,000 is the usual train, but they are occasionally made with 19,200 by having sixteen teeth in escape wheel instead of fifteen. This seems to be an objectionable way of getting the quick train. The teeth of the escape wheel being closer together, a small roller must be used to get the same intersection, and as there is less time for the detent to return there is great danger of mislocking.

For the convenience of getting the seconds hand to jump half-seconds, a 14,400 train is sometimes adopted in pocket chronometers. In this case the escape wheel has twelve teeth, the numbers of the rest of the train remaining the same.

The other parts of the pocket Chronometer Escapement are similar to those of the Marine Chronometer already described.

To Examine the Escapement..—See that the wheel is true and the teeth smooth and perfect, and that the rollers properly fit the staff. See that the end shakes and side shakes are correct. See that the " lights " between the wheel teeth and the edge of the roller are equal on both sides when the wheel is locked. If they are not, the foot of the detent must be knocked a trifle to or from the centre of the roller till the lights are equal. If the light is more than sufficient for clearance, the roller must be warmed to soften the shellac, and the impulse pallet moved out a little. If the light is excessive there will be too much drop on to the locking after the wheel tooth leaves the impulse pallet, and with a large drop there is danger of tripping.

To ensure safe locking the detent should be set on so that when the banking screw is removed, and the locking pallet is free of the wheel teeth, it will just spring in as far as the rim of the wheel.

In pocket chronometer escapements it is especially necessary to see that the face of the locking stone is angled so as to give perceptible draw. Many pocket chronometers fail for want of it.

The gold spring should point to the centre of the roller. Bring the balance round till the discharging pallet touches the gold spring preparatory to unlocking, and notice how far from that point the balance moves before the gold spring drops off the face of the pallet. Then reverse the motion of the balance, and see if the same arc is travelled through from the time the *back* of the pallet touches the gold spring till it releases it. If not, the horn of the detent must be bent to make the action equal.

Bring the discharging pallet on to the gold spring, and let it bend the detent so that the locking stone is as much outside the wheel as it was within when the wheel was locked. The gold spring should then drop off the discharging pallet. Make it to length, sloping off the end from the side on which the pallet falls to unlock, and finish it with great care. The gold spring should be thinned near its fixed end as much as possible, and

G

the detent spring thinned if it is needed. The judgment of the operator must determine the proper strength in both cases. The nose of the detent horn should be nicely flattened and the corners rounded off.

The locking pallet should not be perfectly upright. It should lean a little from the centre of the wheel, and a little towards the foot of the detent, so that the locking takes place at the root of the stone, and then the action of locking and unlocking does not tend so much to buckle the detent. The face of the impulse pallet, too, should be slightly inclined so that it bears on the upper part of the wheel teeth. By this means the impulse pallet will not mark the wheel in the same spot as the locking pallet.

Try if the escape wheel teeth drop safely on the impulse pallet by letting each tooth in succession drop on, and after it has dropped, turn the balance gently backwards ; you can then judge if it is safe by the amount the balance has to be turned back before the tooth leaves the pallet. If some teeth do not get a safe hold, the impulse roller must be twisted round on the arbor to give more drop.

If the escapement is in beat, the balance, when the balance spring is at rest, will have to be turned round an equal distance each way to start the escapement. When the balance spring is in repose, the back of the discharging pallet will be near the gold spring, and if the balance is moved round till the gold spring falls off the back of the pallet and then released, the escapement should start of itself ; and in the other direction also, if the balance is released directly the wheel tooth leaves the face of the impulse pallet, the escapement should go on of itself.

Chronometrical Thermometer.—An instrument devised by the late Astronomer-Royal for magnifying the effect of changes of temperature during chronometer trials. It is an ordinary chronometer, except that the metals composing the compensating laminæ of the balance are reversed, the steel being outside and the brass inside.

Chronoscope.—[*Pendule aux guichets.*—*Die Uhr mit spring-enden. Zahlen.*]—This word, which from its derivation appears to be a more suitable title for the watches generally known as chronograph, is used to denote a clock in which the time is shown by figures presented through holes in the dial.

Chuck.—[*Tasseau.*—*Der Einsatz (Drehstuhl).*]—An adjunct to a lathe, used for holding the work to be operated upon. (See Lathe.)

Circular Error.—[*Erreur circulaire.*—*Der Pendelschwingungs-fehler.*]—In a clock the difference of time caused by variations in the extent of arc described by the pendulum in following a circular instead of a cycloidal path.

Clams.—[*Pinces.—Die Schiebezange.*]—(1.) A kind of pincers kept open by a spring and closed by a sliding thimble, or by being pinched in the jaws of a vice.

(2.) Shields of wood or soft metal placed over the jaws of a vice. (See Vice.)

(3.) A cylindrical piece of boxwood divided by a thick saw-cut for the greater part of its length and closed by a sliding ring, used by jewellers for holding the shank of a ring.

(4.) Jewelled slits through which balance spring wire is drawn to make it of uniform thickness after rolling.

Clepsydra.—[*Clepsydre.—Die Wasseruhr.*]—A machine for indicating time by the passage of water. One of the earliest forms of the Clepsydra, which was in use in Egypt about 200 B.C., is shown in Fig. 1. A supply of water ran through the pipe H into the cone A, and from there dropped into the cylinder E. A conical stopper B regulated the flow, and the superfluous water escaped by the waste pipe I. The Egyptians divided the period between sunrise and sunset into twelve equal hours, so that the conical stopper had to be adjusted each day, and marks for every day in the year and for the particular latitude of the place

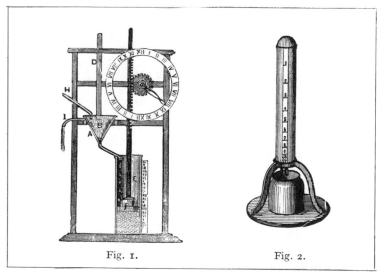

Fig. 1.   Fig. 2.

were cut on the stalk D as a guide to the position of the stopper. A floating piston terminating in a rack served to actuate a pinion, to the arbour of which an hour hand was fixed.

Fig. 2 represents a very simple form of Clepsydra used in the seventeenth century. It is merely a glass vessel, with an orifice at the bottom, filled with as much water as will flow out in exactly twelve hours, figures being placed at the proper distances to denote the successive hours.

Click. — [*Cliquet. — Der Sperrkegel.*] — A detainer. — (See " Ratchet Wheel " and " Detent.")

Side Click Gauge.—To facilitate the selection of side clicks

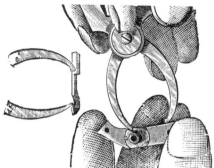

for barrel ratchets, Mr. Mairet has devised a gauge, shaped something like a douzième gauge. The manner of its use will be apparent from the engraving. The distance from the screw hole on the bar to the bottom of the ratchet tooth can be taken and registered to the fifth of a millimetre. Then, by measuring from the screw hole to the hook of likely clicks, one of the accurate distance between these points can be quickly chosen.

Clinker.—(See Red Stuff.)

Clock.—[*Horloge.—Die Uhr ; Wanduhr ; Standuhr.*]—Generally, any mechanism for timekeeping other than a watch or chronometer. Although clockmakers still understand a clock to mean a timekeeper that strikes the hours, the term has commonly a wider significance.

Cleaning and Repairing Clocks.—In taking down clocks prior to cleaning, it is a good plan to mark where the teeth of the wheels engage with the pinion leaves, for if there should be any slight inaccuracy the teeth may not gear so well if altered ; and in striking trains the lifting pins and the run after warning will then be correct when the clock is again put together, that is, assuming they were right beforehand. In many French clocks there is already such a mark, one leaf of the pinion being sloped off and a dot being made on the wheel tooth that corresponds with it. All the parts may be placed in a bath of paraffin, which forms as good a detergent as any, and, as they are taken out, brushed with a moderately hard clock brush ; clean the pivot holes by twirling a pointed peg in each one. Tarnished gilt plates, and polished ones, if not much stained, may be restored by immersing in a cyanide bath (see Gilding). Badly stained polished plates may be repolished with rotten stone used on a willow polisher or soft brush.

For the ordinary run of clocks the following course of examination practised by Mr. Alfred Gray and other good clockmakers is recommended :—After taking the movement from its case, remove the hands, dial, minute cock and bridge, to try the escapement with some power on, and note any faults there. Next remove the cock and pallets, and if it is a spring clock, put a peg between the escape wheel arms to prevent it from running down, and let down the spring. Here sometimes is a difficulty ; if the spring has been set up too far, and the clock is fully wound, it may not be possible to move the barrel arbor sufficiently to get the click out of the racket. In many old clocks there will be found a hole drilled at the bottom of, and between the great wheel teeth, directly over the tail of the click ; so that you can put a key on the fusee square and the point of a fine joint pusher through the hole, release the click and allow the fusee to turn gently back until it is down. Having let down the spring, try all depths and end-shakes and all pivots for wide holes ; if it is a striking clock, do the same with the striking train, paying particular attention to the pallet pinion front pivot to see if it is worn and the rack depth made unsafe thereby—also seeing that none of the rack teeth are bent or broken. Having noted the faults, take the clock to pieces, look over the pivots, and note those that require repolishing. Finally, take out the barrel cover and see to the springs ; if a spring is exhausted or soft, several of the inner coils will be found lying closely round the arbor.

Pallets.—In most cases some repairs will be required to the

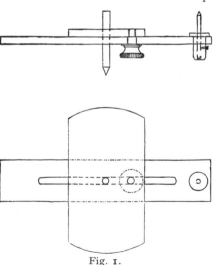

Fig. 1.

pallets, as these nearly always show signs of wear first ; if they are not much cut, the marks can be polished out, and for this purpose a small disc of emery about three inches in diameter, mounted truly on arbor, and run at a high speed in the lathe, will be of great assistance ; finishing off with iron or steel polisher and sharp red stuff. If you have to close the pallets to make the escape correct, see that the pallet arms are not left hard, or you may break them. If the pallets require much alteration or renewal, and a depth tool sufficiently large is not at hand a tool by Mr. Gray, similar to Fig. 1, will answer as a substitute. It consists of a base plate with a slot in the centre ; on this base plate slides another plate at right angles to it, whose position is secured by a set screw and stud passing through the slot. Each of these plates carries a tube with a boss perpendicular to the face of the plate ; in these are fixed either conical centres or hollow bushes, as required, to receive a wheel and pinion or wheel and pallets. After making any alteration in the pallets, it may be necessary to correct the depth ; should it only require a slight alteration, it will be sufficient to knock out the steady pins in the cock and screw it on so that it can be shifted by the fingers until you have got the depth correct, then screw it tight and broach out the steady pin holes, and fit new pins. If much alteration in the depth is required, it may be necessary to put in a new back pallet hole ; this can be made from a piece of hollow stopping broached out and turned true on an arbor. It is not safe to rely on the truth of the stopping unless it is turned on an arbor first. The hole in the plate is now drawn in the direction required with a round file, and opened with a broach from the inside until the stopping enters about half way. Of course, in finishing broaching the hole, you will roughen the extremities to form rivets. Drive the stopping in and rivet it with a round-faced punch from the outside ; reverse it, and resting the stopping on the punch, rivet the inside with the pane of the hammer : remove any excess of brass with the file, chamfer out the oil sink, and stone off any file marks ; finally open the hole for the pivot to the proper size.

Making New Pinions.—Frequently one meets with an escape pinion that has become so badly cut or worn so as to be useless, and if a new one cannot be purchased it will be necessary to make it from pinion wire. In sectoring the pinion wire to the wheel, bear in mind that it will become slightly smaller in filing up. Considerable practice is required to make good-shaped pinions quickly and well. A piece of pinion wire of a slightly greater diameter than the pinion is to be when finished, is cut about one-eighth of an inch longer than required, and the position of the leaves or head marked by two notches with a file. The leaved portion of the wire that is not required is filed down on a filing block, taking care not to remove any of the arbor in so doing ; a centre is then filed at each end true with the arbor, and then projected through a hole in a runner or the cone plate shown under the head of lathe and turned. Get the pinion head quite true by straightening the arbor, if necessary, and turn the arbor and faces of the pinion square and smooth. The pinion is now filed out true, using a hollow-edged bottoming file for the spaces, and a pinion-rounding file for the sides of the leaves. In using the bottoming file the pinion is rested in the device shown under the head of Gallows Tool, and held in the fingers when finishing the leaves. The file marks are taken out with fine emery and oil ; the polishers may be pieces of wainscot oak, about a quarter of an inch thick, five inches broad and six inches long, used *endway* of the grain. One end is planed to a V shape to go between the leaves and the other cut into the grooves by rubbing it on the sharp edges of the pinion itself, which speedily cuts it into grooves to fit. The pinion is rested while polishing in a groove cut in a block of soft deal, which allows it to give to the hand, and keeps it flat. When the file marks are all out, the pinion is ready for hardening. Twist a piece of stout binding wire round it, and cover it with soap ; heat it carefully in a clear fire, and quench it in a pail of water that has been stirred into a whirlpool by an assistant, taking care to dip it vertically. When dried, it is covered with tallow and held over a clear fire until the tallow catches fire ; it is allowed to burn for a moment, and then blown out and left to cool. The leaves are polished out with crocus and oil in the same way that they previously were with emery. Now, if the pinion is put in the centres and tried, it will probably be found to have warped a little in hardening. This is corrected in the following manner : The *rounding* side of the arbor is laid on a soft iron stake, and the *hollow* side stretched by a series of light blows with the *pane* of the hammer, given at regular intervals along the curve. Having got the leaves to run quite true by this means, turn both arbors true, and polish them with the double sticks—these are simply two pieces of thin boxwood, about three-eighths of an inch wide and three inches long, hinged together at one extremity and open at the other ; between these the arbor is pinched with oil and fine emery, and they are traversed from end to end to take out the graver marks. The brass for the collet, to which the wheel is riveted, is now drilled, broached, and turned roughly to shape on an arbor. The position on the pinion arbor is marked with a fine nick, and the collet soldered on with soft solder and a spirit lamp, taking care not to draw the temper of the arbor when doing so. Wash it out with soda and water, and polish the arbor with crocus, turn the collet true, and fit the wheel on. If the pinion face is to be polished, it is now done, the facing tool being a piece of iron about one-sixteenth of an inch thick, with a slit in it to fit over the arbor with slight freedom, used with oilstone dust first, and the sharp red stuff.

Sometimes cut pinions are used for the centres, and then the body of the arbor is sufficiently large to allow the front pivot to be made from the solid arbor ; but if the centre pinions are made from pinion wire in the manner just described, the leaves are beaten together on an anvil to form a solid mass for the froitt pivot. This piece should project sufficiently far through the pivot hole to allow it to be squared to receive the friction spring which drives the motion work. In cases where this pivot is much cut, it may be turned down and have a steel tube soldered on to form a new one ; as these pinions are very long and flexible, some difficulty will be experienced in turning this pivot, unless a back stay is used to support the arbor, and prevent it springing from the graver. A cheaply-made form of backstay is shown under the head of Lathe.

Worn Pinions.—In common clocks, where both third and escape pinions are worn by the wheel teeth, if the pivots are still in good condition the third pinion leaves can be turned back from the outer end rather more than the thickness of the centre wheel, the pivot shoulder also turned back the same distance, the pivot re-made, burnished and shortened. Then the pivot hole in the *front* plate is opened with a broach to about twice its original size, and a stopping with a good large shoulder is turned true on an arbor and riveted into the plate. The thickness of the shoulder of this stopping will depend on the amount the arbor was shortened, and must be such as just to give correct end shake to the pinion. By shifting the third wheel and its pinion thus, a fresh portion of both the third and escape pinions is brought into action, and as good results will be obtained as by putting two new pinions.

Escape Wheel.—Often in old clocks the escape wheel is so much out of truth that anything like close escaping is out of the question, as so much drop has to be given to enable some teeth to escape, that nearly all the power is lost ; in such a case, a new wheel is a necessity, and if you want to get a good *hard* wheel you must make the blank yourself. Take a piece of hard sheet brass, about twice as thick as the wheel is to be when finished, and cut from it a square sufficiently large for your wheel ; then with a hammer with a slightly rounded face, reduce it to nearly the thickness you require. In hammering go regularly over the surface, so that no two consecutive blows fall on the same spot ; and when one side is done turn it over and treat the other in the same way. File one side flat, find the centre and drill a hole nearly as large as required for the collet ; cement it with shellac to a flat-faced chuck in the lathe, and centre it true by the centre hole. Mark with the graver the size of the wheel, and with a narrow cutter remove the corners ; face the blank with the graver and turn it to size, leaving it slightly larger than the old wheel ; knock it off the chuck and reverse it, bringing the turned face next the chuck, turn that face flat and to thickness, and it is ready for cutting. After it is cut, remove any burrs with a fine file, and mark a circle to show the thickness of the rim, and on that circle divide it into the number of arms it is to have ; mark also a smaller circle slightly larger than the collet on which it is to be riveted, draw lines through the divisions in the outer circle and the centre of wheel to mark the centre of the arms. Drill a hole between each of the arms to enable you to enter the file, which to begin with should be a coarse round one, then follow with the crossing file, holding the wheel between a piece of thick card in the vice ; finish by draw-filing the arms and crosses with a very smooth file, followed by a half-round scraper used as when draw-filing. This leaves the surface smooth and ready for the burnisher, of which tool two different shapes will be required, one oval and the other half-round. These tools, when in use, require to be repeatedly cleaned on a piece of leather, and passed over the palm of the hand to prevent tearing up the surface of the metal. The wheel teeth are now polished out with a short-haired brush and fine crocus and oil ; then take out the file marks from both sides of the wheel with water of Ayr stone and oil, and it is ready for riveting it on. The riveting stake for clock work is like the ordinary pinion riveting stake used by watchmakers, only it is in two pieces dividing down the centre of the holes ; if it were in one piece, the pinion head would prevent it passing through a hole of the proper size to fit the collet ; it has two steady pins to ensure its coming together properly. Take a slight chamfer out of the front of the wheel-hole, and roughen the surface of it with a graver, turn the collet down to fit in *tightly*, and rivet it on with a half round punch, taking care to strike light blows and keep the wheel turning while riveting. It is then ready for stoning off and polishing with a flat wood polisher and fine crocus and oil. In crossing out a small delicate wheel, it is a good plan to fasten it with shellac to a flat plate of brass, having a hole in it rather larger than the inside of the rim of the wheel. In this way all danger of bending a tooth of the wheel accidentally is avoided, and the crossing can be finished without removing it from the plate.

Striking Work.—The parts most frequently found to require repair in the striking train of clocks, are the pivots of the upper pinions, especially those of the fly, pin wheel and pallet wheel. If a pivot is only slightly cut, it can be re-turned and polished, and a new hole put in ; but if to entirely remove the marks the pivot would have to be much reduced in diameter. a new pivot is the only resource.

New Pivots.—In putting in a new pivot, the arbor may be centred in the cone plate as shown under the head of Lathe, or in any other suitable tool if that one is not to hand. A short stiff drill should be used, ground to cut in one direction only, rather thin at the point, and for a short distance behind the cutting angles quite parallel. The drill should be left quite hard, or, if a soft arbor is to be drilled, it may be tempered to a light straw colour, and the rest of the shank rather softer. If this is lubricated with either turpentine or benzine, but little difficulty will be found in drilling the arbor ; the hole should be rather deeper than the pivot is long, and in size rather larger than the pivot is to be. A piece of staff steel is now centred, hardened and blazed off, and turned down true to fill the hole, and very slightly tapered (if too taper, the arbor will be split in driving it in) ; when it fits half-way in, draw-file it carefully, and cut it to lengths, filing the outer end off *square*. A few blows of a light hammer will fix it firmly in position, then the extreme end of the pivot can be turned to a centre, through a hole in the lantern runner. The pivot can now be turned down to size, polished, burnished, and the end rounded up. Should the pallet wheel front pivot require repairing, a centre may be cut with the graver in the end of the square, then a male centre can be used, and the pivot turned and polished in the usual manner.

Gathering Pallet and Rack.—In many old clocks, particularly in long case striking-clocks, the rack and gathering pallet are frequently found in very bad condition ; the pallet perhaps fitting the square very badly, thus making its depth with the rack uncertain. In the absence of a proper forging, a pallet may be made from a square bar of steel, thick enough to give the requisite length of boss. Mark the length of the tail of the pallet, and file it down to almost the required thickness ; file also the opposite face of the bar smooth and flat. Mark the position of the hole, and drill it at right angles to the face ; the diameter of the hole will be the same as the small end of the square on the pallet pinion—measuring across the flats, of course. Start the corners of the square in the position required with a good square file ; then take the piece of broken square file of rather a coarse cut, and of the same taper as the square on the pinion ; oil it and drive it in with a few light blows of the hammer, turn the pallet over and knock it out again, turning it a quarter round each time you withdraw it. In a few minutes you can thus form a good square straight hole, and fit it accurately to pinion-square. Put it on an arbor and turn the ends square and to length, see that the tail is at right angles to the hole, also file the boss to form and shape the lip ; this is usually made straight and the back sloped off ; consequently it scrapes the rack teeth with its extreme end only, and wears quickly. As the pallet is in reality a pinion only with one leaf, its durability is increased by curving the face similar to a pinion leaf cut in half. The end of the tail of the pallet should be rounded and finished off smoothly at right angles to its face, its length such that it is well free of the pin in the rack when gathering the last tooth but one, and rests fairly on the pin when the rack is up. If the tail of the pallet were left quite straight, and the end filed off square, there would be a danger of the rack being held up by the pallet, particularly when the pin in the rack is planted lower down than it should be, its proper position being rather above the top of the teeth. The tail of the pallet is therefore curved to just throw the rack off.

If any of the rack teeth are damaged at the points, it may be necessary to slightly top all the teeth and file them up again ; only the backs, or curved sides of the teeth should be filed, finally taking the burr off with the oilstone slip. To make the depth correct again, the rack arm is hammered a little, to stretch it, care must be taken to keep the teeth truly in circle, also to see that they are well free of the boss of the gathering pallet—not only when it is in position resting on the rack pin, but also when it has moved into the position that it would be in when the clock has warned. If the boss of the pallet is not perfectly concentric, it may be just foul of the rack teeth in this position, although free when tried with the pallet resting on the stop pin.

Run.—Clocks are occasionally met with in which the hammer begins to lift as the clock warns, with a lot of useless run *after* the hammer has fallen. This is just the reverse of what should be the case, as the more run *before* the hammer begins to lift, the less probability there will be of the clock failing to strike when the oil gets thick.

Rack Tail.—A frequent source of trouble in some old clocks is the spring tail to the rack ; it is intended to allow the hands to set forward without allowing the clock to strike. If the spring is weak and the rack spring strong, it sometimes gives a little and allows the rack to fall lower than it should, consequently a wrong hour is struck ; an excess of end-shake to the hour-wheel will also cause this fault, if the snail is mounted on the hour-wheel pipe. This is, of course, easily corrected by a thicker collet in front of the minute hand.

Pendulum Suspension Spring.—This in ordinary clocks gets but little attention. The best material is straight lengths of steel, to be obtained from the mainspring maker, of various thicknesses. The chops at the top of the spring are usually made either by folding a piece of brass over to form both sides, or by cutting a slit in a piece of brass of suitable thicknesses, and closing the slit down with the hammer upon the spring until it fits it. A much better plan is to make the chops of two pieces of brass, and rivet them together ; the bottom edges should be slightly rounded off to prevent any chance of the spring breaking at that point, as it sometimes does if the edges are left sharp. Most suspension springs err in being too thick, but it is not always advisable to substitute a much thinner spring, especially should there be but little room for the pendulum to vibrate in, as the arc may be so much increased as to cause the pendulum to strike the sides of the case, rendering it necessary to substitute a lighter weight or a weaker mainspring. The slit in the top of the pendulum is also generally cut with a thin saw, and then closed in ; but there is no certainty of keeping it straight this way, and it is better to file a true slot and fit a slip of brass to fill it up to the proper size, thus keeping the spring true with the rod.

Main Springs.—In selecting a new mainspring it is often not safe to accept the thickness of the old one as a guide. English mainsprings now are made more elastic and of a better surface than formerly ; and with a pin pallet escapement having but little recoil, a stronger main spring may cause the pins to bottom in the escape wheel, or if there is but little room in the case the pendulum may strike the sides. This latter difficulty is also likely to arise with a half-dead escapement. If the old spring is not broken an adjusting rod may be put on the barrel-square and the weight set to balance the spring when it is wound two or three turns. The strength of the new spring can then be compared before putting the clock together.

When a spring of proper length is broken close to the eye it will be sufficient to soften the inner end, punch a fresh hole for the hook, and carefully bend round another eye.

Cheap locks.—American and other cheap clocks, that will not pay for taking down, often have the balance or pendulum removed, and are then wound up and immersed in a paraffin bath, to run down. I do not recommend this plan, but merely state the fact that it is practised.

Clock Watch.—[*Montre à grande sonnerie.—Die Selbstschlag-Taschenuhr.*]—A watch that strikes the hours in passing, as distinguished from a repeater which strikes the hours at any time on putting special mechanism in motion.

Club Tooth.—[*Dent à talon.—Der Kolbenzahn.*]—The form of tooth mostly used for lever escape wheels of watches and which has part of the impulse angle on the tooth. (See Lever Escapement.)

Cock.—[*Coq.—Der Kloben.*]—A bracket other than a hang-down bracket.—The balance cock of a watch is the bracket which supports the upper end of the balance staff. Escape cock is the bracket that supports the upper end of the escape wheel and pallet staff arbors. In clocks, the pendulum cock is the bracket supporting the pendulum.

Cole, James Ferguson (born 1799 ; died 1880), an able watchmaker and expert springer. He devoted much attention to the lever escapement, of which he devised several forms, and was for some time Vice-President of the Horological Institute.

Collet.—[*Collet.*—*Der Putzen.*]—Part of a cylindrical piece of metal which is of superior diameter to the rest. A collar ; a flange.

(2) A small piece of metal fitting friction tight to the balance staff of a watch or chronometer, and which is pierced to receive the inner or lower end of the balance spring. Sometimes the collet is slit from the outside to the hole, to give it a better grip on the balance staff.

(3) The underside of a brilliant.

(4) That part of a ring in which the stone is set.

Compensation Balance.—[*Balancier compensé.*—*Die Kompensationsunruh.*]—A watch or chronometer balance in which the centre of gyration is caused to approach or recede from the centre of motion in different temperatures, so as to compensate for the effect of such variation not only on the balance itself, but upon the balance spring.

Berthoud, in 1773, tabulated the effect of temperature upon one of his marine watches. He reckoned that in passing from 32° to 92° (Fahr.) it lost per diem by—

Expansion of the balance...	62 secs.
The loss of spring's elastic force	312 ,,
Elongation of the spring	19 ,,
	393 or 6m. 33s.

Doubtless Berthoud's observation was correct as far as the total amount of the temperature error goes, but there appears to be no warrant for assuming that a part of the loss was due to elongation of the spring. The thickness and the width of the spring would be increased in precisely the same proportion as the length, and as the strength of a spring varies as the cube of its thickness, the spring would be absolutely stronger for a rise of temperature if the relative dimensions only were considered.

Sir G. B. Airy, by experiment in 1859, showed that a chronometer with a plain uncompensated brass balance lost on its rate 6.11 secs. in 24 hours for each degree of increase of temperature.

To counteract this effect of change of temperature, chronometers and fine watches are furnished with a balance which expands and contracts with heat and cold. The halves of the rim are free at one end and fixed at the other to the central arm, which is of steel. The inner part of the rim is of steel, and the outer part, which is of brass twice the thickness of the inner, is melted on to the steel. As brass expands more than steel, the effects of an increase of temperature is that the brass in its struggle to expand bends the rim inwards, thus practically reducing the size of the balance. With a decrease of temperature the action is reversed. The action, which is very small at the fixed ends of the rim, increases towards the free ends, where it is greatest. In a marine chronometer there is one large weight at about the middle of each half rim, which is shifted to or from the fixed end, according as the compensation is found on trial to be less or more than is desired. In pocket chronometers and watches a number of holes are drilled and tapped in the rim, and the compensation is varied by shifting screws with large heads from one hole to another, or by substituting a heavier or lighter screw. In the marine balance there are two screws with heavy nuts on opposite sides of the rim, close to the central arm, for bringing the chronometer to time. These nuts are slit, as shown in the drawing, to clasp the screw spring-tight and so avoid backlash. In watch balances there are four such screws placed at equal distances round the rim. These, of course, are not touched for temperature adjustment.

Fig. 1 shows the Marine Chronometer Balance. A compensation watch balance is shown in Fig. 2. In all but the finest

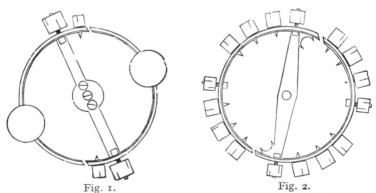

Fig. 1.   Fig. 2.

Marine Chronometer Balance.   Compensation Balance for Watches.

work the quarter mean-time screws are not fitted with nuts, but made with heavy heads, and screwed into the balance from the outside like the compensation screws. It will be observed that the cuts in the rims of the balances are not radial. The object of cutting them at an angle, as shown, is that the free end of the rim may be stopped from bending unduly towards the centre when the balance is roughly handled.

Before using a compensation balance it is the practice of good adjusters to spin it close to the flame of a lamp, so as to subject it to a higher temperature than it is ever likely to meet with in use. Mr. Arthur Webb raises the temperature till the ends of the rim butt at the notch. The balance is then placed on a cold plate, and afterwards tested for poise. If necessary the balance is trued, and the operation repeated till the balance after heating is found to be in poise, or is rejected. Mr. Walsh tells me that to get a low temperature after heating he immerses the balance in ether. Mr. Arthur Webb suggests that compensation balances would be more certain and permanent in their action if they were hardened and tempered after the brass is melted on to the steel. It appears probable that uncertainty of action may sometimes be traced to careless hammering of the brass, which is better compressed by rolling than by hammering.

Sir G. B. Airy has demonstrated that the loss in heat from the weakening of the balance spring is uniformly in proportion to the increase of temperature. But the compensation balance fails to meet the temperature error exactly ; the rims expand a little too much with decrease of temperature, and with increase of temperature the contraction of the rims is insufficient ; consequently a watch or chronometer can be correctly adjusted for temperature at two points only. A marine chronometer is usually adjusted at 45° and 90°, unless special adjustment is ordered to suit particularly hot or cold climates ; pocket watches at about 50° and 85°. In this range there would be what is called a middle temperature error of about 2 secs. in 24 hours with a steel balance spring. The amount of the middle temperature error cannot be absolutely predicated, for in low temperatures when the balance is larger in diameter, the arc of vibration is less than in high temperatures when the balance is smaller, and consequently its time of vibration is affected by the isochronism, or otherwise, of the balance spring. And advantage is sometimes taken of this circumstance to lessen the middle temperature error by leaving the piece fast in the short arcs. To avoid middle temperature error in marine chronometers, various forms of compensation balances have been devised, and numberless additions or auxiliaries have been attached to the ordinary form of balance for the same purpose. (See Auxiliary and Middle Temperature error.)

Compensation Adjustment.—A hot and cold chamber are required for the temperature adjustment. The " oven " is a box made of sheet copper or zinc, generally with a water jacket to the bottom, the exterior of which is heated by means of a gas jet. There is sometimes an automatic gas governor for keeping the temperature uniform. (See Oven.) The oven is furnished with a glass door, and, of course, a thermometer. The " ice-box " (which may often be dispensed with in winter) is also a metal chamber, with a receptacle for ice round the sides, jacketed all over with a non-conductor. The adjustment for temperature is made after observations of the alteration of rate in the two extremes at which it is decided to expose the piece. A short exposure to the temperature, or single observation, cannot be taken as a reliable indication of the effect, for the unnatural connection of the metals composing the rim of the balance requires time to settle.

The usual course is to place the piece to be tested, after its rate has been carefully noted, into the oven. After twenty-four hours, the rate is again noted. Say it has gained on its rate 8 seconds. It is them removed to the ice-box, and subjected to the other extreme temperature for 24 hours. At the end of that period a comparison shows that it has lost on its rate 7 seconds. Although the alteration in the two extremes is not equal, there is sufficient evidence that the balance is over-compensated. In the oven, the rims bending too far inwards reduced the effective diameter of the balance too much, and caused a consequent gain. In the ice-box the rims expanded too much, and as a consequence the piece went slower. If it is a marine chronometer under trial, the weights have to be shifted a little towards the fixed ends of the rim. They must be shifted equally, or the balance will be thrown out of poise, and it is well to see that the slots in the weights are easy and do not grip the rim. If a watch is being tested, two opposite screws must be shifted towards the fixed end, care being taken not to screw them too tight for the rim.

The piece is then again subjected to the extremes of temperature, and as the compensation adjustment gets closer the piece is taken from the ice-box and placed a second time in the oven for verification before the alteration is made. As the trial proceeds the piece is allowed to remain more than 24 hours in each extreme, oftentimes a week. When the compensation is perfect, the balance is poised, the piece (if it is a watch) is adjusted for positions and finally brought to time. (See Balance Spring.)

Compensated Pendulum.—[*Pendule compensé.*—*Das Kompensationspendel.*]—A pendulum constructed so that the distance between the point of suspension and the centre of oscillation remains constant during the variations of temperature. (See Pendulum.)

Compensation Curb.—[*Compensateur.*—*Der Kompensations-bogen Rückerszeiger.*]—A laminated bar composed of brass and steel, fixed at one end and free at the other ; the free end carries the curb pins that regulate the acting length of a watch or chronometer balance spring. The compensation curb was invented by Harrison, but though often modified and re-invented, it is never used now. Bréguet devised a form in which only one curb pin was affected by the compensating laminæ ; this pin was caused to approach and recede from the other, thus varying the play of the spring between the pins.

Conical Pendulum.—[*Pendule conique.*—*Das konische Pendel.*]—A pendulum whose bob moves in a circle. Conical pendulums are not much used except for equatorial clocks or other purposes where continuous motion is desired. One revolution of a conical pendulum is performed in the same time that a vibrating pendulum, whose length is equal to the vertical height of a conical pendulum makes two vibrations. If extra impulse is given to a conical pendulum, the circle described is enlarged, the vertical height lessened, and the time of its revolution decreased. There is generally a spade attached to and revolving with the pendulum bob, so arranged that as the circle of the pendulum is enlarged the spade dips deeper into a vessel containing glycerine or some other resisting medium, and by thus offering increased opposition to the progress of the pendulim, brings it back to its normal position.

Conical Pivot.—[*Pivot conique.*—*Der konische Zapfen.*]—The form of pivot used in English watches when the holes have end stones. The name does not correctly express the shape of the pivot itself, which should be straight ; it is only the shoulder that can be at all conical. (See Polishing.)

Consular Case.—Usually understood to be the ordinary double-bottom watch case when fitted with a high-rounded glass. The original consular cases were single-bottomed and applied to watches wound through a hole in the dial. They took the place of the old pair cases and were so named in honour of Napoleon Bonaparte, at that time Consul of France.

Contrate Wheel.—[*Roue de champ.*—*Das Kronrad.*]—The fourth wheel in a watch having a verge escapement. A form of wheel used to transmit motion from an arbor to another which stands at right angles to the first. In a contrate wheel the teeth, instead of sticking out from the periphery, stand up at right angles to the plane of the wheel.

Conversion. — [*Conversion.* — *Umarbeitung.*] — A converted watch is one in which an escapement of a different kind has been substituted for the original one. A great number of verge watches have lever escapements fitted to them. Duplex, and

H

occasionally chronometer escapements, are also discarded for levers. The operation of converting is spoken of as making a conversion.

Many conversions have the fourth wheel too large with correspondingly coarse teeth, the escape pinion being in some cases larger than the third. A fourth wheel about three-fourths the size of the old contrate wheel will generally be right. Pallets of 6-size to a 10, 12, or 14-sized movement will do. Balance a little under half the diameter of top-plate, if the old verge balance is not used. The old jewel holes should not be used even if sound, as they are mostly too large, a hindrance to a good vibration. A new bar to carry the escapement is a cleaner job than using the old potance. This may be either screwed to the top plate, in which case the pallets and wheel can be put together first, or to the pillar plate near the barrel and the fourth arbor. The third arbor is cut off near the pinion face and a pivot formed. When the pinion leaves are very long they must be turned back and refaced, or there may be too little freedom for the pallets and wheel.

Suitable numbers for the fourth wheel and escape pinion can be selected from the Table of Lever Trains in " Train " if the verge has a seconds train, that is, if the contrate wheel rotates once in a minute. For non-seconds train the following rule may be observed—

Crown Wheel of 15.—When the verge has a crown wheel of 15 teeth and an escape pinion of either 7 or 8, use a pinion of the same number and a fourth wheel with the same number of teeth as the contrate wheel.

If the escape pinion of verge has 6 leaves use a pinion of 7 and add a sixth to the number of contrate wheel for fourth-wheel teeth. For instance, contrate wheel 54 and pinion of 6 ; use pinion of 7 and fourth-wheel 63.

Crown Wheel of 13.—When the verge has an escape pinion of 6, use an escape pinion of 7 and the number of contrate wheel teeth for fourth wheel. (The effect will be to take a ninetieth part from the original number of vibrations, which in a train not exceeding 18,000 will never be more than 198.)

When the verge has an escape pinion of 7, use an escape pinion of 8 and the number of contrate wheel teeth for fourth wheel. (This will add a one-hundredth part to the original number of vibrations.)

Crown Wheel of 11.—When the verge has an escape pinion of 6, use an escape pinion of 8 and one less than the number of contrate wheel teeth for fourth wheel.

When the verge has an escape pinion of 7, use a pinion of 8 and a fourth wheel with one-sixth less teeth than the contrate wheel.

Crown Wheel of 9.—The few verge trains with crown wheel of nine have escape pinions of 6. In converting use an escape pinion of 8 and a fourth wheel with one-fifth less teeth than the contrate wheel.

Full directions for applying the balance spring are given under the head of "Balance Spring," but before selecting the spring it is necessary to ascertain the number of vibrations.

Multiply together 30 and the numbers of centre, third, and fourth wheels.

Also multiply together the numbers of third, fourth, and escape pinions.

Divide the first product by the second, and the result will be the number of vibrations per hour.

For example : Take a watch with centre wheel 80, third 72, fourth 50, and pinions of 8. Here $30 \times 80 \times 72 \times 50 = 8,640,000$ and $8 \times 8 \times 8 = 512$. Then $8,640,000 \div 512 = 16,875$, the number of vibrations per hour. The number of vibrations per minute can, of course, be obtained by dividing this number by 60.

It is also useful in bringing the watch to time to know the number of seconds in which the fourth wheel rotates.

Multiply together 3,600 and the numbers of the third and fourth pinions.

Also multiply together the numbers of centre and third wheels.

Divide the first product by the second, and the result will be the time of one rotation of fourth wheel in seconds.

Taking the preceding example :—$3,600 \times 8 \times 8 = 230,400$, and $80 \times 72 = 5,760$. Then $230,400 \div 5,760 = 40$ secs.

The above rules render the calculation of conversions so simple that Tables of Trains appear unnecessary ; but for the few who still desire them they are given at the end of the book.

Coomb.—[*Limon de pierre noire.—Der Wassersteinschmutz.*] —The paste obtained by rubbing together two pieces of blue-stone with oil. It is used for polishing brass. (See Polishing.)

Copper.—[*Cuivre.—Das Kupfer.*]

Corn Tongs.—[*Brucelles pour joyaux.—Die Kornzange.*]— Tweezers with the gripping points formed to resemble the shell of a barleycorn. They are used by jewellers for picking up stones, &c.

Cremaillere.—[*Crémaillère.—Der Aufziehrechen Repititions-uhr.*]—The winding rack of a repeating watch.

Crescent, Passing Hollow.—[*Passage de fourchette.—Der Ausschnitt in der Rolle, Ankerhemmung.*]—The hollow formed in the roller of a lever escapement to permit of the passing of the safety pin.

Crocus.—(See Red-Stuff.)

Crossed Out Wheel, Pierced Wheel.—[*Roue croisée.—Das*

Schenkelrad.]—A wheel with arms.   The radius of a circle divides the circumference into six equal parts, and the marks so placed would denote the position of the arms for wheels with six arms. For wheels with three arms every other mark may be taken.

Crown Wheel.—[*Roue de recontre.—Das Gangrad einer Spindeluhr.*]—The escape wheel of the verge escapement.

Crutch.—[*Fourchette de pendule.—Die Pendelgabel.*]—A wire fixed to the pallet staff arbor of a clock.   The free end of it communicates impulse to the pendulum.   It either passes into a longitudinal slit in the pendulum rod or is formed into two fingers to embrace it.   The pendulum rod is sometimes fitted with a flat piece of brass to work in the crutch.   Should this be very closely fitted and the pendulum spring a little twisted, the clock will stop, a fault occasionally overlooked.   Where the crutch is in contact with the pendulum rod it should be VERY SLIGHTLY oiled.

Crystal Case.—[*Boîte à glace plate.—Uhr mit Flachglas.*]—A watch case constructed for a flat glass.   The bezel not jointed gives the best fit.   The bevel of the glass should not be carried too far into the centre, or the figures become indistinct ; nor should the flange of the bezel be too narrow, or it may help to cause the same fault.

Cumming, Alexander (born at Edinburgh, about 1732, died at Pentonville, 1814), author of an excellent treatise on clock work, which was published in 1766.   He kept a shop in Fleet Street which after his death was occupied by his nephew, John Grant.   Among the fine and curious clocks at Buckingham Palace is one Cumming made for George III., which registers the height of the barometer every day throughout the year.   He had £2,000 for the clock, and £200 a year for looking after it.

Curb Pins.—[*Goupilles de raquette.—Die Rückerstifte.*]—Two small pins embracing the balance spring of a watch near its attachment to the stud.   The time of vibration of the balance is altered according as these pins are shifted by means of the index to or from the point of attachment of the spring.   If they are moved towards the attachment, the amount of spring practically in use is lengthened, and the watch loses ; if shifted the other way the contrary effect is produced.

Cycloid.—[*Cycloide.—Die Cycloide.*]—A curve generated by a point in the circumference of a circle rolling along a straight line.   The correct path for a pendulum to secure uniformity in the time of its vibration through arcs of different extent.

Cycloidal Cheeks.—[*Guides cycloidaux.—Cycloidische Backen.*]—Pieces of metal placed on each side of a pendulum suspension cord or spring to lead the pendulum through a cycloid. Not used now.   (See Huyghens.)

Cylinder Escapement, Horizontal Escapement.—[*Echappement à cylindre.—Die Zylinderhemmung.*]—(Patented by
Tompion in 1695 and subsequently perfected by Graham.) The
balance with this escapement is mounted on a hollow cylinder
large enough in the bore to admit a tooth of the escape wheel.
Nearly one-half of the cylinder is cut away where the teeth enter,
and impulse is given to the balance by the teeth, which are
wedge-shaped, rubbing against the edge of the cylinder as they
enter and leave. The teeth of the Verge Escapement lie in a
vertical plane in the plan of a watch, and the term horizontal,
therefore, fairly distinguished the Cylinder Escapement when it
is introduced, but now that all the escapements in general use
answer to the title, " Cylinder Escapement " appears to be the
more suitable definition.

The Cylinder is essentially a frictional as distinguished from
a detached escapement. It performs fairly well, and is just
suited for the lower grades of watches. The vibrations of the
balance are not so much affected by inequality in the force
transmitted and other faults if the escapement is a frictional one,
and the work comparatively coarse, as when a highly detached
escapement and very fine pivots are used. It is certainly remarkable that English watchmakers should have been so baffled by a
constructional difficulty as to throw aside the cylinder escapement. Mudge and other eminent English makers used hard
brass for the escape wheel, and, occasionally, ruby for the
cylinder, but without overcoming the tendency to cutting and
excessive wear of the acting surfaces. It remained for the Swiss
to bring the problem to a successful issue by making both wheel
and cylinder of steel, and hardening them. The production of
the cylinder escapement is now monopolised by the Swiss and
the French, who, with the aid of machinery, manufacture the
escape wheels and cylinders for an almost incredibly low price.

Action of the Escapement.—Fig. 1 is a plan of the cylinder
escapement, in which the point of a tooth of the escape wheel
is pressing against the outside of the shell of the cylinder. As
the cylinder, on which the balance is mounted, moves round in
the direction of the arrow, the wedge-shaped tooth of the escape
wheel pushes into the cylinder, thereby giving it impulse. The
tooth cannot escape at the other side of the cylinder, for the shell
of the cylinder at this point is rather more than half a circle ; but
its point rests against the inner side of the shell till the balance
completes its vibration and returns, when the tooth which was
inside the cylinder escapes, and the point of the succeeding tooth
is caught on the outside of the shell. The teeth rise on stalks
from the body of the escape wheel, and the cylinder is cut away
just below the acting part of the exit side, leaving only one-

CYLINDER ESCAPEMENT.

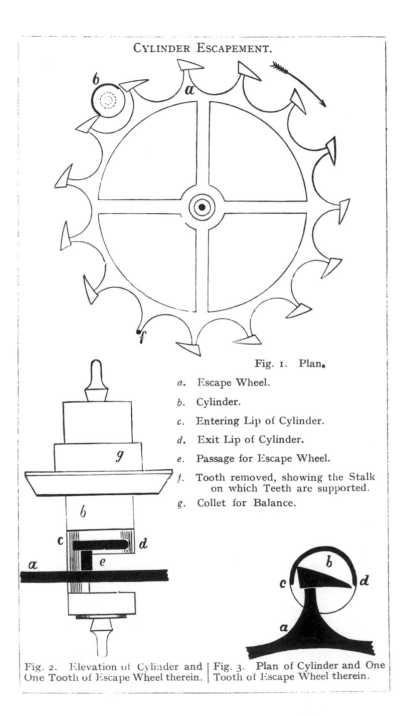

Fig. 1. Plan.

a. Escape Wheel.

b. Cylinder.

c. Entering Lip of Cylinder.

d. Exit Lip of Cylinder.

e. Passage for Escape Wheel.

f. Tooth removed, showing the Stalk on which Teeth are supported.

g. Collet for Balance.

Fig. 2. Elevation of Cylinder and One Tooth of Escape Wheel therein.

Fig. 3. Plan of Cylinder and One Tooth of Escape Wheel therein.

fourth of a circle in order to allow as much vibration as possible. This will be seen very plainly on examining Fig. 2, which is an elevation of the cylinder to an enlarged scale.

Proportion of the Escapement.—The escape wheel has fifteen teeth formed to give impulse to the cylinder during from 20° to 40° of its vibration each way. Lower angles are, as a rule, used with large than with small-sized watches, but to secure the best result either extreme must be avoided. In an escapement with very slight inclines to the wheel teeth, the first part of the tooth does no work, as the tooth drops on to the lip of the cylinder some distance up the plane. On the other hand, a very steep tooth is almost sure to set in action as the oil thickens. The diameter of the cylinder, its thickness, and the length of the wheel teeth are all co-related. The size of the cylinder with relation to the wheel also varies somewhat with the angle of impulse, a very high angle requiring a slightly larger cylinder than a low one. If a cylinder of average thickness is desired for an escapement with medium impulse, its external diameter may be made equal to the extreme diameter of the escape wheel × .115.

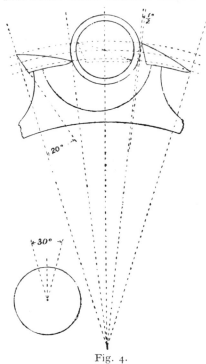

Fig. 4.

Then to set out an escapement, if a lift of say 30° be decided on, a circle on which the points of the teeth will fall is drawn within one representing the extreme diameter of the escape wheel at a distance from it equal to 30° of the circumference of the cylinder. Midway between these two circles the cylinder is planted. (See Fig. 4.) If the point of one tooth is shown resting on the cylinder, a space of half a degree should be allowed for freedom between the opposite side of the cylinder and the heel of the next tooth. From the heel of one tooth to the heel of the next = 24° of the circumference of the wheel ($\frac{360}{15}$—24), and from the point of one tooth to the point of the next also = 24°, so that the teeth may now be drawn. They are extended within the innermost dotted circle to give them a little extra substance, and their tips are rounded a little, leaving the points of the impulse planes the most advanced. The backs of the teeth diverge from a radial line from 12° to 30°, to give the cylinder clearance ; a high-angled tooth requiring to be cut back more than a low one. A curve whose radius is about two-thirds that of the wheel is suitable for rounding the impulse planes of the teeth. The internal diameter of the cylinder should be such as to allow a little freedom for the tooth. The acting part of the shell of the cylinder should be a trifle less than seven-twelfths of a whole circle, with the entering and exit lips rounded as shown in the large plan, Fig. 3, the former both ways, and the latter from the inside only. This rounding of the lips of the cylinder adds a little to the impulse beyond what would be given by the angle on the wheel teeth alone. The diameter of the escape wheel is usually half that of the balance, rather under than over.

Examination of the Escapement.—See that cylinder and wheel are perfectly upright. Remove the balance spring, and put the cylinder and cock in their places. Then with a little power on, and a wedge of cork under the balance to check its motion, try if all the escape wheel teeth have sufficient drop, both inside and out. If, with the cylinder planted the correct depth, there is sufficient drop inside the cylinder and none with-out, the cylinder is too large ; if the reverse fault is apparent the cylinder is too small. If some of the teeth only are without necessary freedom, make a hole in thin sheet brass of such a size that one of the teeth that has proper shake will just enter. Use this as a gauge to shorten the full teeth by. For this purpose use either steel and oilstone dust or a sapphire file, polish well with bell-metal and red stuff, and finish with a burnisher. Be careful to operate on the noses of the teeth only, and round them both ways so that a mere point is in contact with the cylinder. If the inside drop is right, and there is no outside drop with any of the teeth, the cylinder may be changed for one a little smaller or for one of the same inside diameter but thinner. Or the wheel may be changed for one a little LARGER, but in this case be sure the larger wheel will clear the fourth pinion. And with insuffi-cient drop inside changing the wheel for one a little smaller will often be more expeditious than removing the cylinder.

If the teeth of the escape wheel are too high or too low in passing the opening of the cylinder, the wheel should be placed on a cylinder of soft brass or zinc small enough to go inside the teeth, with a hole through it and with a slightly concave face. A hollow punch is placed over the middle of the wheel while it is resting on the concave face of the brass or zinc cylinder, and one or two light taps with a hammer will bend the wheel sufficiently. In fact, care must be taken not to overdo it. It rarely happens that the wheel is free neither of the top nor bottom plug, but should this be the case, sufficient clearance may be obtained by deepening the opening with a steel polisher and oilstone dust or with a sapphire file. A cylinder with too high an opening is bad, for the oil is drawn away from the teeth of the escape wheel.

If a cylinder pivot is bent, it may very readily be straightened if a *bouchon* of a proper size is placed over it to get a leverage.

When the balance spring is at rest, the balance should have to be moved an equal amount each way before a tooth escapes. By gently pressing against the fourth wheel with a peg this may be tried. There is a dot on the balance and three dots on the plate to assist in estimating the amount of lift. When the balance spring is at rest, the dot on the balance should be opposite to the centre dot on the plate. The escapement will then be in beat, that is, provided the dots are properly placed, which should be tested. Turn the balance from its point of rest till a tooth just drops, and note the position of the dot on the balance with reference to one of the outer dots on the plate. Turn the balance in the opposite direction till a tooth drops again, and if the dot on the balance is then in the same position with reference to the other outer dot, the escapement will be in beat. The two outer dots should mark the extent of the lifting, and the dot on the balance would then be coincident with them as the teeth dropped when tried this way ; but the dots may be a little too wide or too close, and it will, therefore, be sufficient if the dot on the balance bears the same *relative* position to them as just explained ; but if it is found that the lift is unequal from the point of rest, the balance spring collet must be shifted in the direction of the least lift till the lift is equal. A new mark should then be made on the balance opposite to the central dot on the plate.

When the balance is at rest, the banking pin in the balance should be opposite to the banking stud in the cock, so as to give equal vibration on both sides. This is important for the following reason. The banking pin allows nearly a turn of vibration, and the shell of the cylinder is but little over half a turn, so that as the outside of the shell gets round towards the centre of the escape wheel, when a tooth is at rest outside of the cylinder, the point of a tooth may escape over the exit lip and jamb the cylinder unless the vibration is pretty equally divided. When the banking is properly adjusted, and a tooth is at rest inside the cylinder, bring the balance round till the banking pin is against the stud ; there should then be perceptible shake between the cylinder and the plane of the escape wheel. If there is no shake the wheel may be freed by taking a little off the edge of the passage of the cylinder where it fouls the wheel, by means of a sapphire file, or a larger banking pin may be substituted at the judgment of the operator. See that the banking pin and stud are perfectly dry and clean before leaving them : a sticky banking often stops a watch. Cylinder watches and timepieces after going for a few months sometimes increase their vibrations so much as to persistently bank. To meet this fault a weaker mainspring may be used, or a larger balance, or a wheel with a smaller angle of impulse. The quickest way is to *very slightly* top the wheel by holding a piece of Arkansas stone against the teeth, afterwards polishing with boxwood and red stuff. So little taken off the wheel in this way as to be hardly perceptible will have great effect.

Broaching Cylinder Escape Wheels.—A holder like the sketch is useful for gripping cylinder escape wheels while they

are being broached. The face of the lantern is crossed out to admit the arms of the wheel. There are four pillars, and there is a projection at the end of each pillar, very much like the tooth of a cylinder escape wheel, so as to form what is called a bayonet joint. The arms of the wheel enter the cruciform slits ; the wheel is turned as far as the pillars will allow, and then when the screw is tightened, the wheel is gripped between the screw and the projections. Before the hole is broached a bit of wire should be inserted in the hole, and a flame directed on the other end of the wire. Sufficient heat will pass along the wire to lower the temper of the hole.

Running in a Cylinder.—Many different methods of procedure are adopted for running in a cylinder. The following, recommended by Schoof, is, from his experience, worthy of consideration. Remove balance cock and chariot, and screw a brass plate of this shape, in place of the latter, so that one of the holes is in the position of bottom jewel. Select a suitable cylinder and put it in the frame with the arbor going through the hole, in the piece of brass upon which the body of cylinder rests. Let the tooth of escape wheel pass into the cylinder, which turn round till the plane of wheel is in the passage. If the cylinder is too high, turn away the bottom of the body. Mark exact height for balance just above the bar of escape wheel. Reduce the lower arbor of cylinder to the thickness of the chariot. Screw chariot and cock to their places, take whole length with a pinion gauge or other measurer, and make the cylinder to it by shortening the upper arbor. To give strength to the cylinder fill it with shellac, when the proper height for wheel teeth has been taken.

A useful tool, by M. Marches', used for getting the distance between the lower end stone and the passage when replacing a broken cylinder, is shown in Fig. 7. The larger view to the right hand is sufficiently explanatory of its application. (See also " Cylinder Height Tool.")

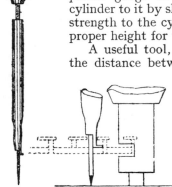

Fig. 7.

New Cylinder Plug. — A slight recess to just admit the cylinder should be turned in a brass stake that has a hole through it large enough for the plug to pass, and a tap on a knee punch, shaped as shown, should remove the old plug. But if the plug is tight, the body of the shell may be stretched slightly, by laying it on a hollowed stake and giving a few taps round its circumference with a hammer. The new plug may be gauged with a Micrometer or Registering Callipers. It should be perfectly parallel. If taper, it is almost sure to split the shell. The pivot is roughed down before the plug is inserted in the shell. The shell is rested on a foot punch, like the second in the sketch, while pressing the plug into its place.

Cylinder Gauge, Jacot Gauge.—[*Filière pour cylindres.—Das Zylindermass.*]—A steel plate having two tapered slits. The full diameter of a cylinder for a cylinder escapement having been noted in one slit, the corresponding number in the other slit gives the size of the great opening, or the proper distance from the back to the lips, which should be .58 of the diameter.

Cylinder Height Tool.—[*Outil pour hauteurs de cylindres.— Das Zylinderhöhenmass.*]—A gauge used for taking measurements when replacing a cylinder.

Fig. 1 is the most usual pattern. It consists of a jaw fixed to a brass tube, and a similar jaw free to move up and down. The brass tube is slit to receive the extremity of the movable jaw, which is fixed to a nut fitted into the tube. The nut terminates in a long steel pivot small enough to enter a watch jewel hole. When the jaws are closed the pivot is just flush with the nose of the tool, consequently the distance it is projected beyond the nose of the tool is exactly represented by the distance of the jaws apart. The screw with the milled head at the top is for opening and closing the gauge, and the set screw at the side is for securing a measurement.

Fig. 2 is a new and improved tool. It has a divided base for striding over a cylinder cock, and there is a pump centre at the bottom, which is kept out by a spiral spring in the arbor, and may be pushed back by the finger and secured in any position by a set screw at the side. To take the extreme length of a cylinder with Fig. 2, the jaws are closed by means of the screw at the top, and the cylinder cock having been removed, the tool is rested on the watch plate, the pump-centre released, and allowed to enter

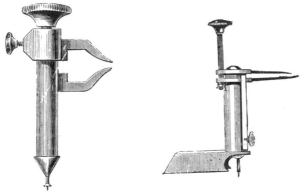

Fig. 1.   Fig. 2.

the bottom jewel hole and rest on the end stone. It is then secured by the set screw. Then the tool is taken away and by means of the top screw the jaws opened something wider than the length of the cylinder. The cylinder cock, with index and end stone removed, is placed in position, the tool is again rested on the watch plate, and the top screw turned till the point of the pump centre is just about to enter the top jewel hole. The distance between the jaws will then be the extreme length of the cylinder required. This example will be sufficient to indicate the method of taking other heights.

Cylinder Plugs.—[*Tampons de cylindre.*—*Die Zylinderspunde.*] —Plugs fitting into the top and bottom of the cylinder of a cylinder escapement, at the extremities of which the pivots are formed.

Dead Beat Escapement.—[*Echappement à repos.*—*Die Grahamhemmung.*]—An escapement in which the escape wheel does not recoil. Generally a dead beat escapement is understood to mean the Graham escapement used in regulators and fine clocks.

The Dead Beat or "Graham" Escapement.—(Invented by George Graham at the beginning of the eighteenth century.)

For regulators and other clocks with seconds pendulum this escapement, which is shown on page 126, is the one most generally approved. The only defect inherent in its construction is that the thickening of the oil on the pallet will affect the rate of the clock after it has been going some time. Notwithstanding this it has held its own against all other escapements, on account of its simplicity and certainty of action. The pallets of the Graham escapement were formerly made to embrace fifteen teeth of the wheel, and until recently ten, but now many escapements are made as shown in the drawing, with the pallets embracing but eight. This reduces the length of the impulse plane and the length of run on the dead face for a given arc of vibration, and consequently the relative effect of the thickening of the oil. The angle of impulse is kept small for the same reason. There is not much gained by making the pallets embrace a less number of teeth than eight, for the shake in the pivot holes and inaccuracies of work cannot be reduced in the same ratio, and are therefore greater in proportion. This involves larger angles and more drop. It is purely a practical question, and has been decided by the adoption of eight teeth as a good mean for regulators and fine clocks where the shakes are small. For large clocks of a rougher character, ten teeth are a good number for the pallets to embrace.

To Set Out the Escapement.—Draw a circle representing the escape wheel to any convenient size, and, assuming the wheel to have 30 teeth and the pallets are to embrace eight of them, set off on each side of a centre line, by means of a protractor, 45°. Lines drawn from the centre of the escape wheel through these points will pass through the centre of the impulse faces of the pallets ; thus, 360 (number of degrees in the whole circle) divided by 30 (proposed number of teeth) = 12, which is the number of degrees between one tooth and the next. Between eight teeth there are seven such spaces and 12 × 7 = 84, and 84 + 6 (half of one space) = 90, the number of degrees between the centres of the pallets. The proper position for the pallet staff centre will be indicated by the intersection of tangents to the wheel circle drawn from the centres of the pallets. But it happens that a tangent of 45° = the radius, and, therefore, the practical method adopted is to make the pallet arms from the staff hole to the centre of impulse face equal to the radius of the escape wheel. If we take the radius of wheel to be =1, it will be found that with the pallet arms this length, the height of the pallet staff hole from the centre of the wheel will be 1·41, and the horizontal distance between the impulse faces of the pallets will be 1 41 also.

The width of each pallet is equal to half the distance between one tooth and the next, less drop, which need not be more than half a degree if the escape wheel teeth are made thin as they should be. The dead faces of the pallets are curves struck from the pallet staff hole The escaping arc = two degrees, is divided into $1\frac{1}{2}°$ of impulse and $\frac{1}{2}°$ of rest. $1\frac{1}{2}°$ of impulse is quite enough

Dead Beat or Graham "Escapement."

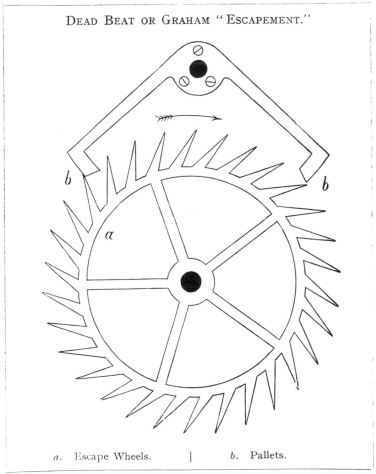

a. Escape Wheels.   *b.* Pallets.

if the escapement is properly made, and if increased beyond 2°, it will be at the cost of the timekeeping properties of the clock from the effect of the thickening of the oil already referred to.

From the centre of the wheel set off two radial lines barely 3° on each side of the radial lines already drawn to mark the centre of the pallets. Then strike the curved dead faces of the pallets just touching the radial lines last drawn.

Now from the pallet centre draw lines through the spot where the curved locking face of each pallet cuts the wheel circle. If you look at the engraving you will see that a wheel tooth is resting on the left-hand pallet. The amount of this rest is $\frac{1}{2}°$, as already stated. Mark off this $\frac{1}{2}°$, which gives the position of the locking corner of the pallet, and then set off another line $1\frac{1}{2}°$ below it, which will mark the spot for the other corner of the pallet. On the right-hand pallet, the line already drawn marks the extreme corner, and it is only necessary, in order to get the locking corner, to set off a line $1\frac{1}{2}°$ ABOVE it.

The wheel teeth diverge from a radial line about 10°, so that their tips only touch the dead faces of the pallets. The tips may be shown each about a quarter of a degree in width.

For escaping over ten teeth, the distance between the centre of the wheel and the centre of the pallet staff should be equal to the diameter of the wheel; with this exception the preceding directions are applicable for setting out.

The wheel is of hard hammered brass, and for regulators is made from an inch and a half to two inches diameter, and very light. The pallets are usually of steel, nicely fitted to the arbor, and in addition, screwed to a collet thereon as shown. In the best clocks the acting faces are jewelled. Sometimes the pallet arms are cast of brass, and the pallets formed of solid jewels. Many good clockmakers put two banking pins in the plate, one on each side of the crutch, to prevent the pallets from being jammed into the wheel by careless handling.

The Graham escapement requires a heavy pendulum, especially if the train is comparatively rough. The clock weight must be sufficient to overcome increased resistance arising from inaccuracy of work; consequently, when the train runs freely, so much extra pressure is thrown upon the dead faces of the pallets that a light pendulum has not enough energy to unlock, and the clock stops.

For clocks with shorter than half-seconds pendulums the pallets are generally made " half-dead," that is, the rests, instead of being curves struck from the pallet staff hole, are struck with a larger radius so as to give a slight recoil to the wheel. Whatever the length of pendulum, it is better to strike the locking faces with a radius rather too large than to err in the opposite direction.

Dead Smooth.—[*Lime très fine.*—*Feile mit feinstem Hieb.*]— The cut of the finest kind of file.

Decimal Fractions.—[*Fractions Décimales.*—*Decimalbrüche.*] —Fractions expressed by tenths. As I have used Decimal

Fractions throughout this book, a few explanatory remarks thereon may not be unacceptable to the younger readers.

Ciphers to the right hand of decimals cause no difference in their value. ·5 or ·50 or ·500 is of the same value. ·5 represents $\frac{5}{10}$, ·50 represents $\frac{50}{100}$, and ·500 represents $\frac{500}{1000}$, and each of them is equal to $\frac{1}{2}$. But every cipher placed on the left hand of decimals diminishes their value tenfold ; thus, ·3, ·03, and ·003 represent respectively three-tenths, three-hundredths, and three-thousandths.

In the addition of decimals, or whole numbers and decimals, the lines must be arranged so that the decimal points are all directly under one another. The addition will then be proceeded with as in whole numbers.

In the subtraction of decimals, or decimals and whole numbers, arrange the lines as in addition.

In the multiplication of decimals, or decimals and whole numbers, arrange the figures as with whole numbers and point off from the right hand of the product as many decimals as there are in the multiplier and multiplicand together. If there are not enough figures in the product to do this, add the requisite number of ciphers to the LEFT hand of the product.

In the division of decimals, or whole numbers and decimals, the quotient must contain as many decimals as the dividend has more than the divisor. For instance, if there are two decimal figures in the divisor and three in the dividend, cut off one figure at the right hand of the quotient as a decimal. If there are not enough figures in the quotient to carry out this rule, add the requisite number of ciphers to the LEFT hand of the quotient.

To reduce a vulgar fraction to a decimal, add a cipher or ciphers to the enumerator, and divide by the denominator ; the quotient will be the decimal required.

EXAMPLE.—Reduce $\frac{7}{8}$ to a decimal,

$$8)7\cdot000$$

·875 the decimal required.

When figures in the quotient repeat continually, they are called a repetend, and the last figure should be marked with a dash to distinguish it from a terminal decimal. Sometimes two, three, or more figures will repeat in regular order ; such figures are termed a circulate, and may be indicated in the same manner as a repetend.

Demi Hunter, Cut Hunter.—[*Guichet.*—*Die Halbsavonnette Uhr.*]—A watch case in which a glass of about half the diameter of the hunting cover is let into it.

Denison, E. B. (born 1816, died 1905) —In 1851 Mr. Denison, a barrister holding a good position at the parliamentary bar, was requested by the Government to draw up, in conjunction with the Astronomer-Royal, a specification for the construction of a large clock for the Northern Tower of the Houses of Parliament, then in course of erection. Vulliamy and other leading clockmakers who were invited to tender for the work, all demurred to a stipulation that the clock should be guaranteed to perform within a margin of a minute a week, which they declared to be too small. Mr. Denison would not yield, and the clockmakers were equally firm. Eventually it was decided to entrust the work to Mr. Dent, who was to make a clock from designs to be furnished by Mr. Denison. Mr. Denison's temerity was justified by his success. The Westminster clock turned out to be the finest timekeeper of any public clock in the world. The Double Three-Legged Gravity Escapement was invented for it, besides a new maintaining power and a novel arrangement for letting off the hours to satisfy another of the conditions, which required the first blow of the hour to be given within a second of true time. Mr. Denison was elected President of the British Horological Institute in 1868, and succeeded his father as baronet, taking the title of Sir Edmund Beckett in 1874. In 1886 he was called to the House of Lords under the title of Baron Grimthorpe.

Dennison, Aaron L.—Born at Brunswick, Maine, U.S.A., 1812, died at Birmingham, England, 1895 ; pioneer of American Watchmaking.

Dent, E. J. (born 1790 ; died 1853).—He worked as a finisher of repeating watches till 1830, when he joined J. R. Arnold in partnership at 84, Strand. During the ten years they were together the business was greatly extended, chiefly through the energy and ability of Dent, who, after the partnership expired, established himself at first at 82, then at 61, Strand, and also at 33, Cockspur Street. Dent accepted the stipulation laid down by Mr. Denison on behalf of the Government, that the Westminster Great Clock should be guaranteed to give exact time within a minute a week, and thus secured the contract for making it, on the understanding that it was to be designed by Denison.

Depthing Tool.—[*Compas aux engrenages.*—*Der Eingriff-zirkel.*]—A tool in which a wheel and pinion can be arranged at their proper working depth, and the distance apart of their centres transferred to the plate in which the pivots of their arbors are to run.

Accuracy of construction is absolutely essential in the depthing tool, and before venturing to use a new one it should be tested. The centres should be turned end for end and transposed, ascertaining after each change if there is any deviation in a circle described by the points ; also if the points when they meet exactly coincide. If possible a comparison should be made with an approved tool by trying in both a large and also a small wheel and pinion. The adjusting screw had better be removed so as to see that the joint works smoothly and that the spring has perfect control over it. If the joint is stiff and appears to be dirty, the joint pin may be taken out and the joint thoroughly cleaned.

A steel cone piece to slide on one centre, and a brass trumpet-piece to slide on the opposite one, are useful for unmounted wheels, &c.

In use the depthing tool is held in the left hand, with the adjusting screw for opening and closing the tool pointing to the right. After making sure that the points on the ends of the pinion are true, it is carefully placed face uppermost in the centres on the left. The tool is opened sufficiently for the teeth

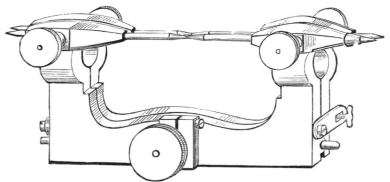

Depthing Tool.

of the wheel to clear the teeth of the pinion, and the wheel is then put in. By means of the regulating screw the teeth of the wheel and pinion may be brought into contact, taking the precaution to feel with a finger of the right hand that they do not butt while coming together, or a broken pivot may be the result. When the teeth are in contact, both the sliding centres of either the wheel or pinion may be loosened and pushed up by the bottom centre or down by the top centre, as may be required, in order to get the faces of the teeth level so as better to observe the depth. The faces being level, and the bottom centre of the pinion tightened, the top centre may be loosened and pressed against the board until the pinion runs sufficiently friction-tight for the shake of the wheel teeth in it to be felt, and then secured with the nut. Now the tool is held up to the light, the wheel turned by the forefinger, and while looking through the wheel and pinion they are brought into depth by means of the adjusting screw. If the pinion has straight cut leaves the shoulder or beginning of the rounding of a wheel tooth should come into contact with the shoulder of a pinion leaf when the two make parallel lines as in Fig. 2 ; but if it is what movement-makers call a bay-leaf pinion, the pinion line will be angled a little towards the centre when the shoulders come into contact. It is important to see that the engaging contact is at the shoulders, for if it is on the roundings of the teeth a butting action ensues. The shake is then to be tried. It is not sufficient to see that there is freedom in one position of the wheel and pinion ; the pinion should be moved round by minute por-

Fig. 2.

tions, so that it can be ascertained whether there is proper shake from the time of first contact till the leaf leaves the wheel.

Butting is generally indicative of a pinion too large. If the excess is not very great, it may be got over by slightly increasing the depth ; but a large pinion is a bad fault, for with it far too

Fig. 3.

much of the action is before the line of centres. Pinions of less than ten leaves may with advantage be rather under-sized, so as to bring the action more after the line of centres. A pinion must not, however, be so small that the wheel tooth drops from leaf to leaf, although if this fault is not of large extent, it may be corrected by altering the shape of the wheel teeth so that the rounding or shoulder is carried more towards the root of the tooth without shortening the tooth. (See Fig. 3.) The original shaped tooth is shown on the left, and the altered shape on the right. The running of an undersized pinion will be also be improved if the depth is made a trifle shallow. With a full-sized pinion and an engagement too deep, a correction may be made by widening the spaces of the wheel and so shortening the rounding of the teeth. (See Rounding-up Tool.)

Being satisfied that the depth is correct, that there is proper shake and no butting, the depth may be marked off. For this purpose the binding nuts are loosened, and the wheel and pinion taken out of the tool. Then, while one centre is kept tight, the tool is held upright with the tight centre in the hole from which the depth is to be marked, and the loose centre is brought down until it touches the plate. Some caution must be exercised, for with a jewelled hole it occasionally happens that a beginner cracks the jewel in getting the centre to height. Great judgment and care are required to ensure the perfect uprightness of the tool, and if it is out in the slightest degree the depth will not be marked off as it was adjusted in the tool. The centres should fixed, and as they rest on the plate the tool observed critically all round ; if the smallest deviation from the upright can be detected, the tool should be re-set till it is correct. When it is right, a portion of a circle is lightly marked across the line where the wheel or pinion is to be planted. Watch plates are marked by holding the tool still and turning the plate. A drilling centre is then marked on the depth line with a pointed drilling or chamfering tool. This operation requires considerable skill, as does also the drilling off : for however carefully the depthing may have been done, the slightest deviation by the drill, either in centring or drilling, will turn a well-marked depth into a bad-action one. The drilling of the hole perfectly upright, too, is essential for a correct depth. Sometimes a table tool is used to ensure straight holes, or they are drilled in the mandril, but a skilled workman will drill them in the ordinary way to a certainty. A beginner should certainly use the mandril. It is better after centring the plate and catching a centre by means of a graver, to reverse the plate in the mandril and drill from the original mark on the other side, for, although the plate may be set true with the pump centre, it is liable to be drawn a little in fixing, and in a small watch extreme accuracy is desirable. Having drilled the bottom hole, a round broach should be passed through it to remove all burr, and a long peg cut to fit the hole. The rest of the mandril is then brought up almost close to the plate so as to support the peg, and keep it from turning. Now, as the mandril is rotated, the slightest want of truth in the hole will be indicated by a movement of the free end of the peg, and the plate must be shifted till the free end of the peg has no motion whatever. Then the upper plate can be fixed, and the top hole drilled and its uprightness tested in the same manner.

Blunt centres of a depth tool are almost certain to get out of truth if sharpened with an oilstone slip. The best plan is to get a large brass runner with a lantern end, having a hole large enough to allow nearly all the conical point of the runner to come through. Then with a brass screw ferrule on the runner and a male centre in the opposite end of the turns, the point may be turned true with the graver.

Derham, William (born, 1657 ; died, and buried in Upminster, in 1735), a clergyman, author of *The Artificial Clockmaker*.

Detached Escapement.—[*Echappement libre.—Die freie Hemmung.*]—An escapement in which the balance or pendulum is, during a considerable portion of its vibration, free or detached from the train.

Detent.—[*Détente.—Die Gangfeder des Chronometers.*]—A term generally used by watchmakers to indicate a detainer. The detent of a chronometer escapement is the piece of steel carrying the stone which detains or locks the escape wheel. The fusee detent of a watch is the pawl or click which takes into the steel ratchet wheel.

De Vick, Henry.—About 1364 he made for Charles V. of

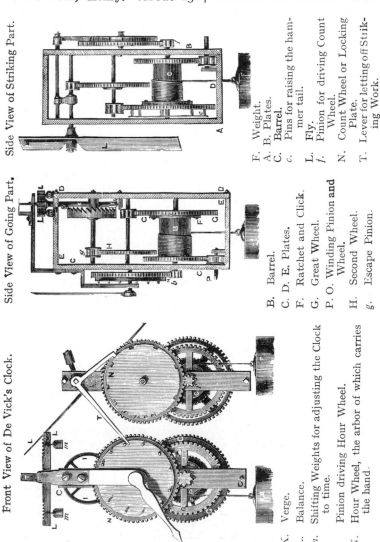

Side View of Striking Part.

F. Weight.
A. B. Plates.
C. Barrel.
c. Pins for raising the hammer tail.
L. Fly.
f. Pinion for driving Count Wheel.
N. Count Wheel or Locking Plate.
T. Lever for letting off Striking Work.

Side View of Going Part.

B. Barrel.
C. D. E. Plates.
F. Ratchet and Click.
G. Great Wheel.
P. O. Winding Pinion and Wheel.
H. Second Wheel.
g. Escape Pinion.

Front View of De Vick's Clock.

K. Verge.
L. Balance.
m. Shifting Weights for adjusting the Clock to time.

Pinion driving Hour Wheel.

N. Hour Wheel, the arbor of which carries the hand.

France the first turret clock of which we have reliable record. It had iron frames and wheels,'and, except that it was controlled by a balance instead of a pendulum, bears remarkable resemblance to the turret clocks of a few years ago. The drawings (p. 133) of the front and side views of De Vick's clock will be of interest.

Dial.—[*Cadran.*—*Das Zifferblatt.*]—(1). The divided plate in a watch or clock used to indicate the motion of the hands. (2). A spring timepiece with a large dial.

Exterior dials of public clocks are generally made too small and with too much decoration, which obscures the position of the hands.

A very good rule is that the diameter of the dial should be one foot for every ten feet the dial is from the ground. Sir Edmund Beckett suggested that dots should be substituted for the figures, so as to render the position of the hands clearer. The same authority advocated a concave form for the dials of public clocks, as with them the parallax of the hands with the dials may be reduced to a minimum, and the hands are more sheltered from the wind. But exposed dials, if concave or with a sunk centre for the hour hand, afford an objectionable lodgment for snow. Illuminated dials for turret clocks are, as a rule, made of opalescent glass. The gas jets behind the dials are turned very low in the daytime by a pin in a wheel, turning once in 24 hours which is driven by the clock. Another pin in the same wheel turns the gas up again at night. A number of holes are drilled in the circumference of the wheel, and according as the nights shorten or lengthen, the man who winds the clock shifts the pins, so that the gas may be turned up at dusk and down as the sun rises.

To Refix a Seconds Piece in an enamel dial, adjust it to position and place three drops of gum on the back of the dial at its junction with the seconds piece. Then gently heat the dial sufficiently to cause white sealing-wax from a pencil-shaped piece applied round the joint to flow into it. When the dial is cool, carefully remove any superfluous wax.

A Name may be removed from an enamel dial by gently rubbing it with a little fine diamantine on the point of the finger.

To Drill or Broach Holes in Enamel Dials.—Use a flat-ended drill or a conical broach of copper into which diamond powder has been hammered. A graver, kept moistened with turpentine, is sometimes used.

An emery countersink as shown in the sketch will be found useful for trimming the edges of holes in dials.

Tarnished Metal Dials.—A good plan is to gently heat the dial and dip in diluted nitric acid, but this must not be adopted for dials with painted figures, for the figures would be destroyed. The following process is not so fatal to the figures, but even with the greatest care it is impossible in some cases to preserve them. Dissolve ½oz. cyanide of potassium in a quart of hot water, and add 2oz. strong liquor ammonia and ½oz. spirits of wine (these two may have been mixed previously). Dip the dials, whether silver, gold, or gilt, in it for a few seconds ; quickly plunge them in warm water ; brush carefully with soap ; rinse and dry in hot box-wood dust.

French Dead Silvering that is discoloured may be restored, and will keep its colour longer than if re-silvered, by brushing it with a very soft brush moistened with cream of tartar and water mixed to the consistency of cream. If the figures are waxed in, no danger need be apprehended, but if painted great care is required to avoid brushing then off.

Silvering Watch and Clock Dials.—Dissolve a stick of nitrate of silver in half-a-pint of rain water ; add two or three tablespoonfuls of common salt, which will at once precipitate the silver in the form of a thick, white curd, called chloride of silver. Let the chloride settle until the liquid is clear ; pour off the water, taking care not to lose any chloride ; add more water, thoroughly stir and again pour off, repeating till no trace of salt or acid can be perceived by the taste. After draining off the water, add to the chloride about two heaped tablespoonfuls each of salt and cream of tartar, and mix thoroughly into a paste, which, when not in use, must not be exposed to the light. To silver a surface of engraved brass, wash the surface clean with a stiff brush and soap. Heat it enough to melt black sealing-wax, which rub on with a stick of wax until the engraving is entirely filled, care being taken not to burn the wax. With a piece of flat pumice-stone and some pulverised pumice-stone and plenty of water, grind off the wax until the brass is exposed in every part, the stoning being constantly in one direction. Finish by laying an even and straight grain across the brass with blue or water of Ayr stone. Take a small quantity of pulverised pumice-stone on the hand, and slightly rub in the same direction, which tends to make an even grain ; the hands must be entirely free from soap or grease. Rinse the brass thoroughly, and before it dries lay it on a clean board, and gently rub the surface with fine salt, using a small wad of clean muslin. When the surface is thoroughly covered with salt, put upon the wad of cloth, done up with a smooth surface, a sufficient quantity of the paste, say to a dial three inches in diameter a piece of the size of a marble, which rub evenly and quickly over the entire surface.

The brass will assume a greyish, streaked appearance ; add quickly to the cloth cream of tartar moistened with water into a thin paste ; continue rubbing until all is evenly whitened. Rinse quickly under a copious stream of water ; and in order to dry rapidly, dip into water as hot as can be borne by the hands, and when heated, holding the brass by the edges, shake off as much of the water as possible, and remove any remaining drops with a clean, dry cloth. The brass should then be heated gently over an alcohol lamp, until the wax glistens without melting, when it may be quickly covered with a thin coat of hard spirit varnish, laid cn with a broad camel's hair brush. For watch dials the varnish may be thinned with spirits of wine.

Dial Feet.—[*Pieds de cadran.—Die Zifferblattfüsse.*]—Short pieces of wire soldered to the back of a watch dial, which fit into corresponding holes in the pillar plate and keep the dial in its proper position. On the other side of the pillar plate small pins are passed through the holes in the dial feet. In repeating watches, where there would not be room for the feet, the dial is often formed with a thin ring to fit over the pillar plate. English repeaters mostly have short flat feet secured by screws in the edge of the pillar plate. Some manufacturers snap their watch dials into a very thin brass rim, which in its turn is snapped into a groove turned in the pillar plate near its edge, a method of attachment that appears in many ways preferable to the feet. In common watches, pins falling out of the dial feet is a fruitful source of trouble. The pins are often too taper, or the hole in the feet too large. The holes should be under a third of the thickness of the foot in size and exactly in the middle of it. When the pin is in position it should just rest on the plate, neither above nor below the surface. A piece of steel hooked to pass over the foot will remove a pin that is pushed in too far. If a dial foot is burst at the hole, a better plan than removing the foot is to encircle it with a piece of tubing soldered to the copper of the dial. The hole in the pillar plate can be opened to suit the tubing, and a new pin fitted with the assurance that the position of the foot has not been altered.

To replace a dial foot, prepare a piece of copper wire with an enlarged surface where it is to be attached to the dial. Tin this surface, scrape away the enamel for its reception by means of a graver moistened with turpentine, tin the copper, place the wire in position and gently heat with a blowpipe.

The subjoined holder is a useful help in soldering a new foot to a dial. It consists of a circular plate with a conical tit in the centre, which is kept down by a spring. On lifting the spring a dial can be slipped underneath ; if the tit is allowed to enter the central hole, it will press the dial sufficiently tight to hold it firmly. Then there is a spring arm fixed by a set screw to the shank of the tool. This may be adjusted to suit the dial feet, and when right will hold a foot perfectly upright and correct to

Holder for Dial Feet, &c.

position, while the soldering operation is going on. It occurs to me that there are many other operations besides soldering dial feet, in which this double spring holder will be available. It is like lending the watchmaker an extra hand.

Diamantine.—[*Diamantine.—Die Diamantine.*]—A preparation of crystallised boron much esteemed as a polishing powder for steel work. " Rubytine," a somewhat similar polishing medium for arbors and pivots, is quicker than diamantine, but does not yield so good a gloss.

Diamond Mill.—[*Meule chargée de diamant.—Die Diamant-schleifscheibe.*]—For cutting and polishing ruby pallets and other hard stones, discs charged with diamond powder and rotated at a high speed are used. The cutting mills are of soft copper about an inch and a half in diameter, into which diamond powder of a coarseness suited to the work has been rolled or hammered. Polishing mills are usually of ivory or tortoiseshell, and very fine diamond powder is used loose instead of being beaten into the mill. Vegetable ivory is now generally preferred. Being slightly porous it takes the diamond powder better and polishes quicker. The diamond powder for charging the mills is graded by pouring it into a vessel containing olive oil and allowing it to settle. The finer diamond powder is then poured off with the oil, and the coarser remains at the bottom of the vessel.

Diamond and Sapphire Files.—[*Limes en diamant et saphire.—Die Steinfeilen.*]—A diamond file is formed of a strip of copper with diamond powder hammered into it. A sapphire file, which is useful for operating on garnet and other soft stones, is made of a piece of sapphire which is flattened down on a diamond mill and shellaced to a brass handle. The file is rendered coarser or finer according to the coarseness or fineness of the mill upon which it is flattened. A file too coarse used for the corners of garnet pallets or other fine edges would be liable to chip the stone. The corners of garnet pallets may be reduced with a strip of copper and diamantine if a sapphire file is not to hand.

Dipleidoscope.—[*Dipléidoscope.—Das Dipleidoscop.*]—A reflecting meridian instrument invented by Mr. J. M. Bloxham

(Patent No. 9793 A.D. 1843), which consists of three plates of glass placed together so as to form a hollow prism. Two of the plates are blackened at the back, and each of them then reflects the rays of the sun on to the third plate. The dipleidoscope is fixed so that these reflections coincide at solar noon. The advantages of the dipleidoscope over the ordinary forms of sun-dials are that the passage of the sun over the meridian is indicated with greater exactness, and the reflections may be discerned in weather too cloudy to see any shadow on the sun dial.

Dividing Plate.—[*Plateforme divisée.—Die Teilscheibe.*]—The circular brass plate in a wheel-cutting engine, in which holes are drilled as a register for the proper division of the wheel teeth.

Dog Screw.—[*Vis clef.—Die Schlüsselschraube.*]—A screw with an eccentric head, or with one side of the head taken off, used for attaching a watch movement to a dome case.

Dog Watch.—A period of two hours which occurs twice in the nautical day, the first beginning at 4 o'clock and the second at 6 o'clock p.m. (See Time.)

Dome.—[*Cuvette.—Der Staubmantel.*]—The inner case immediately covering the movement of a watch is called a dome when it snaps on to the band of the case. A case so made is spoken of as a dome case.

Double-bead.—[*Double-filet.—Doppelrand (Gehäuse).*]—A form of watch case in which there are two beads or projections running round outside the cover.

Double Bottom.—[*Boîte à double fond.—Gehäuse mit doppeltem Boden.*]—The form of watch case used in most full-plate watches. The inner bottom is in one piece with the band of the case, and the movement is attached to the case by means of a bolt and joint. The movement of a watch with a double bottomed case can only be got at from the front of the watch. The bezel has first to be opened and the bolt pushed back, when the movement is free to turn on the joint and fold out the case.

Double Roller Escapement.—[*Echappement à double rouleau.—Die Ankerhemmung mit Doppelrolle.*]—A variety of the Lever Escapement, in which a separate roller is used for the guard action. (See Lever Escapement.)

Double Sunk Dial.—[*Cadran à centre rapporté.—Eingesetztes Stunden und Sekundenblatt.*]—A dial with recesses for the hour hand and seconds hand.

Douzieme.—[*Douzième.—Das Zwölftel.*]—A unit of measure often used by watchmakers. The $\frac{1}{12}$ of a line = $\frac{1}{144}$ of an inch.

Douzieme Gauge.—[*Outil aux Douzièmes.—Das Zwölftelmass.*]—A measuring instrument consisting of two limbs hinged together with the point of attachment closer to one end. The

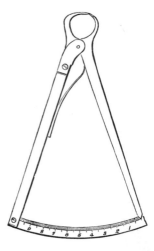

short end of the tool terminates in two jaws which embrace the work to be gauged. Fixed to one of the long limbs is a scale, and the other limb terminates in a pointer, so that the size of any object in the jaws is exaggerated on the scale. The name Douzieme gauge applies only to those instruments that register twelfths of a line. In some gauges of this form the inch is divided into tenths and hundredths, and in others the millimetre is adopted as the unit. Occasionally there is a double scale representing inches and millimetres, with a pointer on the other limb for each scale. A slotted arm and screw are sometimes added to fix the limbs in order to retain a measurement. This, of course, adds to the weight of the tool.

Dovetail.—[*Queue d'aigle.—Der Schwalbenschwanz.*]—Under-cut. A groove made to receive a slide in the form of a fan. In a verge escapement the slide carrying the inner pivot of the escape wheel arbor. In full-plate lever watches the slide is jewelled and supports the bottom pivot of the balance staff.

Draw.—[*Tirage.—Der Anzug (Ankerhemmung).*]—In a lever escapement, the locking faces, instead of being curves struck from the centre of motion of the pallets, are cut back at an angle which is called the " draw," or " the angle of draw." The fronts of the escape wheel teeth are also set an angle so that the teeth bear on their points only, and by this means the lever is drawn to the banking pins so that the guard pin may be free of the roller. In the chronometer escapement, the locking face of the stone and the teeth of the escape wheel are angled in the same way, to keep the detent pressed against the banking, and prevent accidental unlocking through a shake or jar.

Draw Plate.—[*Filière à tirer.—Das Zieheisen.*]—A tool for reducing wire. Holes with rounded mouths are formed in steel or in jewel, through which a wire is drawn.

Drill.—[*Foret.—Der Bohrer.*]—A tool for cutting round holes in hard substances.

Drilling.—[*Perçage.—Bohren.*]—Drilling tools may be divided into two classes, those in which the object to be drilled rotates while the drill is stationary, and those in which the work is fixed and the drill rotates. It may be conceded that as a rule greater accuracy is possible with the object to be pierced rotating ; on the other hand, it is often more convenient, and quicker, to keep the work stationary. The mandril and lathe are examples of the former ; as also is Boley's excellent horizontal driller in Fig. 1.

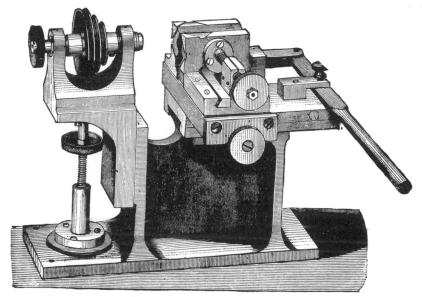

Fig. 1.

This is furnished with a well-designed vice for holding the object to be pierced. It has long angular jaws, so that a piece of wire can be instantly gripped with the assurance that it is in line with the drilling spindle. There is a circular sink on the face of the vice, and another on the top for holding discs to be perforated either through the face or the edge. Having clasped the work in the vice, the centre of the drill can be brought to any spot by means of a screw to adjust the height of the head stock, and a slide rest to traverse the vice. The vertical driller shown in Fig. 2 serves to illustrate the second class. Here the drills are held in a split chuck ; the slide worked by a cross handle at the back is convenient, so is the adjustable stop at the top. There is a lug cast on the bottom of the tool for holding it in the vice. The uprighting tool generally used by escapement makers, and the ordinary drill stock, are also examples of the second class. Occasionally, for special purposes, the work and the drill are both rotated. The flat, pointed drill, like Fig. 3, in which the two cutting edges form about a right angle, is suited for brass or other metals of moderate hardness. For drilling with a bow, where the motion is reversed, the cutting edge of the drill had better be bevelled from both sides, but where continuous motion can be secured, the cutting edge should be close to the face of the drill, and for soft metal may even be turned up a little in advance of the face. Drills for tempered steel require to be more obtuse, and are generally rounded, or spoon-shaped, as it is called, like Fig. 4, though sometimes flat-ended, chisel-shaped drills are preferred. For making square sinks to receive screw heads, and the like, a pin drill is used (Fig. 5). Oil is used as a lubricant when drilling wrought iron and steel, and for very hard steel, turpentine may be substituted with advantage. For long holes of small diameter a twisted drill is desirable, so as to clear out the shavings and avoid unnecessary heating. On an emergency, a twisted drill can be quickly formed from a piece of flat steel, twisted while red hot by nipping one end in a vice. For long holes of large diameter

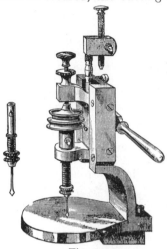

Fig. 2.

nothing beats a half-round drill. Care must be taken to keep the cutting edge of the chisel-shaped, or half-round drills, exactly in the centre of the hole, and before using such a cutter the hole must be bored a little distance to the correct size, so as to form a guide for it, and if a small hole can be first drilled for the whole length, so much the better. Half-round cutters should be turned on the rounding

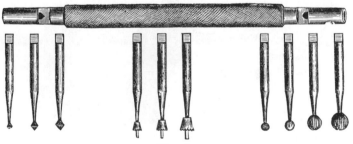

Fig. 6.

part. It is almost impossible to ensure their truth if they are filed up, and if they are not true they will not prove satisfactory in working.

There is now to be obtained a very handy holder with the set of ten sinking tools, as shown in Fig. 6, for forming sinks of various sizes and shapes.

New Pivot.—One of the best and simplest devices for centring and drilling up a small staff or pinion to receive a new pivot is shown in Figs. 7 and 8. There is a steel runner accurately fitted to the turns and pierced throughout its length. The

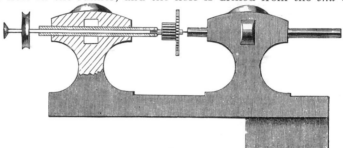

hole is tapped at one end to receive a bush with a trumpet-mouthed hole. The arbor to be drilled for a new pivot is placed

Fig. 7.

in the turns with the shoulder to be drilled resting in the trumpet-mouthed runner, and the other end centred exactly opposite in an ordinary runner. A drill stock, having on it a ferrule, is fitted to the hole in the runner, and the hole is drilled from the *end* of

Fig. 8.

the turns. Fig. 8 shows the arrangement clearly. Two or three of the little trumpet-mouthed bushes should be provided for different-sized pivots. By adopting this method the arbor is centred at once, and the hole is bound to be true throughout its length if the tool is correctly made. Instead of adapting the turns to this purpose, the special staff drilling tool, shown in Fig. 9,

Fig. 9.

may be used. After hardening the drill by flaring it in

the air, it should be let down all but the extreme cutting edge. For this purpose, catch the point in a pair of pliers, and then hold the drill in the flame of a lamp till the temper is lowered sufficiently, when plunge it into oil. The pliers preserve the point quite hard.

Swiss drills, sold in sets, and sized by the millimetre gauge, from ·10 to 1·1 mills, are very convenient. They are quite hard enough for drilling *foreign* pinions, and have a broad chisel point. A drill cutting both ways requires more pressure behind it than when formed with a more acute cutting angle to cut one way only. Use turpentine as a lubricator.

Drilling tools having runners with conical mouths pierced for the reception of the drill, are made, the work being rotated, and the end to be operated on supported in a conical hole of suitable size. But the way first described, in which the work is stationary, offers advantages when dealing with very delicate arbors. Both methods, and most other kinds of small drilling, are provided for in several lathes, which will be found fully illustrated under the head of Lathe. A pinion or staff too hard to drill may often be sufficiently lowered without discolouring by boiling it in linseed oil. If not, the extreme end of it may be held in hot pliers. A pair of tongs having circular noses with holes in the centre, and lined with asbestos, are useful for holding a balance or wheel to preserve it from the heat while the staff or pinion on which it is mounted is being lowered.

Drill Stock.—[*Porte foret.—Der Einsatzbohrer.*]—A piece of steel, pierced at one end for the reception of a drill, and having a ferrule mounted on it. The best kind for ordinary use has the nose made

Fig. 1.

after the manner of a self-centring chuck, so that the drill is sure to be gripped concentrically, as in Fig. 1.

The form of drill stock shown in Fig. 2 will be found useful for many purposes. The drill holder rotates within the

Fig. 2.

loose body, and the latter may be pinched in the vice or fitted to the turns ; making, in either case, a very steady drilling tool. The pulley is loosely fixed to the drill-holder, so that, in working, the strain on the drill is not so great as with the ordinary drill stock having a ferrule attached rigidly, and as a consequence, the annoyance of a broken drill is often avoided.

Drop.—[*Chute.—Der Abfall.*]—In an escapement, the space through which the escape wheel moves without doing work.

Drum.—[*Tambour.—Die Trommel.*]—A small French clock enclosed in a brass cylindrical case. The distinguishing feature of most of these clocks is that they never go for long together.

(2) The barrel on which the driving cord in turret clocks is wound also answers to the name of drum.

"Drum" Escapement.—The escapement used in French Drum Clocks is a continual source of trouble to English clock-

jobbers. It receives impulse at every other vibration only. The clocks have going barrels, and the idea of the escapement appears to be that by providing a long frictional rest on one of the pallets the extra pressure of the escape wheel tooth, when the mainspring is fully wound, will be sufficient to prevent any considerable increase in the arc of vibration of the pendulum. But the clocks often stop from deficiency of power when the spring is nearly down, and stop when they are fully wound because the small and light pendulum has not energy enough to unlock the pallet. The best that can be done is to alter the resting pallet to the form indicated by the dotted lines in the figure, and see that the wheel teeth and pallets are well polished.

The hole for the back pivot of the pallet staff, even though it is already free, may be broached out considerably with advantage.

Duplex Escapement.—[*Echappement duplex.—Die Duplex-hemmung.*]—(Invented about 1780. Credited to Dutertre, Tyrer, and others.) A watch escapement in which the escape wheel has two sets of teeth. One set lock the wheel by pressing on the balance staff, and the other set standing up from the face of the wheel give impulse to the balance.

This, like the Chronometer, is a single beat escapement, that is, it receives impulse at every other vibration only. It is shown on the next page. The escapement has two sets of teeth. Those farthest from the centre lock the wheel by pressing on a hollow ruby cylinder or roller fitted round a reduced part of the balance staff, and planted so that it intercepts the path of the teeth. There is a notch in the ruby roller, and a tooth passes every time the balance, in its excursion in the opposite direction to that in which the wheel moves, brings this notch past the point of the tooth resting on the roller. When the tooth leaves the notch, the impulse finger, fixed to the balance staff, receives a blow from one of the impulse teeth of the wheel. The impulse teeth are not in the same plane as the body of the wheel, but stand up from it so as to meet the impulse finger. There is no action in the return vibration. In the figure, the detaining roller travelling in the direction of the arrow is just allowing a locking tooth of the wheel to escape from the notch, and the pallet is sufficiently in front of the tooth from which it will receive impulse to ensure a safe intersection.

DUPLEX ESCAPEMENT.

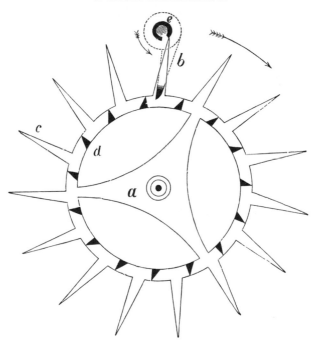

a.	Escape Wheel.	*d.*	Impulse Teeth.
b.	Impulse Pallet.	*e.*	Ruby Roller.
c.	Locking Teeth.		

The balance is never detached, but the roller on which the wheel teeth rest is very small and highly polished, so that there is but little friction from this cause, and the alteration in its amount is, therefore, not of such consequence as might be imagined. A very usual proportion is for the diameter of the roller to be one-fifth of the diameter of the largest part of the escape wheel, which it intersects from 30° to 80°, measured from

K

the centre of the roller. The impulse teeth should have considerable drop on to the pallet. 8° are not an unusual amount. The escape wheel is made as light as possible, of hard hammered brass of very fine quality. The points of the impulse teeth are usually two-thirds the distance of the points of the locking teeth from the centre of the wheel. The impulse pallet is sometimes jewelled.

The staff requires to be planted with great exactness, and one of the most frequent causes of derangement of the Duplex Escapement is the wearing of the balance pivots. In such cases, the pivots have been repolished, new holes, or, at all events, a new bottom hole should be put in. See also that the point of each locking tooth is smooth and nicely rounded, and that every impulse tooth falls safely on the pallet; if some are shallow, twist the impulse pallet round so as to give more drop. Or if the roller depth is also shallow, carefully make the teeth of equal length by topping, and then, supposing it to be a full-plate watch, very slightly tap the cock and the potance towards the wheel until the escapement is made safe. In a three-quarter plate the recess for the jewel setting may be scraped away on one side and rubbed over on the other. The extra amount of intersection of the impulse pallet in the path of the wheel teeth thus made can be easily corrected by polishing off the surplus amount, if any.

It is of the utmost consequence in this escapement that all the jewel holes should fit accurately, and that the balance staff should have very little end shake, otherwise the pivots will be found to wear away very quickly.

A loose roller is occasionally the cause of stoppage. The staff and roller should be carefully cleaned from oil which would prevent the shellac from sticking, and if the staff is polished where the roller fits it may be grayed for the same reason. Then warm the roller and fix with shellac.

It sometimes happens that the impulse pallet, in running past, just catches on the impulse tooth, and when the balance leans towards the escape wheel, the continual recurrence of this causes the vibration to fall off, and gradually stops the watch. If the locking teeth are already the right depth, the fault should be corrected by polishing a very little off the corner of the pallet with a bell-metal polisher if the pallet is of steel, or with an ivory polisher and the finest diamond powder if it is jewelled. But the greatest care must be taken not to overdo it.

A small drop of oil should be applied to the notch and nowhere else, except to the pivots.

When the escapement is in beat, the notch in the roller is between the locking tooth resting on it and the line of centres, or a little nearer the latter ; out of beat is a frequent cause of stoppage.

The idea of this escapement is seductive ; and at one time it was considered an excellent arrangement, but it has proved to be quite unreliable. The best proportion of its parts and the finest work are insufficient to prevent it setting. On the introduction of the Lever it declined, and is rarely made now.

Duplex Hook.—[*Levée de duplex.—Der Duplexfinger.*]—The impulse pallet of the Duplex Escapement.

Duplex Roller.—[*Rouleau de duplex.—Die Duplexerolle.*]—A hollow cylinder of ruby, slit throughout its length, fitted on the balance staff of the Duplex Escapement, against which the long teeth of the escape wheel rest except during the impulse, when they pass through the notch or slit.

Dust Cup.—[*Calotte.—Die Staubhülse.*]—A guard fitted round the fusee arbors of watches and chronometers to exclude dirt. There are two varieties of cups—" saucer " and " balance-wheel "—the former, shaped like a saucer, is generally of gold and is used in three-quarter plate watches ; the latter is somewhat similar in form to the Verge balance wheel.

Dust Pipe.—[*Chapeau.—Das Staubhütchen.*]—A cup-shaped shield round the set hands arbors of watches and chronometers for preventing the entry of dirt into the movement. In the best English watches the cup is solid with the arbor. In marine chronometers the pipe is screwed to the plate.

Earnshaw, Thomas, born at Ashton-under-Lyne in 1749. He invented the spring detent escapement and the compensation balance, both substantially as now used in chronometers. Earnshaw was in business in Holborn, near the turning now called Southampton Row ; and died at Chenies Street in 1829.

Eccentric Arbor.—[*Arbre excentrique.—Der excentrische Dreh-stift.*]—An arbor used chiefly for holding barrel or fusee arbors which may require to be turned after the centres are cut off. It is made in two parts. One part consists of a centre and ferrule ; the other part is split, and has a square hole to receive the square of a barrel arbor or fusee, and generally two screws to nip it, as shown.

There is a disc on the end where it meets the ferrule. The holes in the disc are large, so that when the three screws are loosened, the two parts of the eccentric arbor may be shifted with relation to each other till the barrel arbor or other work runs true.

Eiffe, James Sweetman (born 1800, died 1880).—A clever chronometer maker, who for some time carried on business in Lombard Street. He invented a compensation balance very similar to that patented by Molyneux.

Electric Clocks.—[*Horloges électriques.*—*Elektrische Uhren.*] —Electric clocks may be divided into four classes : (1) Clocks in which electricity is used to impel the pendulum and turn the hands so that periodical winding of a spring or weight is dispensed with. (2) Clocks that are driven by a weight or spring and wound in the usual way, but in which the vibrations of the

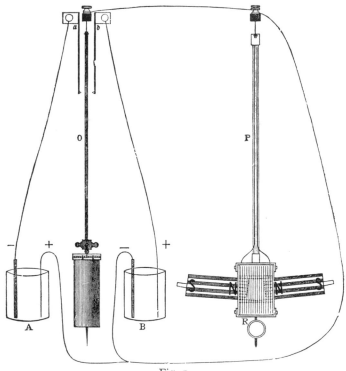

Fig. 1.

pendulum are controlled by currents transmitted automatically from a standard timekeeper. (3) Clocks wound by electricity ; and (4) clocks in which the mechanism is quite uncontrolled, but the proper position of the hands is ensured by periodical electric currents from a standard.

Alexander Bain, in 1843, invented (patent 9754) a system of

driving clocks and keeping a number of them, in circuit, going to time by an electric current applied to the pendulum.

For his pendulum bob he used a helix of wire, each turn insulated from the preceding, the wire ending in two insulated springs by which the pendulum was suspended. Two bundles of permanent magnets, having their similar poles slightly separated in the centre, were fixed at one end to the clock case, so that the pendulum bob was free to oscillate over them. The pendulum was caused to complete the galvanic circuit by a sliding bar resting on the battery terminals acted on by its vibration, and thus rendered automatic.

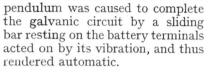

Charles Shepherd, Sir Charles Wheatstone, Dr. Hipp of Neuchatel, F. J. Ritchie, and many others, also elaborated methods for attaining the same end.

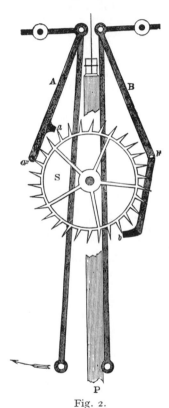

Figs. 1 and 2 refer to Ritchie's arrangement of electrically-driven clocks. Each of the secondary clocks is provided with a pendulum, P, similar to Bain's, vibrating over two bundles of magnetic bars SN-NS, Fig. 1. The master clock is furnished with slight springs, *a* and *b*, one on each side of the pendulum, o, which are attached, one to the copper terminal plate of battery A, and the other to the zinc terminal of battery B, the other poles, + and —, of each battery passing by the line wire, or through the earth, to one of the suspension springs of the controlled pendulum, P, thence down the rod and around the ball, R, to the second suspension spring, and by the line wire to the pendulum rod of the master clock. The wheel-work is carried forward by a kind of gravity escapement which locks the wheel, so that no force can carry forward the wheel beyond the stop, and no power less than the weight of the gravity arms will drive it backwards. Fig. 2 represents an escape-wheel, S, acted on by two gravity arms, A and B, one on each side. The pendulum, P, is moving in the direction of the arrow. The right hand arm *b* is resting on a tooth, pressing it forward by its weight.

Fig. 2.

The tooth, however, resting on the stop, $a2$, resists its action until it is raised out of its position by the vibration of the pendulum, when the wheel is moved forward till a tooth is caught by the stop, $b1$, on the arm B, which locks it till relieved by the reverse swing of the pendulum. The action is, of course, repeated with succeeding teeth of the wheel.

It is clear that if the current fail in clocks of the first class, not only do they cease to register, but on the resumption of the current they start with the error accumulated during its cessation. R. L. Jones, therefore, proposed to make the secondary clocks each complete in itself, and furnished with a pendulum practically similar to P in Fig. 1. A standard clock with the same length of pendulum as those controlled, was caused to transmit currents which accelerated or retarded the motion of the controlled pendulums, according as their vibrations were a little too slow or too fast. This system works very well in Greenwich Observatory, where

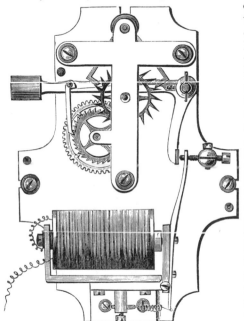

Fig. 3.

the circuit is short, and constant attention is paid to the electric apparatus; but the late C. V. Walker, after many years' trial with a clock at London Bridge, controlled from Greenwich, pronounced it to be impracticable. If the current failed long enough for the controlled pendulum to get one beat out, the controlling apparatus had no means of setting the clock right again. Nevertheless, others appear to have met with greater success. Professor Becker designed a system for about 200 street-corner dials in Glasgow. In this case the Standard Greenwich mean-time clock at the Observatory controls about eight weight regulators by means of an alternating current passing every second through the Observatory clock and the Ritchie pendulums of these regulators. The regulators stationed in the various districts of the city form centres for the further distribution of time. The regulator makes automatically a contact each half minute, and the current closed by this contact works a number of relays stationed in the same room with it. Each relay again is the centre of a separate street circuit with 8 to 12 dials. For the sake of testing the regulators and the dials, each regulator makes every hour at a particular minute a contact which lasts about $\frac{3}{4}$ of a minute. Further, one dial of each circuit is placed at the station of the circuit, and all those at the same station make a contact at the same minute as their regulator does. The contact springs are connected in series, and one end is joined to the Observatory circuit, the other to earth. Therefore, by joining at the Observatory an electric bell to battery and earth, the ringing of the bell at the proper time shows that the regulator, and at least one dial of the various circuits, are correct.

In Pond's electric clock a motor attached to the frame winds the weight or spring once an hour, or after other periodical intervals, the motor being started by an arm, which, carried round by the train, completes the electric circuit on arriving at a certain point ; when the barrel has made nearly a rotation a projection from it strikes the arm away, breaking the contact and stopping the motor. The motor is actuated by one or more Leclanche or other cells. This invention, though regarded favourably when introduced a few years ago, does not seem to have been largely adopted in England.

Fig. 3 shows the self-winding electric clock of Messrs. F. Hope-Jones and G. B. Bowell. The pendulum is driven by an escapement in the usual way. The necessary turning force is applied to the escape-wheel by means of a weighted lever, which in falling through a small arc communicates its power through a click and ratchet to a wheel gearing with the escape-wheel. As the weighted lever falls in its action of driving the pendulum, its lower limb meets the contact screw in the tail of the armature, and the circuit of the electro-magnet and of a group of dials in series with it is closed by the contact of these two parts. The weight is then thrown up and its momentum causes a quick break as the insulated screw comes into action, and the armature is stopped by the poles of the magnet. This instrument has, therefore, the two essentials of a good switch—a severe rub and thrust and a quick break. The switch consists of two moving parts, the armature which drives and the weighted lever which is driven, so that the entire energy required to keep the pendulum swinging is mechanically transmitted through the surfaces of the contact at each operation.

The construction of a well-thought-out electrically-driven

regulator, by Mr. T. J. Murday, with provision for driving subsidiary dials, will be gathered from the diagram below. On the left hand are front and side views of the upper part of the pendulum rod. Compensation for temperature changes is effected

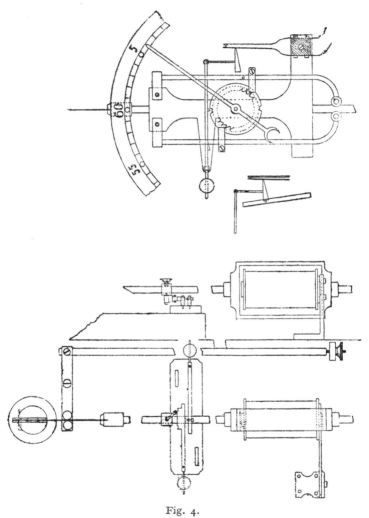

Fig. 4.

by means of the expansion or contraction of a zinc rod, which actuates a lever carrying two pins, and the arrangement is such that the effective part of the suspension spring is lengthened or shortened exactly to the same amount as the steel rod of the pendulum expands or contracts through changes of temperature.

About twelve inches below the suspension, an ebonite block carrying the contact levers is fixed. A trailing piece or " toggle " with a roller is suspended from the side of the pendulum so as to sweep over a raised part or block on the upper lever. So long as the arc of oscillation of the pendulum is sufficient to carry this roller right off the block at each end, no contact is made and no impulse received by the pendulum. On the arc of oscillation falling off slightly, the roller fails to clear the end of the block and falls back into a groove ; this action depresses the upper contact lever into contact with the lower one, causing a current to flow through the electro-magnet, which gives the pendulum a fresh impulse. By this intermittent contact action, the arc of oscillation of the pendulum is maintained constant within very narrow limits—no matter whether the battery be weak or strong. The driving magnet may be placed at the bottom of the case and act on an armature fixed beneath the bob on the pendulum rod ; or it may take the form of a single bobbin fixed by a bracket on the back of the case, so as to allow a frame-shaped armature (as shown in drawing) to swing over it. In practice this armature frame is inserted in the pendulum rod a few inches above the bob.

The movement operated by this pendulum is shown on the right hand of the diagram. It consists of a centre seconds action with two gravity levers acting on the escape wheel. These levers are raised alternately by the swing of the pendulum, and in falling back give impulse to the escape wheel, and through this to the usual train of motion work for the minute and hour hands.

The mechanism for transmitting half-minute currents to operate secondary dials is also shown on the movement diagram. A disc fixed on the escape wheel arbor is cut away at two points on its edge. A lever rests lightly on the edge of this disc until one of the parts cut away comes under a projection on the lever ; this allows the lever to drop, and with it a block which comes between a pin on one of the gravity levers and the contact springs. As the whole weight of the pendulum is pressing this lever outwards at the moment, a firm, steady contact is obtained —absolutely free from the danger of sending " double impulses," and of a much more certain action than is obtained by the usual " pin contacts " attached to weight-driven regulators for this purpose.

Mr. T. J. Murday's mechanism for actuating the hands of

secondary dials is shown in Fig. 5 : *a* is a wheel fixed on the minute wheel spindle, and gearing with a pinion *b* on the spindle of the ratchet wheel *c*. This wheel *c* has large well-cut teeth, so

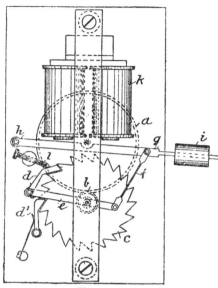

that clicks *d*, *d'* may engage deeply and safely therein. The spindle of the wheel *c* forms the fulcrum of a rocking lever *e*, which carries the click *d* engaging with *c*. The other end of this lever is connected by a rod *f* with an armature *g*, which is pivoted at *h*, the other end of the armature carrying a weight *i*. When the electro-magnet *h* is energised by the current transmitted every half minute by the regulator, the armature *g* is drawn up to the poles, causing the lever *e* to move through an arc sufficient to place the click *d* in another tooth of the wheel *c*. In this

position the pivoted end of the click *d* lies just behind the detent click *d'*, thus preventing *d* from rising out of the tooth. This means that the wheel *c* is firmly locked and unable to move either way. When the current ceases the armature *g* falls away, and this movement being transmitted to the lever *e*, the wheel *c* by means of the click *d* is carried round a distance of one tooth. The click *d* then comes against a stop *l*, so arranged that *d* cannot rise out of the tooth. The wheel *c* is thus locked firmly in this position also ; it cannot " race " forward owing to the locking action of *d* just explained, and the detent click *d'* as securely holds it from moving backwards. The weight *i* must be sufficiently heavy to drive the movement safely.

Fig. 6 illustrates the " Synchronome " combination of a gravity impelled pendulum and automatic switch as devised by Mr. F. Hope-Jones for the Synchronome Company.

The weighted lever (A) is normally at rest and is released once every half-minute by a click (B) on the pendulum (D) which in obedience to one shallow tooth in the wheel (C) removes the vertical support (F) and allows the wheel (H) to give impulse to the pendulum through the pallet (J) as the lever (A) falls. When the lever meets the armature (E) it is thrown up again by the magnet (G).

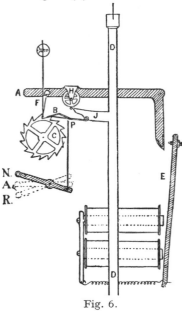

Fig. 6.

There are many advantages in the form of the switch illustrated. The weighted lever is let down by the pendulum with a clear and steady movement straight into contact, and the stroke of the two moving parts of the switch is considerably increased. The gradual failure of current which occurs when primary batteries are used is revealed by a great increase in the duration of contact, so that every instrument in circuit becomes a battery indicator. The wide difference in duration of contacts can be employed to ring a warning bell, so adjusted that it will not respond to short contacts. Such failure of current will not stop the clock, because the latent energy of the pendulum will assist the electro-magnet (G) in lifting the weighted lever (A) by means of the wheel (H), which it will find in its path on its return to the left. When eventually the current is insufficient even to complete the work of raising the lever which the pendulum has thus begun, and the pendulum consequently stops, it will stop on open circuit.

With this form of switch the dials can be readily set to time. In the illustration a little switch-handle (N) will be observed on the left-hand side, which carries a pin (P) underneath the driving click. The handle is shown in its normal position, and then the click releases the gravity arm once every half-minute, but when it is turned to the mid-way position (A, accelerate) the lever will be released every two seconds, thus setting forward the dials automatically at the rate of 15 minutes per minute, while when it is moved into the lowest position (R, retard) the click is out of action altogether, and the pendulum swings free. Alterations of a few seconds slow or fast can be effected by moving the wheel forward or backward by hand, each tooth representing two seconds.

It will be observed that the pendulum is entirely free save for the rotation of the idle guide wheel (c). The surface of the pallet (J) fixed to the pendulum forms an arc of a circle whose centre is coincident with the centre of suspension. The pallet may therefore run so close under the little wheel (H) on the gravity arm that the drop of the latter is imperceptible, and the impulses will be uniform in strength and position in spite of variation of arc or point at which the release occurs.

Bain devised a simple hand-setting arrangement. An electro magnet was placed behind the dial of the clock to be corrected. The armature of the magnet carried a rod ending in a **V**-shaped fork immediately above the figure XII. A pin projecting back from the minute hand was so placed as to clear the arms of the **V** when the armature was at rest, and forced to occupy the angle of the **V**, when, by the attraction of the magnet, the armature and rod were drawn upwards at the completion of each hour.

In London the Standard Time Company adopted a system invented by John A. Lund, of setting clocks right by sending an electric current from a master clock which causes two clips to

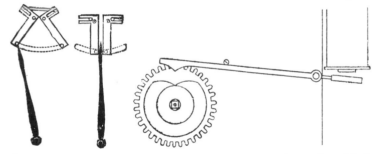

close on the minute hand of each clock, so that if the hand is in advance of or behind the twelve on the dial the clips adjust it to its proper position.

Recently the hand-closing clips have been abandoned in favour of a cam attached to the minute wheel into which an adjusting arm drops exactly at each hour. The hand-setting arrangement and the cam action are both shown in the subjoined cut. This company has for synchronising clocks throughout London thirty circuits, covering an area of sixteen square miles. Each clock under control periodically signals, between the hours, its time back to the central station. In this way an erratic timepiece, which otherwise might get out of control of the synchroniser, may be at once located and seen to.

Ellicott, John, born 1700, died 1772.—He wrote one or two pamphlets on the construction of pendulums, and was the inventor of a compensation pendulum in which the bob rests on the longer ends of two levers, of which the shorter ends are depressed by the superior expansion of a brass bar attached to the pendulum rod. This device has not proved to be of practical value. Ellicott's shop was in Sweeting's Alley, Cornhill, close to the old Royal Exchange.

Emery.—[*Emeri.*—*Der Schmirgel.*]—Corundum mixed with iron and other impurities. It is graded to various degrees of fineness, and used for grading or abrading metallic surfaces. The chief kinds required by watchmakers are No. 46, called grit or corn emery, used for sharpening cutting burnisher ; No. 70, grit, used for smooth burnishers ; flour emery, a finer kind also used for smooth burnishers ; and washed or double-washed emery, used for snailing and giving a grey polish to steel work.

Emery File.—[*Pierre à émeri.*—*Die Schmirgelfeîle.*]—A solid stick of emery used as a file.

Emery Stick.—[*Lime à émeri.*—*Die Schmirgelpapierfeile.*]—A stick of wood round which emery paper is glued.

Emery Wheel.—[*Roue à émeri.*—*Das Schmirgelrad.*]—A wheel composed of emery and some uniting material, used for abrading metals which are held against its rapidly revolving surface. The best wheel with a fine cut for surfacing metal is an American one, in which vulcanite is the cementing medium. In another good kind silica is used. Either of these may be run in water if desired. For occasional use, and for grinding milling cutters and the like, a wheel may be made of a wooden body, to which emery is attached with glue. The glue must become thoroughly hard before the wheel is fit for use. To work effectively the circumferential velocity of an emery wheel must not be less than 2,000 feet per minute. For watchmakers' tools an emery wheel is preferable to a grindstone, by reason of its even texture.

End Stone.—[*Pierre de contre pivot.*—*Der Deckstein.*]—A small disc of jewel on which a watch pivot rests. In most English watches all the escapement pivots run on end stones which are screwed in the plate or cock at the end of the jewel hole. The arbors of the train are prevented from moving endways by the fact that the pivot is of lesser diameter than the rest of the arbor. It passes easily through the whole in which it works ; the body of the arbor cannot follow, but is caught by the projecting shoulder. The holes for the train pivots are termed " thorough holes," whether they are jewelled or not, so as to distinguish them from holes provided with end stones which are called capped jewels.

Engine Turning.—[*Guillochis.—Das Guilloche.*]—The wavy circular curves cut into the outside of watches for decoration.

Epicycloid.—[*Epicycloïde.—Die Epicycloide.*]—A curve generated by a point in the circumference of a circle as it rolls upon another circle. The usual shape of the teeth of a wheel that drives. (See Wheels and Pinions.)

Equation of Time.—[*Equation.—Zeitgleichung.*]—Adding to or subtracting from the true solar time the amount necessary to obtain equal or mean time. (See Time.)

Equatorial Clock.—[*Horloge équatoriale.—Das Uhrwerk eines Aquatoriums.*]—A clock for driving an equatorial telescope. It has generally a conical pendulum or other device for rendering its motion continuous.

Escapement.—[*Echappement.—Die Hemmung Der Gang.*]—The device in a watch or clock by which the motion of the train is checked, and the energy of the weight or mainspring is communicated to the pendulum or balance. (See Anchor, Chronometer, Cylinder, Dead Beat, Duplex, Double Three-Legged, Drum, Lever, Pin Pallet, Pin Wheel and Verge.)

Escape Pinion.—[*Pignon d'échappement.—Das Gangradstrieb.*]—The pinion on the escape wheel arbor.

Escape Wheel.—[*Roue d'échappement.—Das Gangrad.*]—The wheel that permits the escapement of the train.

Escaping Arc.—Twice the angular distance a pendulum has to be moved from its point of rest, in order to allow a tooth of the escape wheel to pass from one pallet to the other.

Eye Glass.— *Loupe.—Die Lupe.*]—The usual form of watchmaker's glass is a convex lens an inch in diameter mounted in horn. Though sometimes extra strong glasses are used for special purposes, the focus for general work ranges from two to

four inches. Some workmen find the muscular exertion of supporting the glass irksome, and attach it to a wire, held in the mouth or behind the ear, or to a light spring coiled round the head. Eye glasses may now be obtained mounted in cork for lightness. Holes are often drilled through the mounting to prevent the glass being dulled by the collection of moisture on it, though some object to the holes, as tending to derange the sight. Apart from the accumulation of moisture on the lens, more than one glass should be at hand in case the eye tires ; glasses of different foci rest the eye. There is a very superior glass called the " Dracip," well polished, and giving a wide flat field. The frame is made of vulcanite, and is very light.

The focus of each glass is legibly stamped on the frame as shown in the subjoined figure.

Young watchmakers should be taught to use the glass to either eye. It is sometimes an advantage in working to be able to do so, and the idle eye would benefit by more use. Mr. Brundell Carter and other oculists declare that the constant use of a glass is not injurious, and watchmakers find the eye to which the glass is applied to be the stronger of the two.

Watch jewellers use a glass with double lenses half an inch in diameter, and with a very short focus.

Facio, Nicholas, introduced watch jewelling (Patent, No. 371, May, 1704).

Ferguson, James, born 1710 ; died, and buried in Marylebone Churchyard, 1776. Astronomer and mechanician.

Ferguson's Clock.—This clock, shown in Figs. 1 and 2, contains only three wheels and two pinions. The hours are engraved on a plate fitting friction-tight on the great wheel arbor ; the minute hand is attached to the centre wheel arbor, and a thin plate divided into 180 equal parts is fitted on the escape wheel arbor, and shows the seconds through a slit in the

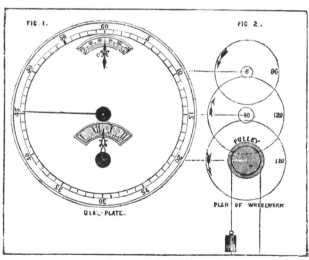

Ferguson's Clock.

dial. The clock has a seconds pendulum. The number of teeth in the escape wheel is higher than is desirable, and the weight of the thin plate or ring in the escape wheel arbor is objectionable, though it might now be made of aluminium, vulcanite, or other very light material.

Ferrule.—[*Cuivrot.*—*Die Schraubenrolle.*]—The small wheel or pulley round which is coiled the string of a bow when it is used to give a rotary motion to a piece of work. A screw ferrule (Fig. 1) has a projecting lug, which is split, and furnished with two screws to enable it to grip the work. In pivoting arbors, where there is not room for a screw ferrule, a ferrule of bone roughed on one side with a file and waxed to the wheel is often used. For this purpose beeswax alone is generally too soft, and the practice is to melt it and mix with it a very little resin.

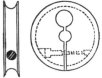

Fig. 1.

Spring Ferrule.—This is a convenient form of ferrule where there is but little room. As shown in Fig. 2, a saw slit is made across the ferrule to meet a hole drilled through it. The larger this hole is, the more springy and better the ferrule will be. But if it is desired that the centre hole should be large enough to take the body of the pinion, then the outer hole must be reduced in proportion, but instead of being circular it can be carried farther round the circumference in the shape of a crescent,

Fig. 2.

A ferrule is useful when it is desired to repair the upper pivot. or turn back the shoulder of the escape pinion of a horizontal escapement without removing the wheel, is shown in Fig. 3. A convex surface is presented to the back of the wheel, and a concave cover very thin in the middle being screwed on from the front of the wheel, bends back the teeth.

Sliding Carrier.—This useful adjunct, if not supplied with a Jacot tool, may with advantage be fitted to it. It is often handier than the screw ferrule, and saves time when used instead of waxing or shellacing. A small steel plug or arbor is fitted to one of the centres as shown on Fig. 4. The ferrule of steel runs on a collet of hard brass, and is kept in its place by a small washer. The collet is pierced to move freely on the steel arbor, and its projecting end slit and then pinched together so as to grip the arbor sufficiently tight to remain in position when in use,

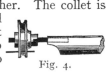

Fig. 3.

Fig. 4.

and yet not so tight but that it may be moved to and fro without trouble. Holes may be made at convenient positions in the ferrule to receive the carrier pin. The shake between the crossings is not objectionable with small-sized wheels, but for large and heavy balances, etc., two pins and a larger ferrule may be used. To compensate for the room taken up by the projecting end of the collet, a little is sometimes taken off the boss of the Jacot tool.

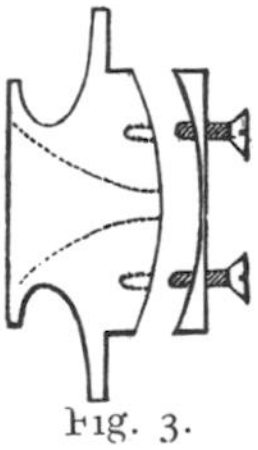

Carrier and Sliding Centre.—Fig. 5 is a most useful arrangement of ferrule and carrier, by Mr. Arthur Webb, suitable either for the Jacot tool or the turns. The ferrule runs on

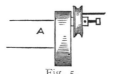

Fig. 5.

an arbor screwed to a disc which is fixed to an ordinary runner (A). The arbor of the ferrule is pierced to receive a very small runner, as shown, which has a head of brass in which the work is centred. This little runner may be drawn further out or pushed further into the arbor as required, and is kept in position by a set screw tapped into the head of the arbor.

Squire's Adjustable Runner, with Ferrule.—Fig. 6 represents an adjustable runner combined with driving pulley, by Studley Squire. At the rear end is fitted a main runner made in various sizes to suit different turns. At the fore end is a dovetail slide, which is adjustable with an up and down action, by means of the nutted screw at the bottom. The dovetail carries a working runner, which is clamped by the screw at the top and is reversible, having a centre of different size at each end.

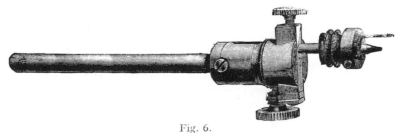

Fig. 6.

Similar runners are made with safety centres. The pulley can be thrown into, or out of, engagement with the work at will, being kept in position upon the runner by means of an ingenious device fitted spring tight upon the runner and provided with a slot in which works a collar projecting from the rear end of the pulley. The pulley has a crank-shaped driving pin, adjustable for driving the smallest escape wheel or the largest balance.

The nicest adjustments of the back centre can be obtained in an instant by a turn of the adjusting screw below the dovetail. This is of particular advantage when working in small Swiss escape pinions, polishing pivots in beds, etc. In turning hollows also, many to whom the fitting of a new pinion or cylinder is not a frequent occurrence will find it a convenience to be able to adjust the work in an oblique position so as to see into the bottom of the hollow.

This tool can further be utilised with advantage in drilling

L

pinions or staffs. It is next to impossible to drill a pinion upright if the wheel be held in the figures, a slight movement of which suffices to break the drill off in the hole. For a job of this kind a lantern runner of the larger size is placed in the turns with the adjustable runner on the opposite side. Supposing it to be a Swiss top 3rd pivot that is required, the pinion is first centred for drilling, then the lower or sound end of the pinion is placed in a hole of the lantern runner, with the wheel resting flat against the face of the lantern. The adjustable runner is now screwed to the level of the pinion, and the drill placed between the two. This adjustment takes but a moment, while it secures a hole perfectly upright, and prevents the annoyance of getting broken drills out of holes.

The adjustable runner is not confined to use in working with *eccentric* runners. For turning with *concentric* ones the dovetail can be lowered so that the working runner takes up a perfectly central position. It is scarcely necessary to point out the advantages of a driving pulley. All the repairs to the pivots of a cylinder or pinion can frequently be done in less time than is required to put on and take off a ferrule, particularly of the wax or cement kind.

Ferrule for Balances.—M. Martinet, in the *Revue Chronométrique*, describes a ferrule for balances, like Fig. 7. It is made of steel, and the balance is held firm by what is really a bayonet joint. The face of the ferrule is recessed or bored out, leaving a thin rim. *a*, *b*, and *c* are three slots cut in the face of the rim, and continued round a portion of the periphery in a wedge-like form to receive the arms of the balance. The balance is placed in and twisted round, the wedge-like slots gripping it firmly. Care must be taken to cut the slots round the periphery in the

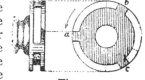

Fig. 7.

right direction, so that the pressure of the tool when operating on the balance tightens it in the slots, and to cut the slots round the periphery pretty near the face, so that in gripping the balance the wedge-like slots allow of a little springing. A ferrule on this principle for compensation balances would, of course, have but two slots instead of three. There are now to

be obtained the set of ferrules as shown in the above sketch They comprise ferrules for balances with slots for two and three arms ; for third and fourth wheels with five arms, and for escape wheels.

File.—[*Lime.—Die Feile.*]—An instrument with a serrated surface for reducing metal and other hard substances by abrasion. (See Bastard, Bull's Foot, Dead, Smooth, Diamond, Pillar, Pivot, Potance, Rat Tail, Ridged-back, Safe-edged, Slitting.)

A new file should never be used on steel or iron, but be reserved for brass till its first keenness has worn off. A file apparently worn out may be revived by placing it out of doors to rust, then immersing it in a strong solution of cyanide of potassium and afterwards well brushing it with file cleaner and oil. File handles should be fitted on perfectly straight with the file ; they may be burnt out with the heated tang of an old file or other piece of steel of the right shape.

File Cleaner.—A piece of carding wire attached to a flat piece of wood forms an invaluable file cleaner. Should there be a bit of metal so tightly jammed between the serrations as to

resist the action of the cleaner, it may be removed with the flattened end of a bit of brass wire. Some files show a disposition to persistently clog. After cleaning out, these should be either rubbed with chalk or slightly oiled, which will in a great measure prevent the accumulation of fixed lodgments.

Filing Block.—This is a hardened steel cylindrical block with different sized grooves for supporting wire while it is being

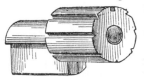

filed. One end of the block is reduced and held in a split socket. While the tail of the split socket is loose between the jaws of a vice, the cylinder may be freely rotated to bring the desired groove uppermost. When the vice is tightened the split socket grips the cylinder, keeping it fast. For rapidly making perfectly straight pins, filing click or mainspring wire, and similar purposes, this filing block is far superior to the box-wood or ivory rest.

Guide for Filing and Adjusting Square Stems.—This useful appliance is the invention of M. Bonierbale. Its function will be readily understood from the following description of the engraving : —A is a round bar flattened on one side, carrying a fixed tail-stock, on the top of which is a hard steel plate C. This steel plate is raised or lowered by means of a sunkhead shouldered screw in the centre, and steadied by the other three screws shown. Variations of level may be obtained by these screws, and any size of parallel or taper square produced. B is a poppet head sliding on the bar carrying the dividend plate F, and a

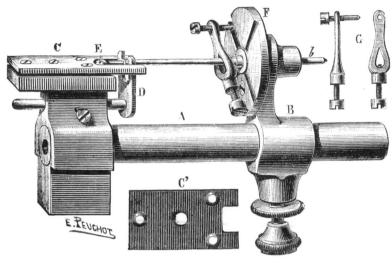

female centre *b*. D is a guide to regulate the length of a square. E is a small circular plate acting as a bearing for the work. This plate is intended to be filed flush with the work, and at the same time. G is a carrier with a screw acting as stop, when placed in the grooves of plate F.

Flirt.—[*Fouet.—Die Anlaufwippe.*]—A lever or other device for causing sudden movement of mechanism. In an independent centre-seconds watch the revolving arm on the last arbor of the independent train, which by taking into the escape pinion regulates the motion of the independent seconds hands. In some forms of chimes clocks, a lever actuated by a spring and discharged every quarter of an hour to knock up the quarter rack hook.

Fly.—[*Volant.—Der Windfang.*]—A two-bladed fan used in clocks to preserve uniformity in the time which elapses between each stroke of the hammer. When the striking train is discharged it would run with increasing speed but for the fly, whose progress is opposed by the air with a force proportional to the square of the velocity of the fly. With a fly too small it will be noticed that when say twelve o'clock is being struck, the last few blows will be delivered much quicker than the first.

Fly Cutter.—[*Fraise à crochet.—Die Zahnfräse. Hakenfräse.*]

—A cutter with a single cutting edge, used for cutting the teeth of brass wheels. This is the usual form, but latterly fly cutters are often made double. A piece of steel fitted to the cutter holder so as to project equally on each side is turned to the form the cutter is desired to be. The steel is thinned on opposite sides till the faces are just coincident with the centre of the holder, and after being filed back from the edge to give requisite clearance is hardened and tempered.

Fly Pinion.—[*Pignon de volant.*—*Das Windfangstrieb.*]—The pinion in a clock which carries the fly. In a repeating watch the last pinion of the running train when no escapement is used to control it.

Fly Spring.—[*Levier.*—*Die Springfeder.*]—A spring that causes the outer cover of a watch case to fly open. (See Case Springs.)

Follower.—In a verge watch, the plug which, fitted to the stud or counter potance in the top plate, carries the outer pivot of the escape wheel arbor.

(2) In a pair of toothed wheels, the one that is driven.

Fourth Wheel.—[*Roue de champ.*—*Das Sekundenrad.*]—The wheel in a watch that drives the escape pinion, and to the arbor of which the seconds hand is attached.

Fraise.—[*Fraise.*—*Die Fräse.*]—A cutter. (See Ingold Fraises.)

Frame.—[*Cage.*—*Das Gestell.*]—The plates of a watch or clock that support the pivots of the train.

Free Spring.—[*Spirale sans raquette.*—*Die Spiralfeder ohne Rücker.*]—A balance spring uncontrolled by curb pins. In marine chronometers, pocket chronometers, and generally fine watches in which an overcoil spring is used, the spring and balance are proportioned to give exact time as nearly as possible, and curb pins are dispensed with. Any future alteration in the rate, if such be needed, has then to be made by an expert.

Friction.—[*Friction.*—*Die Reibung.*]—Friction is defined as the resistance experienced when one hard body is rubbed upon another, caused by the tendency of the asperities which exist on all surfaces, however highly polished, to interlock, together with the natural attraction which bodies have for each other, and possibly some electric action. The force necessary to overcome friction varies directly as the weight or pressure with which the bodies are kept in contact, and is independent of the amount of surface over which the weight or pressure is spread. If the surfaces are too small in proportion to the pressure, they will be rapidly worn away ; and if the disproportion be very great, one or both of the surfaces will be destroyed by abrasion. With most substances used for the acting surfaces of machines this abrading

action would very quickly develop itself were the surfaces allowed
to come into absolute contact, therefore a film of some lubricant
is interposed, which of itself has a retarding influence ; but in
machines of any size or weight the amount is very small, compared with the friction proper—the attrition of the metallic surfaces. The viscidity of the unguent employed is also proportioned to some extent to the weight ; but it is possible to have
the weight so small that, even with the most fluid unguent, the
adhesion of the unguent is far greater in amount than the attrition of the metallic surfaces. In the balance staff pivots of a
watch, for instance, there can hardly be said to be friction at all ;
nearly the whole of the resistance arises from the adhesion of the
oil, the amount of which is exactly proportionate to the extent of
the surfaces in contact, instead of to the weight ; and as the
fluidity of the oil cannot be kept constant, owing to the action of
the metallic surfaces and the atmosphere, the varying sum of
the resistance is a source of great perplexity to watchmakers.
The surfaces in contact are, therefore, wisely made very small in
cases where the greater part of the resistance arises from the
adhesion of the oil. Towards the barrel, where the pressure is
greater, larger surfaces are desirable, and thicker oil should be
used. When very thin oil, which is proper for the escapement,
is used to the centre and fusee pivots, one often hears complaints
of these pivots becoming dry, and cutting. In clocks the barrel
and train pivot holes are often improperly reduced to nearly the
same lengths as the verge and escape holes. Where extra surface is desired for the bearings of arbors, &c., it should be
obtained by increasing the length and not the diameter, for if the
diameter of a pivot or bearing of any kind is doubled, the resistance to motion is doubled also, because the revolving surface is
then twice the distance from the centre of motion.

The coefficient of friction is a fraction representing the proportion of the weight or pressure on two surfaces in contact that
will be required to overcome the friction between them.

Frictional Escapement.—[*Echappement à friction.—Hemmung
mit reibender Ruhe.*]—An escapement in which the balance is
never free from the escapement. The Cylinder, Verge and
Duplex are the best known examples of frictional escapements
for watches.

Friction Spring.—[*Ressort de Minuterie.—Die Minutenrohr-
feder.*]—In house clocks a bow-shaped spring squared at its
centre on to the centre arbor. Its ends press against the face of
the cannon wheel and so drive the hands. (See Motion Work.)

Frodsham, Charles (born 1810, died 1871), a skilful and
successful watchmaker, who succeeded J. R. Arnold, at 84, Strand.
He conducted many experiments with a view of elucidating the

principles underlying the action of the compensation balance and the balance spring, and wrote several valuable technical works. He was for some time a Vice-President of the Horological Institute.

Frosting.—[*Adouci mat.*—*Der Mattschliff.*]—The grey surface produced on steel work for watches, &c. (2) The granular, or " matted" surface given to brass pieces prior to gilding.

Frosting Steel Keyless Work.—After the work has been prepared with a surface free from scratches, it is rubbed with a short backward and forward motion on a small glass slab with a thickish paste of oilstone dust and sweet oil. Before mixing this paste, look over the pounded oilstone with a very strong magnifying glass, and carefully remove all the black atoms, which, if left, would inevitably scratch the work. It may afterwards be rubbed in the same manner on unglazed note-paper; very fine oilstone dust, and oil. The work is cleaned and finished by rubbing in a circular direction with pith, or instead of rubbing with pith the work may be carefully breaded and immersed in benzine.

Frosting Watch-Caps, Plates, &c.—2½ nitric acid, 2 muriatic acid, full strength. Dip the articles for a few seconds ; rinse in water ; scratch-brush with a circular motion, then gild.

Full Plate.—[*Deux platines.*—*Das Vollplatinenwerk.*]—A full-plate watch has a top plate (*i.e.* the bottom plate of the movement looking from the dial) of a circular form, and the balance is above this plate.

Fusee.—[*Fusée.*—*Die Schnecke.*]—A spirally grooved pulley of varying diameter interposed between the barrel and the centre pinion of a watch or clock for the purpose of converting the varying force of the mainspring into a constant pressure at the centre pinion. A chain or gut is coiled round the fusee when the mainspring is wound. The first coil is on the larger end of the fusee, and the last on the smaller end, from whence the chain is attached to the barrel. As the mainspring runs down,

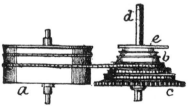

Mainspring barrel and fusee.
a, Mainspring barrel ; *b*, fusee ; *c*, great wheel ; *d*, winding square ; *e*, snail-shaped flange.

the barrel rotates and coils on to itself the chain or gut from the fusee ; but while the mainspring when fully wound turns the fusee by uncoiling the chain or gut from the smallest part of the fusee it gets the advantage of the larger radius when its energy becomes lessened. The fusee is supposed to have been invented by Jacob Zech, at Prague, about 1525. By means of an adjusting

rod the uniformity of the force transmitted by the fusee is tested.
In watches and marine chronometers the fusee is always fitted
with a chain. In clocks gut is often used. In modern watches
and clocks the fusee is furnished with maintaining power to drive

WATCH FUSEES.

Great Wheel.	Turns in Fusee to Centre Pinion of 10.	Turns in Fusee to Centre Pinion of 12.	Great Wheel.	Turns in Fusee to Centre Pinion of 10.	Turns in Fusee to Centre Pinion of 12.	Great Wheel.	Turns in Fusee to Centre Pinion of 10.	Turns in Fusee to Centre Pinion of 12.
40	7·5	9	56	5·35	6·52	72	4·16	5
42	7·14	8·57	58	5·17	6·21	74	4·05	4·87
44	6·81	8·3	60	5	6	75	4	4·8
46	6·52	7·83	62	4·83	5·81	76	3·94	4·76
48	6·25	7·5	64	4·68	5·66	78	3·84	4·61
50	6	7·21	66	5·54	5·45	80	3·75	4·54
52	5·76	6·97	68	4·41	5·35	84	3·57	4·28
54	5·5	6·6	70	4·28	5·14			

the train while the fusee is being turned backwards during the
process of winding (see Maintaining Power). Such fusees are
called going fusees, to distinguish them from the plain fusee
in old verge watches.

The calculation for the turns of fusee is so simple as scarcely
to call for any remark. The centre wheel of a watch turning
once in an hour, and the turns on fusee being calculated for 30
hours, it is evident that the number of turns in any case is
inversely proportionate to the number of times the teeth of
centre pinion is contained in main wheel.

EXAMPLE.—With great wheel of 75, and centre pinion of 10 ;
75 ÷ 10 = 7.5 and 30 ÷ 7.5 = 4, which is the number of turns
given in the table above.

NOTE.—The proportion of the top to the bottom diameter of a fusee can
only be determined when the force of the spring at these two points is known ;
but in all cases the force at the top must bear the same proportion to the
bottom diameter as the force at the bottom bears to the top diameter. (See
Adjusting Rod.)

Repairing Watch Fusee Top Pivot.—First file up and
re-polish the square, taking off the corners sufficiently to prevent
them standing above the pivot when it is re-polished. Put the
square into an eccentric arbor and get the fusee quite true. Now
put a screw ferrule on to the fusee back arbor, and place the
whole piece in the turns with the eccentric in front, using the

bow on the ferrule at back.   If the pivot is much cut it should be turned slightly with the point of the graver.   Polish first with steel and coarse stuff, afterwards with bell-metal and fine-stuff, and finish with the glossing burnisher.—H. B.

To Put in a Watch Fusee Top Hole.—Put the pillar plate in the mandril and peg the bottom hole true, then turn out the top hole to the required size for stopping.   The stopping (a hollow one) should be small, and no longer than just sufficient to form the rivet.   If there be danger of bending the plate, the stopping should be softened slightly (the hammering will reharden it), and the ends turned hollow to facilitate the rivetting.   The top hole is now to be turned to nearly right size for the pivot, testing it frequently for truth with the peg, as much broaching is especially to be avoided.   In finishing the stopping use POLISHED cutters, take off the corners of the hole, and polish the cup or chamfer for the oil with peg and red stuff.   The same procedure is to be followed with ¾-plate fusee, and it will be found best to finish the stopping in fusee piece before screwing the steel on to the brass.   Be careful to give the fusee but little end shake ; if it be at all excessive the stop work and the maintaining work will become uncertain, and either or both may fail.—H. B.

Fusee Cap.—[*Crochet de fusée.—Der Stellungshaken (Schnecke)*.] —A thin steel plate on the smaller end of the fusee.   It has a projecting nose, and as the last coil of the chain or gut is wound on to the fusee, it pushes the stop finger in the path of the nose of the fusee cap.

Fusee Engine.—[*Outil à tailler les fusées.—Die Schnecken-schneid maschine.*]—A kind of screw-cutting lathe used for cutting the grooves on fusees.

Fusee Hollow.—[*Sillon de fusée.—Die Schneckenzapfenunter-drehung.*]—The hollow cut behind the shoulder of the fusee upper pivot to prevent the overflow of the oil.

Fusee Piece.—[*Chapeau pour fusée.—Messingfutter für den Schneckenzapfen.*]—The circular plug screwed to the top plate in which the upper pivot of the fusee works.

Fusee Sink.—[*Noyure de fusée.—Die Schneckenausdrehung.*]— The sink cut in the top plate of a watch to give space for the fusee.

Gallows Tool.—[*Support pour pignons (outil).—Der Trieb-halter(Werkzeug).*] --A tool in which a pinion is placed by clockmakers when the leaves or bottoms are to be filed.   The head is rivetted very loosely on to the shank, so as to give to the motion of the hand in filing.

Gathering Pallet, Tumbler.—[*Collecteur.*—*Der Schöpfer.*]— A revolving finger that in striking clocks and repeating watches moves the rack one tooth for each blow struck.

Gauge.—[*Outil à mesurer.*—*Das Mass.*]—A measuring instrument. (See Callipers, Cylinder, Cylinder-Height, Douzième, Glass-Height, Micrometer, Pinion, Pinion-Height, Pivot, Pump-Cylinder, Slide.)

Gilding.—[*Dorage.*—*Die Vergoldung.*]—The coating of inferior metal with gold. There are two methods of depositing the gold ; the water or mercurial process, and the electro.

Water Gilding.—The operation of water gilding is as follows : Six parts of mercury are heated with one of gold, and the resulting amalgam squeezed, so as to separate superfluous mercury. Thus nearly four of the six parts of mercury will be squeezed out, and a mass of the consistence of butter left. The object to be gilt is rubbed over with a solution of subnitrate of mercury ; thus it becomes covered with a superficial layer of that metal. And now the amalgam is applied, and will at once attach itself to the mercurial surface. It is then washed, and gently heated over some burning charcoal, and the amalgam kept uniformly brushed over the surface by means of a soft brush. The heat is still kept up, until the surface assumes a dull yellow colour, when it is removed and polished by a wheel brush, kept moist by dilute vinegar. A mixture of beeswax and verdigris is then applied ; the latter having an affinity for mercury, removes any which is still left on. Lastly, the article is burnished, washed with dilute nitric acid, and afterwards with water, and dried. The whole operation is most noxious to the health of those who practise it.

Electro Gilding.—In large gilding establishments the current is now obtained by means of dynamo-electric machines ; but for the guidance of those who occasionally require to gild small articles, I have been favoured by Mr. Willis with the following concise directions, and if they are implicitly followed a good result will be ensured.

The Battery.—Procure a Smee's battery, say quart size, fill up the cell to within two inches of the top with 1 part of sulphuric acid to 12 parts of water ; let it stand till the mixture gets cold ; place the battery in it, and it is then fit for use. If you want to lessen the power, raise the battery out of the cell as to have less surface exposed to the action of the acid.

Place in the binding screws two copper wires of convenient length, ·125 inch diameter ; to the one in connection with the platinized silver plate hang a piece of fine gold to supply the gold solution or bath with gold ; to the one in connection with the zinc plates hang the articles to be gilded.

The Solution.—Dissolve 1 oz. of cyanide of potassium in 1 pint of distilled water in an earthenware or glass vessel ; hang the anode (the piece of gold) so as to dip into the solution, and on the other wire hang a piece of copper wire also dipping into the solution. If kept gently warmed, the solution in about three hours will be charged with gold and fit for use.

Preparation of the Articles.—If the articles are not perfectly clean the gold will not be deposited evenly, and the solution will be spoilt. Therefore carefully clean them in a bath of diluted acid or by well brushing them with a weak solution of cyanide.

Whatever surface is desired must be put on before gilding.

Articles that have been soft soldered should have the solder touched with a weak solution of sulphate of copper, which will deposit a thin coating of copper, to which the gold will take freely.

To Gild.—Hang the articles on the wire in connection with the zinc ; if more than one at a time, see that they do not touch each other. Hang the anode to dip in the solution, and in from half a minute to two minutes as they will be gilt ; if not thick enough, wash them and repeat. Keep the solution warm while using.

NOTE.—Always take the battery out of the cell when not in use. See that the connections are clean and bright. If the deposit is red or foxey, either you have too much battery power or your solution has too much cyanide or is made too hot. If it is pale, then the reverse. Always use fine gold. When after use the zinc plates are attacked by the acid, they must be remalgamated by placing them in a dish with 1 part of sulphuric acid to 10 parts of water, and brushing mercury over them till they have a silvery appearance ; no part of the plate must be left undone, or the acid will soon attack it. Then wash in clean water and stand on edge to drain off the superfluous mercury.

A complete outfit for occasional gilding and plating with the Bunsen battery, which is equally effective, may now be obtained at the material shops ; as well as requisites for nickel plating.

Tarnished Gilding may be revived by immersing for a few moments in a cyanide bath consisting of one ounce of cyanide of potassium, one ounce of lump ammonia to half-a-pint of water, and rinsing at once in hot water. The articles may then be dipped in spirits of wine, and dried in boxwood dust. The cyanide solution should be kept in a glass stoppered bottle, and used with care, as it emits a poisonous vapour.

Greasy articles should be cleaned in benzine before dipping in the cyanide.

A good recipe for renovating all kinds of gilt or polished work, which is used by French cleaners of zinc clock cases, and by many of them regarded as secret, is as follows :—

Cyanide of Potassium	...	...	...	5 parts.
Bicarbonate of Soda	...	...	...	1 ,,
Sulphuric Ether	...	...	...	$2\frac{1}{2}$,,
Alcohol	...	...	...	$2\frac{1}{2}$,,
Distilled Water	...	...	...	25 ,,

The articles remain in the bath a few seconds and are then rinsed with water, and afterwards with alcohol, and dried in boxwood dust.

Glass.—[*Verre.*—*Das Glas.*]

The shape and titles of the watch glasses in general use are as follows :—

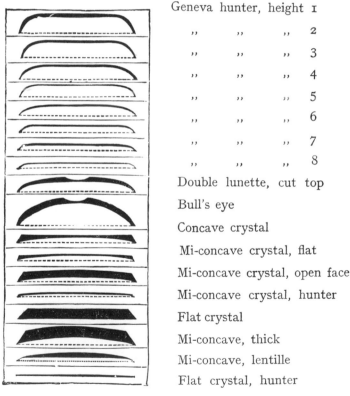

Geneva hunter, height 1
,, ,, ,, 2
,, ,, ,, 3
,, ,, ,, 4
,, ,, ,, 5
,, ,, ,, 6
,, ,, ,, 7
,, ,, ,, 8
Double lunette, cut top
Bull's eye
Concave crystal
Mi-concave crystal, flat
Mi-concave crystal, open face
Mi-concave crystal, hunter
Flat crystal
Mi-concave, thick
Mi-concave, lentille
Flat crystal, hunter

Glass Height Gauge.—This is especially useful in fitting glasses to hunting watches where there is but little spare room.

A plate of brass is formed with a ⅃ shaped base on which two
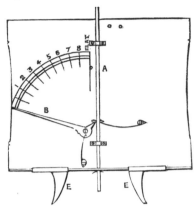
feet (E E) suitable for standing in the bezel of a watch slide. The top of the plate is folded over at the back to a right angle thus ⌐, to serve as a table on which to rest a glass. When the lower end of the sliding rod A is level with the feet, its upper end is a shade above the top of the plate and the pointer B is at " flat," so that when the low end of the rod rests on the cannon pinion or hand of a watch, the projection of the rod above the top of the plate gives the height required for the glass. At the same time the pointer shows on the scale the size corresponding to the labels on Geneva glasses. Mr. Virgo invented this tool, but to Mr. Squire is due the credit of adding the pointer and scale.

Glass Plates.—[*Plattes de glace.—Das Schleifglas.*]—Used for frosting and polishing steel work on, as well as for burnishing endstone settings and other brass pieces, are surfaced with emery. Perfect flatness cannot be ensured in preparing only two plates, for one may be dished and the other rounding, and yet they would touch each other all over. A glance at a ball and socket joint or other similar device will show this clearly. The best plan is to get three pieces of plate glass of a convenient size, and rub them together in rotation with rather coarse emery until a true surface is obtained. Then the plate intended for burnishing is rubbed on the other two with double washed emery, and may afterwards be polished with putty powder. When in use the burnishing plate should be kept scrupulously clean with alcohol and a wash-leather.

Gnomon, Style.—[*Aiguille de cadran solaire.—Der Sonnenuhrzeiger.*]—That part of a sun-dial whose shadow indicates the time.

Going Barrel.—[*Barillet tournant.—Das gezahnte Federhaus.*]—The barrel of a watch or clock round which are teeth for driving the train direct without the intervention of a fusee.

Going Fusee.—[*Fusée d'entretien.—Die Schnecke mit Gegengesperr.*]—A fusee with maintaining power attached. Old verge watches are still to be met with without maintaining work.

Gold.—[*Or.—Das Gold.*]—Below is the Mint value of gold. The value of the alloy is not taken into account, but in dealing with any considerable quantity of gold alloyed with silver, the value of the latter may be worth consideration. As a rule one-third to one-half of the alloy in gold chains and jewellery is silver.

Value of Gold per Ounce Troy				Value of Gold per Ounce Troy			
Pure Gold	£4	5	0	13 Carat	£2	6	0½
22 Carat	3	17	11	12 Carat	2	2	6
18 Carat	3	3	9	9 Carat	1	11	10½
15 Carat	2	13	1½	8 Carat...	1	8	4

See also " Hall Mark."

Mystery Gold.—This substance, resembling 9-carat gold and resisting the action of nitric acid, was found by Mr. F. Lowe to consist of silver 2·48, platinum 32·02, and copper 65·5 parts.

Gold Spring, Passing Spring.—[*Ressort d'or.*—*Die Goldfeder. Die Auslösungs feder (Chronometerhemmung).*]—A very thin gold spring attached to the detent of a chronometer escapement, through which the escape wheel is unlocked by a pallet when the balance moves in one direction, but which yields and allows the pallet to pass when the balance moves the opposite way.

Gong.—[*Timbre.*—*Die Tonfeder.*]—A piece of steel wire bent into the form of a volute, used for striking the hours on in quarter and carriage clocks when a deep tone is required. The hammer should be faced with leather slightly burnt to harden it.

Gongs.—Steel wires on which the hammers strike in repeating watches. They are fixed at one end to the plate, and run round inside the band of the case.

Graham, George, born at Kirlinton, Cumberland, 1673, died 1751, and buried in Westminster Abbey. He was apprenticed to one Aske, and on completion of his apprenticeship, became assistant to Tompion. About 1700 he perfected the Cylinder Escapement which Tompion had previously patented. He invented the Dead Beat Escapement for clocks and the mercurial pendulum. For some years prior to his death he kept a shop opposite the " Bolt and Tun " in Fleet Street.

Graham Escapement.—See Dead Beat.

Graver.—[*Burin.*—*Der Stichel.*]—A cutting tool formed of a square piece of steel, the cutting end of which is made diamond-shaped. It is the chief turning tool used by watchmakers. In turning, the graver should be presented to the work so that the body of the tool lies nearly at a tangent to the work operated upon. The sketch ap-

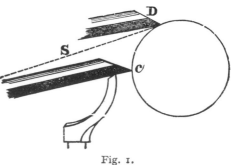

Fig. 1.

pended (Fig. 1) will show what is meant. The point of the
graver at *c* is too low, and the handle too high : the graver
would rub away the metal instead of cutting it. This is
often a favourite position with beginners, who exert so much
pressure against the work that it either flies out of the lathe or
the tool slips under it and breaks it. The point of the graver D
is too high, and the tool is not held at the proper angle, but if it
followed the dotted line s it would cut well ; and though the best
place for the cutting point is between D and C, it had better be
too high than too low. The face of the graver at D is formed at
about the proper angle for ordinary work ; for pinion hollows
and other special purposes a longer
face is often preferred. Gravers
should be proportioned in size to the
work in hand and have handles of
cane or other light material. The
face only, and not the sides of a
graver, should be whetted; after the
operation the face should present a
flat surface, if it is rounding towards
the edge a clean cut will be out of
the question. Fig. 2 shows a tool
by Boley for holding a graver or

Fig. 2.

other cutter while being whetted so as to ensure a flat face
at the desired angle, but such devices are not to be recommended
except for slide rest cutters. Perseverance will give the ability
needed to produce a flat graver by hand.

The very sharp point of the graver should have one light
touch on a smooth stone to make it an almost imperceptible
chisel-shape ; steel will then turn easier, and whether brass or
steel is being manipulated, the graver will last better. When
turning hard steel the graver should touch a different spot with
each stroke of the bow, for the steel is then not so likely to
glaze.

There is a kind of graver to be had under the name of
"Progress," which is particularly suited for turning hard steel,
drilling holes in enamel dials, glass, etc.

Gravity.—[*Gravité.—Die Schwerkraft.*]—The force that draws
all bodies towards the centre of the earth, and by which
a pendulum drawn aside from its point of rest is impelled to
vibrate. The velocity acquired by a falling body increases
directly as the time occupied in falling ; the vertical space
passed through will be as the square of the time. At the sea-
level in London, a body exposed freely to the action of gravity
falls 16·095 feet in the first second, and at the end of that time
has acquired a velocity of 32·19 feet per second. It must be

remembered that it started with no velocity, and that the velocity is gradually and uniformly accelerated, therefore, if falling for two seconds, it would have passed through 64·38 feet, and at the end of that period be travelling with a velocity of 64·38 feet per second. g is used by mathematicians to represent the acceleration in the velocity of a body exposed to the action of gravity, which varies in different latitudes, but as before stated is 32·19 feet per second in the latitude of London. The velocity acquired by a body in falling through a given height is the same whether it fall vertically or descend through an inclined plane. It would take the same time to fall vertically through the diameter of a circle and through any chord terminating in the extremity of the diameter, and the velocity acquired by descending chords of the same circle is as the length of the chord.

Gravity Escapement.—[*Echappement à gravité.*—*Die Schwerkrafthemmung.*]—An escapement in which impulse is given to the pendulum by a weight falling through a constant distance.

For large turret and other clocks which have to move a number of heavy hands exposed to wind and snow, the Graham and similar escapements are not perfectly adapted. The driving weight of the clock must be sufficient to move the hands under the most adverse circumstances. Then at times, when the wind and snow assist the hands in their motion, the whole of the superfluous power is thrown on the escapement, and accurate performance cannot be expected. The custom has been to use for such clocks a remontoire of some kind, that is, an arrangement by which the train, instead of impelling the pendulum direct, winds up a spring. This spring in unwinding administers a constant impulse to the pendulum. A gravity escapement partakes somewhat of this principle. The train raises an arm of certain weight a constant distance, and the weight of this arm in returning impels the pendulum. Until Mr. Denison invented the double three-legged for the great clock at the Houses of Parliament gravity escapements were regarded with suspicion as having a tendency to trip.

Double Three-Legged Gravity Escapement.—Invented by E. B. Denison (Lord Grimthorpe), 1854. This escapement, shown in Fig. 1, consists of two gravity impulse pallets pivoted as nearly as possible in a line with the bending point of the pendulum spring. The locking wheel is made up of two thin plates having three long teeth or "legs" each. These two plates are squared on the arbor a little distance apart, one on each side of the pallets. Between them are three pins which lift the pallets. These pins are generally the bodies of three screws used to connect the locking plates through a three-leaved pinion answers the purpose. In the drawing one of the front legs is resting on a block screwed to the front of the right-hand pallet. This forms the locking. There is a similar block screwed to the back of the left-hand pallet, for the legs of back plate, which is

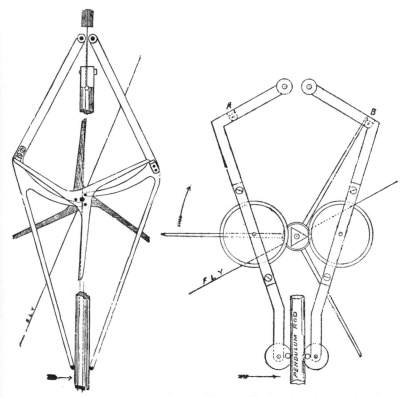

Fig. 1.—Double Three-Legged
Gravity Escapement.

Fig. 2.—Single Three-Legged Gravity
Escapement.

shaded in the drawing, to lock upon. Projecting from each of the pallets is an arm. The tip of the one on the right-hand pallet is just in contact with one of the pins which has lifted the pallet to the position shown. The pendulum is travelling in the direction indicated by the arrow, and the left hand pallet has just given impulse. The pendulum rod in its swing will push the right-hand pallet far enough for the leg of the front locking plate, which is now resting on the block, to escape. Directly it escapes the left-hand pallet is lifted free of the pendulum rod by the lowest of the three pins. After the locking

M

wheel has passed through 60 degrees, a " leg " of the back locking plate is caught by the locking block on the left-hand pallet. There should be a couple of banking pins to stop the pallets from going lower than the left-hand one is shown. This allows the lifting pins to have a little free run before reaching the arm.

As the three-leaved pinion always lifts the pallets the same distance, the pallets in returning give a constant impulse to the pendulum. The friction in unlocking would, of course, vary with the pressure transmitted through the train, but the effect of such variation is found to be practically of no moment. To avoid any jar when the locking leg falls on the block there is a fly, kept by a spring friction-tight on an enlarged portion of the arbor. This fly causes the legs to fall smoothly and dead on the blocks, and thus avoids all danger of tripping.

All the parts are made very light, of steel, with the acting surfaces hardened. The pallets and locking wheels may be cut from an old saw plate, and a three-leaved pinion for lifting obtained by breaking out every alternate leaf from a six-leaved pinion. The distance of the lifting pins from the centre should not be more than $\frac{1}{8}$ or less than $\frac{1}{12}$ of the radius of the locking legs. They should be placed as shown, the one last in action and the one about to lift being vertically under each other ; the lifting is then performed across the line of centres. The distance of the centre of the escape wheel from the pivots of the pallets = the diameter of the escape wheel. The length of the tails of the pallets is immaterial. For symmetry they are generally made as shown. The most frequent mistake in constructing this escapement is that the parts, especially the locking plates, which can hardly be too light, are made too heavy. Lord Grimthorpe suggests that the fly should be made of aluminium for lightness. The beat pins may be of brass or ivory. They, and the pendulum rod where they touch, should be left perfectly dry. If oiled they become sticky, and the action of the escapement will be unsatisfactory.

Single Three-Legged Gravity Escapement.—It will be observed that the double three-legged really acts as an escapement with a three-toothed wheel, the extra wheel being added to spread the impulse pallets more, and so obtain a greater vertical lift for a given horizontal movement. Fig. 2 is a single three-legged arrangement by Dr. Waldo and Professor Lyman. The locking blocks A and B are well placed for ensuring sound lock and easy release. Instead of the usual three pins, there is, for lifting, a triangular steel block which acts against large light friction rollers pivoted one on each pallet. The locking wheel is commendably light. The beat pins are mounted eccentrically for easy adjustment, a good plan often followed.

A *Four-Legged Gravity Escapement* is occasionally used for regulators and other clocks with seconds pendulums. It may, perhaps, when thoroughly well made, and with the locking blocks jewelled, be better than the Graham for such a purpose, as it is free from the error due to thickening of the oil ; but from the small number of teeth in the escape wheel, it requires in the train either very high numbered wheels or an extra wheel and pinion. This is a distinct advantage in a turret clock, because the large amount of power required to drive the leading off rod is thereby more reduced by the time it reaches the escapement. In a regulator the extra pair of wheels and the greatly increased wear of the escape pinion are a drawback sufficient to prevent its general adoption, considering the good performance of the Graham. There is the additional advantage with the Graham that the escape wheel rotates once in a minute and affords a ready means of obtaining the seconds indicator.

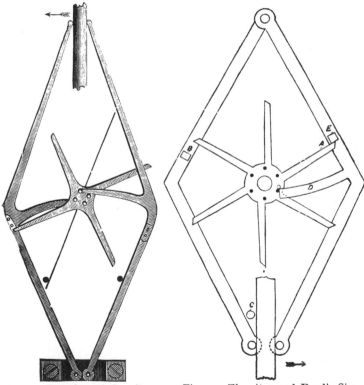

Fig. 3.—Four-legged Gravity Escapement.

Fig. 4.—Thwaites and Reed's Six-legged Gravity Escapement.

For a regulator with a four-legged gravity escapement the escape wheel is usually 3 inches in diameter, and there are 8 lifting pins ; four, as shown, to actuate the left-hand pallet, and equidistant from these are four others projecting from the other side of the wheel to give impulse to the right-hand pallet.

Thwaites & Reed's Six-Legged Gravity Escapement. This modification of Denison's Gravity escapement gives excellent results. Impulse is given to the pendulum at each alternate vibration only. The rotations of the escape wheel are only half what are required with the double three-legged, and a much lighter driving weight will suffice. In the engraving, the pendulum swinging in the direction of the arrow will unlock the tooth A and allow the wheel to move till it is stopped by the block B on the neutral arm ; by this time the lifting pin which had been in contact with the arm D is carried to the position indicated by the outline just above the real pin, which is black ; so that in the return vibration the arm D clears the lifting pin and follows the pendulum, giving it impulse. In its excursion the pendulum moves the neutral arm till the wheel is unlocked from the block B, and the wheel then again takes the position shown in the drawing. The neutral arm simply rises and falls, giving no impulse ; but when resting against the banking pin C is in the proper position to lock the wheel.

Great Wheel.—[*La première roue.—Das Schneckenrad.*]—The wheel on the fusee arbor which drives the centre pinion. The largest wheel in a watch or clock.

Grimthorpe.—See Denison.

Guard Pin, Safety Pin.—[*Dard.—Der Sicherheitsstift. Das Sicherheitsmesser.*]—A pin in the lever escapement that prevents the pallets leaving the escape wheel when the hands of the watch are turned back. (See Lever Escapement.)

Gymbals.—[*Suspension de chronomètre.—Die Universalaufhäng-ung (Schiffscompass).*]—A universal joint on which marine chronometers are mounted. The object of the gymbals is to keep the chronometer level, notwithstanding that the wooden chronometer box is constantly tilted by the motion of the ship. Invented by Carden (according to Berthoud) to suspend the ship's lamps, and first used for a marine timekeeper by Huyghens.

Hair Spring.—[*Ressort spiral.—Die Spiralfeder.*]—The balance spring of a watch. (See Balance Spring.)

Half Chronometer.—[*Demi chronomètre.—Die Präsicions Taschenuhr.*]—This term originally used to denote watches having an escapement compounded of the lever and chronometer, appears now to be applied to fine lever watches which have been adjusted for temperature and positions.

Half Dead Escapement.—[*Echappement à demi repos.*—*Die Hemmung mit unvollkommener Ruhe.*]—A clock escapement in which there is little recoil.

Half Plate. — [*Demi Platine.* — *Das Halbplatinenwerk.*] — A watch in which the top pivot of the fourth wheel pinion is carried in a cock so as to allow of the use of a larger balance than could otherwise be got in. A sounder jewelling, too, is often secured by cocking the fourth wheel pinion. (See Plate.)

Half-timing.—A preparatory timing of verge watches by letting them run without the balance spring. If the watch ran nearly to half time the balance was considered to be a suitable one.

Hall Marks.—[*Der Feingehaltsstempel.*]—Hall marks are impressed on watch cases, jewellery, and plate after the quality of the metal has been ascertained by assay at certain official Assay Halls. The marking of jewellery is with few exceptions optional. The hall marking of all watch cases of gold or silver made in Great Britian and Ireland is compulsory. The cost is only the actual outlay incurred in assaying and stamping. The hall mark consists of several impressions in separate shields : there are the standard or quality mark ; the mark of the particular office at which the article was assayed ; some character by which the date of marking may be traced, and, if duty is chargeable, the head of the reigning sovereign.

The oldest and most important of the Assay Halls is that presided over by the Goldsmiths' Company of London, which is situated in Foster Lane, just at the back of the Old General Post Office, St. Martin's-le-Grand. The privilege of assaying and marking precious metals was conferred on the company by statute in 1300. The company received a charter of incorporation in 1327, and their powers have been confirmed subsequently by several acts of Parliament.

Many early watch cases, especially silver ones of London make, are met with which have no hall mark, the powers of the company not being so strictly enforced then as now, or the value of the official assay not being so generally recognized.

Repoussé cases, with other artistic wares of a similar character, are exempted from assay.

It was not till 1798 that a lower standard of gold than 22 carat was allowed, 18 carat was then recognized ; in 1854 three further standards, 15, 12 and 9 carat, were introduced.

The standard mark of the London Hall is a lion passant for sterling silver. A lion passant was also the standard mark on 22-carat gold up to 1845.

For gold of 22 carat the standard mark is now a crown, and the figures 22. For 18-carat gold the standard mark is a crown and the figures 18.

For 15-carat gold 15 and 0·625 ⎫ Pure gold being 24 carats, these decimals
  ,, 12 ,, 12 ,, 0·5 ⎬ represent the proportions of pure gold
  ,,   9 ,,   9 ,, 0·375 ⎭ in the article so marked.

The London Hall Mark prior to 1823 was a crowned Leopard's head ; from January 1st, 1823, it was uncrowned ; specimens of both styles are appended.

Date marks of the London Hall, given on pages 184, 185, are, with one or two exceptions, actual reproductions which I have made from watch cases. Specimens of the earliest marks are not to be obtained.

There was a duty on silver articles of sixpence an ounce from 1719 till 1758, but no special duty mark ; in 1784 a similar duty was imposed, and then the head of the reigning sovereign was impressed to denote the payment of duty. The Act came into operation on December 1st, 1784, and at first the head had a curious appearance, being incised, or incuse as it is called, instead of in relief as the other marks were. Cases with the London mark and the letter K, which corresponds to the period from May, 1785, to May, 1786, have the duty head incuse after which the head appears in relief with London marks. The Wardens of the Birmingham Assay Office have a pair of cases with the head incuse, and the Birmingham mark with the letter N, which would denote the period from July, 1786, to July, 1787. In 1804 the duty on silver was increased to 1s. 3d., and on gold to 16s. an ounce. In 1815 a further increase to 1s. 6d. and 17s. 6d. respectively was made, and the duty continued at these amounts till 1890, when it was finally abolished. Watch cases were exempted from duty in 1798.

The maker's mark before 1697 was some emblem selected by him ; in that year it was ordered to be the two first letters of his surname ; since 1739 it has been the initials of the maker's Christian and surnames.

On March 25th, 1697, the quality of standard silver was raised from 11 ozs. 2 dwt. to 11 ozs. 10 dwt. of pure silver in 12 ozs. of plate ; a lion's head erased was then used as the standard mark, and a figure of Britannia as the hall mark ; but on June 1st, 1720, the old standard of 11 ozs. 2 dwts. and the old marks of a lion passant and leopard's head were reverted to, although the higher standard with the figure of Britannia is still occasionally used.

Marks of other Assay Offices.—CHESTER.—Hall mark, a sword between three wheat-sheaves. Prior to 1779 it was three demi-lions and a wheatsheaf on a shield. Standard mark for 18-carat gold, a crown and the figures 18. For silver, a lion passant. Before 1839 a leopard's head in addition. Chester date marks are given on page 186.

BIRMINGHAM.—Hall mark, an anchor in a square frame for gold, and an anchor in a pointed shield for silver. Standard

mark for 18-carat gold, a crown and the figures 18 ; for silver a lion passant. Birmingham date marks are given on page 186.

SHEFFIELD.—A crown (silver is only assayed).

EXETER.—A castle with three towers.

YORK.—Five lions on a cross.

NEWCASTLE.—Three castles.

NORWICH.—A castle and lion passant. (The Norwich Assay office is now closed.)

EDINBURGH has a thistle for the standard mark, and a castle for the hall mark.

GLASGOW has a lion rampant for the standard, and a tree, a fish, and a bell for the hall mark.

DUBLIN has a harp crowned as the standard mark for sterling silver and for 22-carat gold, with the figures 22 added in the latter case; for 20-carat gold, a plume of three feathers and 20 ; for 18-carat gold, unicorn head and 18. The lower qualities of 15, 12, and 9, are marked with the same standard mark as is used
at the London Hall. The Hall mark for Dublin is a figure of Hibernia.

An Act of Parliament in 1904 ordained that foreign plate imported into Great Britian should be assayed and marked. To this end special stamps were devised for the various Assay Offices. At Goldsmiths' Hall, London, besides the quality mark, the sign of the constellation " Leo " is used, at Birmingham a triangle, and at Chester an acorn.

Swiss Hall Marks.—In Switzerland the hall marking of watch cases is only compulsory when the article already bears some indication of the quality gold or silver of which it is made, and even then the English or other recognized hall mark is accepted as a substitute for the Swiss.

GOLD.		SILVER.	
			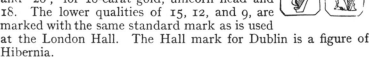
18-C (or 0·755).	14-C (or 0·583.)	Sterling Silver (or 0·935	0·800

The shields represented in the subjoined tables are those used
shield is invariably in the shape of a rectangle, with the

NOTE.—The Date Mark is altered on the 30th of May
till the 30th of May in

a	1678	a	*	A	1716	a	1736	A	1756	a	1776
b	1679	b	1697	B	1717	b	1737	B	1757	b	1777
c	1680	c	1698	C	1718	c	1738	C	1758	c	1778
d	1681	S	1699	D	1719	d	1739	D	1759	d	1779
e	1682	E	1700	E	1720	e	1740	E	1760	e	1780
f	1683	ff	1701	F	1721	f	1741	f	1761	f	1781
g	1684	G	1702	G	1722	g	1742	G	1762	g	1782
h	1685	h	1703	H	1723	h	1743	h	1763	h	1783
i	1686	i	1704	I	1724	i	1744	J	1764	i	1784 †
k	1687	k	1705	K	1725	k	1745	K	1765	k	1785
l	1688	l	1706	L	1726	l	1746	L	1766	l	1786
m	1689	m	1707	M	1727	m	1747	m	1767	m	1787
n	1690	n	1708	N	1728	n	1748	n	1768	n	1788
o	1691	o	1709	O	1729	o	1749	O	1769	o	1789
p	1692	p	1710	P	1730	p	1750	p	1770	p	1790
q	1693	q	1711	Q	1731	q	1751	Q	1771	q	1791
r	1694	r	1712	R	1732	r	1752	R	1772	r	1792
s	1695	s	1713	S	1733	s	1753	S	1773	s	1793
t	1696	t	1714	T	1734	t	1754	T	1774	t	1794
		u	1715	V	1735	u	1755	u	1775	u	1795

* This letter appears to have been used only from March to May, 1697.
Britannia and a lion's head erased was used instead of the
† Watch cases marked between December, 1784, and May, 1798, would

WATCH CASES MARKED AT GOLDSMITHS' HALL, LONDON.

for silver and for 22-carat gold. For lower qualities of gold the corners taken off, like the one surrounding A in 1876.

in each year, lasting from the date indicated in the Table
the following year.

A	1796	a	1816	A	1836	a	1856	A	1876	a	1896
B	1797	b	1817	B	1837	b	1857	B	1877	b	1897
C	1798	c	1818	C	1838	C	1858	C	1878	c	1898
D	1799	d	1819	D	1839	d	1859	D	1879	d	1899
E	1800	e	1820	E	1840	e	1860	E	1880	e	1900
F	1801	f	1821	f	1841	E	1861	F	1881	f	1901
G	1802	g	1822	G	1842	g	1862	G	1882	g	1902
H	1803	h	1823	H	1843	h	1863	H	1883	h	1903
I	1804	i	1824	J	1844	i	1864	I	1884	i	1904
K	1805	k	1825	K	1845	k	1865	K	1885	k	1905
L	1806	l	1826	L	1846	l	1866	L	1886	l	1906
M	1807	m	1827	M	1847	m	1867	M	1887	m	1907
N	1808	n	1828	N	1848	N	1868	N	1888	n	1908
O	1809	O	1829	O	1849	O	1869	O	1889	O	1909
P	1810	p	1830	P	1850	p	1870	P	1890	p	1910
Q	1811	q	1831	Q	1851	q	1871	Q	1891	q	1911
R	1812	r	1832	R	1852	r	1872	R	1892	r	1912
S	1813	S	1833	S	1853	S	1873	S	1893		1913
T	1814	t	1834	T	1854	t	1874	T	1894		1914
U	1815	u	1835	U	1855	u	1875	U	1895		1915

From the 25th March, 1697, to the 1st June, 1720, the figure of
crowned leopard's head and a lion passant, see page 182.
bear an extra stamp representing the head of George III. ; see page 182.

BIRMINGHAM ASSAY OFFICE DATE LETTERS.

NOTE.—The Date Mark is altered on the 1st July of each year, lasting from the Date indicated in the Table till the June following.

Letter	Year	Letter	Year	Letter	Year	Letter	Year	Letter	Year
A	1773	a	1799	A	1825	A	1850	a	1875
B	1774	b	1800	B	1826	B	1851	b	1876
C	1775	c	1801	C	1827	C	1852	c	1877
D	1776	d	1802	D	1828	D	1853	d	1878
E	1777	e	1803	E	1829	E	1854	e	1879
F	1778	f	1804	F	1830	F	1855	f	1880
G	1779	g	1805	G	1831	G	1856	g	1881
H	1780	h	1806	H	1832	H	1857	h	1882
I	1781	i	1807	I	1833	I	1858	i	1883
J	1782	j	1808	K	1834	K	1859	k	1884
K	1783	k	1809	L	1835	L	1860	l	1885
L	1784	l	1810	M	1836	M	1861	m	1886
M	1785	m	1811	N	1837	N	1862	n	1887
N	1786	n	1812	O	1838	O	1863	o	1888
O	1787	o	1813	P	1839	P	1864	p	1889
P	1788	p	1814	Q	1840	Q	1865	q	1890
Q	1789	q	1815	R	1841	R	1866	r	1891
R	1790	r	1816	S	1842	S	1867	s	1892
S	1791	s	1817	T	1843	T	1868	t	1893
T	1792	t	1818	U	1844	U	1869	u	1894
U	1793	u	1819	V	1845	V	1870	v	1895
V	1794	v	1820	W	1846	W	1871	w	1896
W	1795	w	1821	X	1847	X	1872	x	1897
X	1796	x	1822	Y	1848	Y	1873	y	1898
Y	1797	y	1823	Z	1849	Z	1874	}	1899
Z	1798	z	1824					a	1900

CHESTER ASSAY OFFICE DATE LETTERS.

The Date Mark is altered on the 1st July, lasting from the Date indicated in the Table till the end of June in the following year.

Letter	Year	Letter	Year	Letter	Year	Letter	Year	Letter	Year	Letter	Year	Letter	Year	Letter	Year
A	1701	A	1726	A	1752	a	1776	A	1797	A	1818	A	1839	a	1864
B	1702	B	1727	B	1753	b	1777	B	1798	B	1819	B	1840	b	1865
C	1703	C	1728	C	1754	c	1778	C	1799	C	1820	C	1841	c	1866
D	1704	D	1729	D	1755	d	1779	D	1800	D	1821	D	1842	d	1867
E	1705	E	1730	E	1756	e	1780	E	1801	E	1822	E	1843	e	1868
F	1706	F	1731	F	1757	f	1781	F	1802	F	1823	F	1844	f	1869
G	1707	G	1732	G	1758	g	1782	G	1803	G	1824	G	1845	g	1870
H	1708	H	1733	H	1759	h	1783	H	1804	H	1825	H	1846	h	1871
I	1709	I	1734	I	1760	i	1784	I	1805	I	1826	I	1847	i	1872
K	1710	J	1735	J	1761	k	1785	K	1806	K	1827	K	1848	k	1873
L	1711	K	1736	K	1762	l	1786	L	1807	L	1828	L	1849	l	1874
M	1712	L	1737	L	1763	m	1787	M	1808	M	1829	M	1850	m	1875
N	1713	M	1738	M	1764	n	1788	N	1809	N	1830	N	1851	n	1876
O	1714	N	1739	N	1765	o	1789	O	1810	O	1831	O	1852	o	1877
P	1715	O	1740	O	1766	p	1790	P	1811	P	1832	P	1853	p	1878
Q	1716	P	1741	Q	1767	q	1791	Q	1812	Q	1833	Q	1854	q	1879
R	1717	Q	1742	Q	1768	r	1792	R	1813	R	1834	R	1855	r	1880
S	1718	R	1743	R	1769	s	1793	S	1814	S	1835	S	1856	s	1881
T	1719	S	1744	S	1770	t	1794	T	1815	T	1836	T	1857	t	1882
U	1720	T	1745	T	1771	u	1795	U	1816	U	1837	U	1858	u	1883
V	1721	U	1746	U	1772	v	1796	V	1817	V	1838	V	1859	A	1884
W	1722	V	1747	V	1773							W	1860	†	
X	1723	W	1748	W	1774							X	1861		
Y	1724	X	1749	X	1775							Y	1862		
Z	1725	Y	1750									Z	1863		
		Z	1751												

* These two cycles are really script capitals.
† Roman capitals continued as in the cycle beginning with 1818.

Hammer Hardening. — [*Forgeage.* — *Metallhärtung mittelst des Hammers.*]—The only way of hardening brass and copper is by compression. Hammering, if uniformly and skilfully done, answers better than rolling or other methods, but considerable practice is necessary in hammer hardening such a piece of brass as a clock plate, or it will be buckled or distorted out of all shape in the process. Steel of which pallets are made, which is required to be just as hard as will allow of its being worked, is compressed by rolling.

Hammer Rods.—[*Verges de marteaux.*—*Die Hammerleitung (bei Turmuhren).*]—In a turret clock, the vertical rods that connect the movement with the hammers.

Hammer Tail.—[*Queue de marteau.*—*Der Hammerhebel.*]—In a striking clock, a continuation of the hammer stalk that is lifted by the pins in the pin wheel.

Hands.—[*Aiguilles.*—*Die Zeiger.*]—The revolving pointers used to indicate the time on the dials of watches and clocks.

Watch Hands.—The best way of taking off watch and chronometer hands is to insert the points of two small gun-metal levers like miniature crowbars, or steel levers like Fig. 1, under the hour hand socket, one either side ; a piece of thin paper may be interposed between the dial and the hands. When the hands are removed in this way there is no fear of cracking the dial or deforming the hands. The

Fig. 1.

points of the gun-metal levers will occasionally require to be dressed. In some cases there is not room to insert the levers, and then a good plan is to provide a pair of sound pliers without any shake at the joint, with points shaped as Fig. 2, and nicely hardened

and polished. They can be inserted between a very close minute and hour hand. They retain the hand ; do not injure it like the nippers do, and are more under control Fig. 2. than tweezers shaped in the same way.

A very useful tool for holding watch hands while broaching the hole is shown in Fig. 3. It consists of a lantern very similar to those used in the Geneva Screw Head Tool, with an extra long boss for the screw, and with the face of the lantern cut away, as shown in Fig. 3, to suit the hand. There is a knurled lock nut to ensure the hand being held fast when it is once jammed up, and a hole is pierced right through the handle of the tool to permit of the passage of the broach. A pair of sliding tongs with wide jaws, having a series of holes in them to let the bosses of different-sized hands pass through (Fig. 4), also form a good tool for the purpose.

In opening the hole of a watch hand to size, time will be

saved, if, supposing the boss of the old hand is available, a broach be tried in the hole of the old hand, and the distance it will enter marked on it by pushing a slip of cork on the broach in front of the old boss. The new hand may then be broached out to the cork with confidence before trying it on.

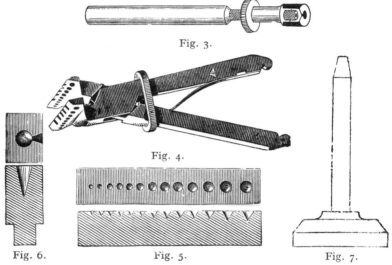

Fig. 3.

Fig. 4.

Fig. 6.   Fig. 5.   Fig. 7.

If it is desired to close the hole of the minute hand, the corner of the boss may be turned off at an angle, leaving the metal thin round the hole, and the hand then placed on a stake and one smart blow given to the boss with a flat surfaced hammer. An hour hand too large may have a slit cut through the boss, which is then sprung together. Some use a stake with a series of conical holes, as in Fig. 5, for placing the boss of the hand in. The hole is then closed by striking the hand ; but even with this stake, if the boss is very stout, the corner must be turned off.

A single stake with the hole a long taper and a slit for the hand is useful (Fig. 6). The hand is placed in boss downwards and struck with a hollow punch, and afterwards reversed and the other side closed.

An hour hand should be of such a length as to just reach the chapters without overlapping them. An hour hand too long is not so readily distinguished from the minute hand, especially if it is comparatively narrow.

It is a common practice, if a seconds hand has a long pipe, to broach it full large, and give a nip to make it grasp a pivot ;

for there is danger of breaking the pivot in removing the hand if a long pipe fits it throughout. A better plan, however, is to broach the long pipe from both ends till it fits. A short pipe will rarely fit too tight.

A piece of brass wire mounted on a foot, as shown in Fig. 7, is a convenience when fitting watch hands, especially where there is a dust-pipe. The top of the wire is slightly hollowed to receive the set square, and with standard like this the dome of the case need not be opened. (See also Riveting Thimble.) Fig. 8 is Haswell''s watch-hand gauge. The proper length for the hour-hand for, say, a 12-size English movement, is from the dot in the centre to the inner curved line

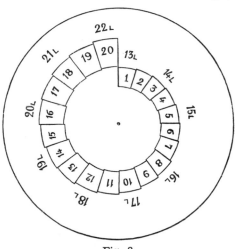

Fig. 8.

by the figure 12, and the proper length of a corresponding minute hand is from the centre to the outer curve. The preceding odd number may be taken as an approximate gauge for chronograph and crystal case hands. A number with *L* added outside the snails indicates the gauge for hands for Swiss watch movements of that number of lignes.

Blackening Clock Hands.—Hands for ships' timepieces or exposed positions, or blued hands that have rusted, may be blackened by holding them over the smoke from a tallow candle and then lightly coating them with lacquer by means of a soft brush.

Hand Vice.—[*Etau à main.—Der Feilkloben.*]—A small vice held in the hand.

Hardening.—[*Trempe—Die Feuerhärtung.*]—Steel is rendered hard by heating it to redness and then cooling it suddenly. Bright steel should not be exposed naked to a fire or flame. It may with advantage be heated in cyanide as described in reference to balance springs on p. 45, or placed in a covered box containing bone dust or animal charcoal in some other form; another plan is to smear soap over the article to be hardened. Water or oil is the medium generally selected for plunging the

190

article in to cool it. Petroleum is recommended if extra hardness is desired. Either mercury or salt water will give great hardness, but the steel is rendered brittle. Oil is the best medium for hardening steel if toughness is desired. (See Hammer hardening.)

Hand Wheel.—[*Roue à main.—Das Handschwungrad.*]—

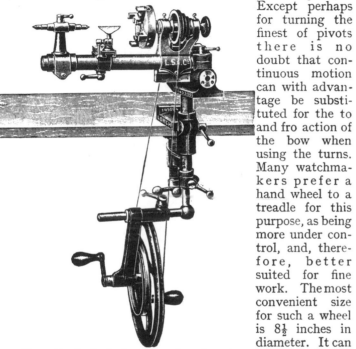

Except perhaps for turning the finest of pivots there is no doubt that continuous motion can with advantage be substituted for the to and fro action of the bow when using the turns. Many watchmakers prefer a hand wheel to a treadle for this purpose, as being more under control, and, therefore, better suited for fine work. The most convenient size for such a wheel is 8½ inches in diameter. It can be easily attached below the bench at a suitable height for the hand. The sketch shows a handwheel attached to the Lorch Lathe. It is so jointed as to be readily adjusted to any required position.

Hanging Barrel..—[*Barillet suspendu.—Das einseitig unterstützte Federhaus.*] A going barrel whose arbor is supported only at the upper end.

Hardy, William.—A clockmaker who flourished at the beginning of the nineteenth century. Among other inventions be devised a compensation balance, the principle of which will be understood from the sketch. Abandoning the cylindrical  laminæ used by Arnold and Earnshaw, he used a straight laminated bar of brass and steel, the brass being underneath. A hole in the centre of the bar served to attach it to the staff, and at each end of the bar was a stalk carrying a spherical weight. These weights could be made more or less active as compensators by screwing them up or down on the stalks which had threads cut on them. By slightly curving the laminated bar upwards or downwards, the weights could be caused to approach or recede from the centre of the balance more or less as desired. At first sight it appeared that the difficulty of the middle temperature error had been overcome. But to obtain sufficient compensation the laminated bar must be so thin and the stalks so long as to leave the balance wanting in rigidity. Nevertheless Hardy's attempt led to the invention of many other balances on the same principle, and therefore deserves to be recorded.

Harrison, John (born, 1693 ; died at his house in Red Lion Square, and buried in Hampstead Churchyard in 1776). The son of a carpenter, he devoted his life to the improvement of timekeepers, and just before his death received from the Government £20,000 for producing a chronometer that would determine the longitude at sea to half a degree. His timekeeper, which is now at Greenwich Observatory, has a vertical escapement driven by a remontoire. It is in a silver case in the shape of a watch case, and is furnished with a going fusee and compensation curb. Harrison is also credited with the invention of the gridiron pendulum, though there is no doubt that Graham had some years before experimented with compensation pendulums composed of metal rods. As nearly as can be ascertained, Harrison invented the going fusee, which has proved to be the most lasting of his conceptions, about 1750.

Hartnup, John.—In 1849 he invented the compensation balance shown in the figure. The rims are composed of brass and steel, as usual, but they are neither upright nor flat, but bevelled, or placed at an angle midway between these two positions. The centre arm *a* is also bimetallic, the brass being uppermost, and connecting the arm with the sections of the rim are two other bimetallic strips *b* *c*, the brass of these being underneath, and the steel on top ; *e* *e* are the weights, and at the ends of the rim the screws for timing and poising.

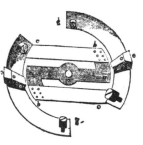

Heart Piece.—[*Cœur.—Das Herz.*]—A heart shaped cam **used in** chronographs to cause the chronograph hand to fly back to zero. First applied by Adolph Nicole in 1862. (Patent No. 1461.)

Hewitt, Thomas.—A chronometer maker who devised many different forms of compensation balances (born 1799, died 1867).

Hindley, Henry (born 1701, died 1771).—He was a clever clockmaker at York, and the inventor of a wheel-cutting engine.

Hollow Fusee.—[*Fusée à carré percé.*—*Die Schnecke mit versenktem Aufziehzapfen.*]—A fusee in which the upper pivot is sunk into the body of the fusee ; the object being to obtain a thinner watch than would otherwise be possible.

Hollow Pinion.—[*Pignon-percé.*—*Das durchbohrte Minutentrieb.*]—A pinion pierced throughout its length, such as the centre pinion in many three-quarter plate watches.

Hooke, Robert (born at Freshwater, Isle of Wight, in 1635 ; died, and buried at St. Helen's, Bishopsgate, in 1703). He invented the balance spring for watches about 1660, and the anchor escapement for clocks a few years later.

Horizontal Escapement.—See Cylinder Escapement.

Horological Institute, Limited (The British)—An Association of Watchmakers founded in 1858 for the purpose of advancing the Horological Arts. It has a commodious building in Northampton Square, Clerkenwell, where classes are conducted for teaching Drawing and Theory. Certificates of competency are issued to repairers after examination. The Institute also publishes monthly an illustrated Horological Journal devoted to technical matters. Members of the Institute pay an annual subscription of a guinea, Associates half a guinea, Fellows 12s. and Student Fellows 6s. It is governed by a Council of about 30 members, Mr. William Barnsdale is the President, Mr. J. F. Cole the Treasurer, and Mr. James Savidge, A.C.I.S., the Secretary, from whom all further particulars may be obtained.

Horse.—[*Chevallet.*—*Der Probierstuhl.*]—A wooden standard for supporting a small clock movement while it is being brought to time. The folding horse shown in Fig. 1 is of metal. Through the arms, *c c*, are holes so that a regulator movement can be secured by the screws that fasten it to the seat board. The sliding bar, *a b*, terminates in a slit with adjusting screw, by means of which a pendulum cock may be clipped. For round movements

Fig. 1.

suitable **wire** clips are passed over the extremities of the arms, *c c,* to keep them immovable a suitable distance apart.

Hour Rack.—[*Rateau d'heures.—Der Stundenrechen.*]—The rack in a striking clock that is moved one tooth for each hour struck. (See Striking Work.)

Hour Wheel.—[*Roue d'heures.—Das Stundenrad.*]—The wheel that carries the hour hand. (See Motion Work.)

Hunter.—[*Montre savonnette.—Die Savonnette Uhr.*]—A watch case that has a metal cover over the dial.

Huyghens, Christian (born 1629, died 1695). A Dutch mathematician, author of *Horologium Oscillatorium,* Paris, 1673.

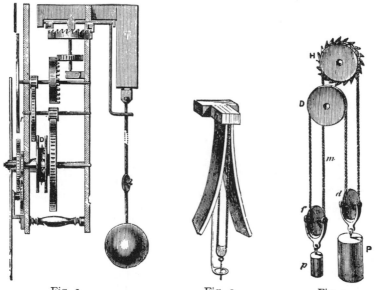

Fig. 1.   Fig. 2.   Fig. 3.

To him is due the credit of first applying the pendulum to clocks.

Huyghens' clock is shown in Fig. 1. The upper part of the pendulum is a double cord hanging between two cycloidal cheeks to give a cycloidal path to the bob. Fig. 2 gives a better idea of this device, which was no doubt of advantage with the long arcs required by the Verge Escapement. Another feature of Huyghens' clock is the maintaining power. P (Fig. 3) is the driving weight, supported by an endless cord passing over the pulley D attached to the great wheel, and also over the pulley H, which is provided with ratchet teeth and pivoted to the inside of

N

the clock case. The cord m is pulled down to wind the clock, and the ratchet wheel H then runs under its click. So that while winding, as in going, one half of P minus one half of p is driving the clock. The pulleys D and H are spiked to prevent slipping of the cord.

When this ingenious maintaining power is applied to a clock with the striking train, the pulley with the ratchet is attached to the great wheel of the striking part, one weight thus serving to drive both trains. A chain is preferable to a cord, owing to the dust which accumulates in the clock through the wearing of the latter. The drawback to the arrangement is that it is not suitable for clocks going more than 30 hours between windings. It is however worth knowing thas a 30-hour striking clock on this plan can be readily converted to an 8-day non-striker by simply disconnecting the striking work.

Hypocycloid.—[*Hypocycloide.—Die Hypocycloide.*]—A curve generated by a point in the circumference of a circle when it is rolled within another circle. The proper shape for the teeth of wheels that are driven by others having epicycloidal addenda. If the tracing circle is half the diameter of the one within which it rolls, the hypocycloid will be a radial line. (See Wheels and Pinions.)

Ice Box.—[*Glacière.—Der Eisschrank.*]—A chamber in which chronometers and fine watches are placed when they are being adjusted for varying temperature. The metal receptacle for the chronometers or watches is surrounded by a space for ice ; the outer vessel has a tap to drain off the water, and is covered with some non-conducting substance.

Great caution is needed in using an ice-box, to preserve the steel work from rust caused by condensation of the air. Mr. Arthur Webb places his chronometers in a brass box with a cover screwing against a waxed joint to exclude the air. Whenever this box is taken out of the ice box, it is allowed to remain unopened for about three hours, and as a further precaution the balance and spring are washed in benzine after a chronometer has been subjected to ice.

Idle Wheel, Intermediate Wheel.—[*Renvoi.—Das Vermittlungsrad.*]—(1) A toothed wheel used to connect two others. An idle wheel introduced causes the follower to rotate in the same direction as the driver, that is, reversing the motion it otherwise would have.

(2) A guide wheel used to change the direction of a driving belt or cord.

Impulse Pin, Ruby Pin.—[*Cheville d'impulsion.—Der Hebestein (Rolle).*]—A pin in the roller of the lever escapement that works into the notch of the lever. On entering the notch it unlocks the escape wheel. It then receives impulse from the lever, and passes out of the notch from the opposite side.

Inclined Plane Clock.—Any watchmaker who wants an attraction for his shop window might do worse than make an inclined plane clock, as shown in the sketch. There is a train of

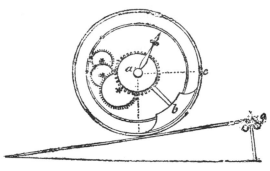

wheels and an escapement as in a watch. The great wheel *a* carries the hand and also the weight *b*. The clock never requires winding. It is every morning simply placed at the top of the inclined plane, down which it gradually rolls during the day, the hand pointing to the hour marked on a dial, which of course covers the mechanism. If the centre arbor goes round once in 12 hours the length of the plane had better be more than twice the circumference of the clock case *c*. Its inclination may be regulated by the screw *g*. The hand might be in the form of a serpent's head, or other grotesque design.

Independent Seconds.—[*Séconde indépendante.—Die springende Secunde.*]—A watch having a seconds hand driven by a special and separate train. The independent seconds hand generally beats full seconds. It is discharged by a push piece running through the pendant, which releases a flirt on the last arbor of the train. This flirt regulates the progress of the hand by taking into a pinion on the arbor of the escape pinion.

Independent Seconds Train.—For 18,000 movement in which the escape pinion has six leaves.

First Wheel	...	80	Pinion of	10
Second do.	...	60	,,	8
Third do.	...	60	,,	8
Fourth do.	...	60	,,	8
Fifth do.	...	48	,,	6

The train is arranged on the pillar plate. The first wheel is on a small separate barrel. The fourth wheel pinion of the independent train carries the seconds hand, and passes through a pipe screwed to the centre of the plate. The cannon pinion fits freely on this pipe, and the motion work is driven as described under the head of "Centre Seconds." The last pinion of the independent train carries a flirt which takes into the leaves of the escape pinion of the usual train. The flirt, and therefore the last pinion of the independent train, thus make six revolutions for each one of the escape pinion. It is essential that the centre wheel of the independent train should rotate once in a minute, and that the flirt should revolve once in a second. If, therefore, with an 18,000 train and escape wheel of fifteen teeth a higher numbered escape pinion is used, a separate pinion of six for the flirt to take into is fixed on the escape wheel arbor.

Index Regulator.—[*Raquette.—Der Rücker.*]—A small lever the shorter end of which carries the curb pins which embrace the balance spring of a watch. The longer end through which it is moved serves also as an indicator of the alteration in the position of the curb pins.

In fitting a Swiss Index it often happens that one is found rather small in the bore. These indexes are flint hard and with the special pliers shown in the sketch they may be readily spilt, thus slightly opening the hole and securing the elastic spring fit of an English Index.

Inertia.—[*Inertie.—Die Trägheit.*]—The property of bodies that impels them to resist being brought into motion if they are at rest, and to resist being brought to rest if they are in motion. Inertia is directly proportionate to the mass of a body. The moment of inertia of a rotating body is its mass multiplied by the square of the distance of its centre of gyration from its centre of motion.

Ingold Fraise.—[*Fraise Ingold.—Die Ingoldfräse.*]—A pinion-shaped cutter used for correcting inaccuracies in the shape of wheel teeth, invented by Ingold, a Swiss watchmaker.

This consists really of a hardened pinion with square, sharp points. The fraise is gradually brought into depth, in a specially arranged depth tool, with a wheel whose teeth are incorrect, and rotated the while by means of a ferrule and bow. The fraises do not supersede the rounding up tool, but may often be used after it with advantage, for if a wheel contain any thick teeth they would not be corrected in the rounding up tool, which also of necessity leaves the teeth slightly hollow. The fraises cut the teeth in the direction they move on the pinion in working, and, therefore, leave a surface which works with the least friction.

A fraise for any particular wheel should be chosen so that when placed upon the wheel, the fraise does not bottom, but just touches the sides, and almost closes over the middle one of the teeth engaged, at the same time just making contact with the teeth right and left. If the fraise chosen is too large, it will cut a jagged and uneven tooth ; and, if too small, will leave a ridge or shoulder on the tooth ; in this, as in everything else, practice makes perfect. As a guide at first, it will be prudent to use the sector to ascertain the most suitable fraise for use ; thus—place the wheel to be operated upon in the sector, and choose a fraise of such a size as will correspond, not to the size indicated by the number of its teeth, but to *two teeth less*.

Involute.—[*Involute.—Die Involute.*]—A curve traced by the end of a string as the string is unwound from a roller. Wheels with involute teeth, though possibly preferable where strength is required, are rarely used for watches and clocks.

Iron.—[*Fer.—Das Eisen.*]—CAST IRON is the crudest kind of manufactured iron. It is brittle, and will not stand much tensile strain, but resists crushing force well. Besides sulphur, phosphorus, silicon, manganese, and other impurities which vary in the produce of different districts, cast iron contains about 3 per cent. of carbon. Cast iron cannot be forged, but is shaped by being poured into moulds when melted.

WROUGHT IRON is obtained by puddling cast iron (*i.e.* stirring it after it is heated to fluidity), whereby many of the impurities are removed, and much of the carbon burnt. The chief distinction in the composition of cast and wrought iron is the amount of carbon, which in the latter is but about ·2 per cent. Though inferior to cast iron for sustaining thrust, wrought iron bears a much greater tensile strain, is tougher, and may be forged into different shapes. Special qualities prepared for electro magnets and the like contain less than ·2 per cent. of carbon. Recently, by the admixture of but ⅛ per cent. of aluminium, castings of wrought iron, ' Mitis castings,' have been produced.

STEEL is pure or nearly pure iron, containing from ·5 to 2 per cent. of carbon, that is, more than wrought and less than cast iron. In the cementing process of steel making, wrought iron, specially selected as free from impurity, is subjected in a closed vessel with charcoal to a temperature of 2000° Fahr. for several days, until the iron is impregnated with the required amount of carbon. By this means " blistered " steel is produced. To obtain " cast " or " pot " steel for tools, blistered steel, containing about 1·5 per cent. of carbon, is cut into small pieces and melted in a covered crucible of Stourbridge clay. For other purposes pieces of blistered steel are cut up and welded together, becoming what is called " shear " steel. Owing to the inventions of Musket, Bessemer, Siemens, and others, mild steel for railway and ship construction, containing about ·5 per cent. of carbon, is made by selecting cast iron, free from sulphur and phosphorus, and blowing the carbon out of it when it is melted. Pure cast iron, sufficiently rich in carbon to give the whole mass the proportion required, is then mixed with it. The presence of sulphur, phosphorus, or silicon makes steel brittle ; manganese is said to render it less liable to become magnetized. For Nickel Steel see p. 24 and the Appendix.

Isochronous. — [*Isochrone.—Zeitgleich.*] — Literally, moving in equal times. A balance spring is said to be isochronous when it causes both the long and the short arcs of the balance to be performed in the same time. The vibrations of a pendulum are called isochronous when they are all of the same duration, irrespective of the extent of the arc through which the pendulum travels. (See Balance Spring.)

Isolator.—[*Isolateur.—Der Isolationshebel.*]—In a minute repeater, a device for keeping the click from contact with the surface piece on the minute snail till the slide in the band of the case is pushed round.

Jacot Tool.—[*Tour Jacot. —Der Zapfenrollierstuhl.*]—A tool used by watch jobbers for renovating pivots by burnishing. One end of the defective arbor is centred, and the pivot to be burnished rests on a bed or slot, of which several of different sizes are formed round the edge of a thin steel disc. For operating on the points of pivots another disc is provided, having different-sized holes round it, through one of which the pivot is projected.

In almost every watch repairing job the pivots require reburnishing, and although some English jobbers stick to the turns it is generally admitted that by means of the Jacot Tool, pivots can be more easily and quickly renovated. The burnisher, being supported parallel to the axis of the pivot, it is impossible to make the pivot smaller at one end than the other, which is very often

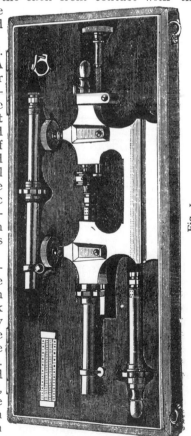

Fig. 1.

the case when pivoting is attempted in the turns by the ordinary run of jobbers.

A new tool should be taken apart and thoroughly cleaned before use, in case of emery or grit having been left in it. To ensure absence of all burr, which would scratch the pivots, the beds should be polished out with boxwood and diamantine. In the very finest tools the beds are jewelled, and then, of course, this precaution is unnecessary. The face of the lantern or disc with holes for rounding up pivots should be polished, and the holes also, by means of a fine wire polisher. Some watchmakers remove the lantern altogether, and substitute one of German silver, as being less likely to scratch the pivots.

A recent improvement in the Jacot Tool is a loose pulley or ferrule and carrier, which may, by turning a screw passing through the body of the tool below the driving runner, be adjusted to suit work of different lengths. This improved tool, with extra runners and a pivot gauge, is shown in Fig. 1 ; at the top right hand corner of the figure is a guard, which is occasionally useful for attaching to the runner, so as to prevent the file or burnisher from running over the shoulder of a pivot and injuring the work.

Fig. 2 is a section of the burnisher used for straight pivots ; Fig. 3 that for conical pivots; Fig. 4 is suited for cutting back shoulders ; and Fig. 5 for rounding off pivots. Burnishers for conical pivots are, of course, varied in form to suit different shaped pivots. The angles shown are adapted for use when the operator works at the Jacot Tool cross-handed as foreign watchmakers do. Many Englishmen work right-handed, and in that case the angles are reversed.

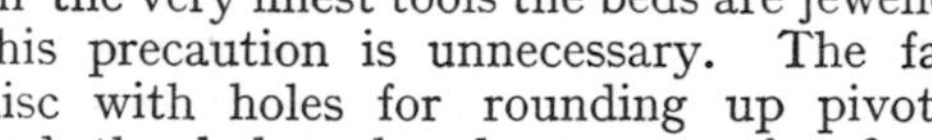

Fig. 2.   Fig. 3.   Fig. 4.   Fig. 5.

The greatest care should be taken of the files and burnishers used with the Jacot Tool. They should have thin and light handles, and be kept in paper cases to avoid contact with other tools on the board, or they will soon become notched. *Thin* oil should be used with the burnisher. Its action is then more sensitive to the touch than when thick oil is used.

The " Universal " pivot file and burnisher, now to be obtained, is very compact, with very light metal handles. One part of the tool is formed for straight, and the other for conical pivots.

Jewelled.—[*Empierré.*—*Mit Steinen versehen.*]—Fitted with precious stones to diminish wear. In a watch all the escape pivots and the fourth wheel pivots usually run in holes made of jewel. The watch is then said to have four pairs of holes jewelled or to be jewelled in eight holes. In addition the acting parts of the pallets and the impulse pin of the escapement are always of hard stone. Sometimes the whole of the holes are jewelled. In the best class of work sapphires or rubies are used, in a lower grade crystal, and in the commonest garnet. In good clocks the pallets and verge holes are jewelled. In thorough holes, such as are used for the train, the bottom jewel hole is usually fitted into a recess hole turned in the plate, and the metal rubbed over in the form of a rivet to secure it. In the upper plate the jewel hole is sometimes fixed in a loose setting, and held in its place by the heads of two screws tapped into the plate close to the recess. In watch escapement holes, where end stones are used, the jewel in a loose setting is fitted into a recessed hole, and upon it is laid the end stone which is also set in metal. The heads of two small screws tapped into the watch plate or cock, as the case may be, serve to secure the jewel hole and the end stone. With screwed jewels the chamfer is often left too flat, so that the pivots enter their holes with difficulty. It should be a steep basin shape.

A jewel, pierced not through its axis of crystallization, sometimes presents a ridgy appearance which no amount of polishing will remove. Such a hole will rapidly cut the pivot working in it. A diamond end stone, whose surface is not coincident with the line of cleavage, will also wear away the end of the pivot in contact therewith. Such a stone is occasionally met with in the balance cock of a marine chronometer.

Ruby, which is really a red variety of the sapphire, though it has a rich appearance, is said to exert a deleterious influence on the oil used for lubrication, and many watchmakers accordingly give the preference to sapphires of a light bluish tint. The relative hardness and other particulars of jewels will be found under the head of Precious Stones.

Screwed Jewel with End Stone.   Jewel Rubbed in.

In making a jewel hole, a suitable stone is selected and flatted to a proper thickness by holding it against a diamond mill which is kept wetted. The operator covers his finger with a wet cloth to preserve the top of his finger from injury and keep a sufficiency of moisture to the stone. If the stone is much out of round, its edge may be nipped with pliers till it is more shapely. It is then shellaced to a small sharp brass chuck, turned on one side with a diamond cutter, polished, and a hole drilled half-way through the stone with a diamond drill. The cutters are fragments of diamond that present a good cutting edge, and the drills are selected from needle-shaped pieces of bort. Both are inserted in handles of annealed brass wire drilled up sufficiently far for the diamond to be inserted large end first ; the end of the brass wire is then squeezed over the diamond with pliers. A very high rate of speed is necessary in turning and drilling the jewels, which are kept moistened with water during the operation. Even with great speed much skill is required to plant the drill at once in the centre of rotation, and if the centre is missed a tit is formed which gives trouble. The stone is reversed on the chuck and the other side brought to shape, drilled, and polished. For polishing, formers of brass to begin and boxwood to finish, corresponding to the shape of the part of the stone under operation, are used with diamond dust and oil. For soft stones the formers are of lead, and used with tripoli and water instead of diamond dust. The hole is polished with pieces of tapered copper wire and finished with a tooth from a tortoise-shell comb. In the American factories the wire is held in a live spindle rapidly rotated in the reverse direction to the jewel, with the advantage of getting over the work quicker, and also securing more perfect roundness of the hole. Flat surfaces of stones are polished on a glass plate with diamond dust and oil, a cork or peg being used to press on and give motion to the jewel. If a jewel is to have a screwed setting, it is sometimes set before the hole is opened.

For examining jewels and other small material kept in bottles, a little glass dish like the sketch is useful. It has a lip for readily returning the contents of the dish.

Jewel Setting.—The plate or cock having been set true in the lathe, a hole is drilled so much less in size than the jewel as to allow of a firm seat for it. The hole is then enlarged to the required depth so as to just admit the jewel, and the seat formed exactly to the shape of it by a tool like Fig. 4, the nose of which is either left flat or curved, according to the contour of the stone. The jewel may be shellaced to the end of a piece of wire for trying it in. A groove is turned round the hole in the plate, to leave a little metal to be rubbed over the stone. The stone is then pressed into its seat with a finely pointed peg, and the brass rubbed carefully over it. This is done with a burnisher shaped preferably like Fig. 5, nicely rounded and polished on a leather buff. While using the burnisher it is occasionally put to the tip of the tongue or drawn over the palm of the hand, for if it is perfectly dry it is apt to tear the brass. If the burnishing is cleanly done with a well-polished tool, a little red stuff on a peg will make a nice finish. The plate or cock is then reversed in the lathe, and the

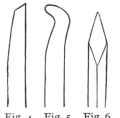

Fig. 4.　Fig. 5.　Fig. 6

metal turned back at an angle, or stripped as it is called. For this purpose a very long-faced graver like Fig. 6 is used. After whetting this graver it is polished with as much care as the finest piece of steel work would be, for on the degree of polish of the graver depends the brilliancy of the stripping.

Jewelling Rest.—Supplied with some of the American lathes is a jewelling rest in which the cutter is carried by one of a pair of sector-like arms. The stone for which the setting is to be bored out is placed between the jaws of the sector twice as far from the joint as the cutter is. When the sector is closed, its arms are in a line with the centre of the lathe. Then if the edge of the cutter is exactly coincident with the lathe centre, and in a line with the face of the arm, it will clearly bore out the setting to fit the stone. Fig. 7 is a view of this ingenious rest. It

Fig. 7

occupies the position of the back centre of the lathe, and would be useful for boring out wheels, sinks in plates, and many other purposes.

To Widen a Jewel Hole.—Chuck the hole in a lathe with cement. Place a spirit lamp underneath to prevent the cement hardening. Hold a pointed bit against the hole while the lathe is running until the hole is true, when remove the lamp. The broach to widen the hole should be made of copper, of the size and shape required, and the point after being oiled should be rolled in diamond dust until it is entirely covered. The diamond dust should then be beaten in with a burnisher, using very light blows so as not to bruise the broach. After the hole is widened as desired, it requires polishing with a broach made of ivory and used with oil and the finest diamond dust, loose (not driven into broach).

Jewel-Setting Rubber.—Watch jobbers often find it handy to have tools to raise a jewel setting in order to replace a damaged

Fig. 8.

hole, and to burnish the setting over again. The points of the one in Fig. 8 are for lifting the setting, the tool being rotated between the finger and thumb, and the points opened by means of the screw as may be required. For rubbing over the setting again, a tool with the points the reverse way is used.

Double Cutter Jewelling Tool.—In this, shown in Fig. 9, are
two cutters follow-
ing each other,
one for the sink
and one for the
lip of a jewel seat.
They can be pre-

Fig. 9.

viously adjusted and tried on a waste piece of brass, with the
assurance that when the plate or cock is operated upon, the
work will be correct.

Jewel Setting Rubbers and Cutters are often attached to the

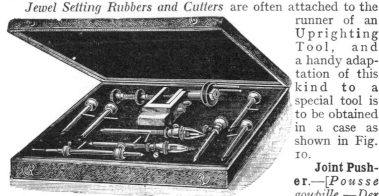

runner of an
U p r i g h t i n g
T o o l,   a n d
a handy adap-
tation of this
kind to a
special tool is
to be obtained
in a case as
shown in Fig.
10.

**Joint Push-
er.**—[*P o u s s e
goupille.—Der
Scharnierstift-*

Fig. 10.

drücker.]—A round piece of tempered steel, used generally in a
wooden handle, for forcing small pins into or out of position.   A
tight pin is started much easier by impact than by pressure, and
a good plan is to have a punch or joint pusher with a square body
which is clasped in the vice, not tightly, but just so as to hold it
in the position.   If then the work is held in one hand against the
point of the joint pusher, the other hand is at liberty to give the
head of the joint pusher a slight tap with a hammer.

Jumper.—[*Sautoir.—Der Sternkegel (Schlaguhren).*]—A click
in the shape of an obtuse wedge, which is pressed into the space
between two teeth of a star wheel by a spring.   A tooth of the
star wheel in passing pushes the click back till it gets just past
the point of the wedge, when the latter jumps the wheels forward
and falls into the next space.

Jurgensen, Urban (born 1776, died 1830).—An eminent
Danish watchmaker.   Author of *The Higher Horological Art* and
Principes de la Mesure du Temps.—Jurgensen was associated with
the leading men of his day.   He experimented with compensation
balances made of brass and platinum, and strongly advocated
the use of gold springs for marine chronometers.   He made

many excellent chronometers for the Danish navy, and very successful metallic thermometers.

Kendall, Larcum.—A watchmaker who in 1769 manufactured a duplicate of John Harrison's celebrated prize chronometer.

Kew Observatory.—The trial for Class A Kew certificates is divided into eight periods of five days each : 1st, pendant up ; 2nd, pendant right ; 3rd, pendant left ; 4th, dial up in refrigerator (about 40° Fahr.) ; 5th, dial up in room (about 65° Fahr.) ; 6th, dial up in oven (about 90° Fahr.) ; 7th, dial down in room ; 8th, pendant up. All but periods 4 and 6 are at the ordinary temperature of the room. The 4th, 5th, 6th, and 7th periods are each extended one day, and on the first day of each the watch is not rated. To obtain the certificate, the daily rate must not exceed 10 secs, ; the *mean difference* of daily rate during each period must not exceed 2 secs. ; the difference of mean daily rate between pendant up and dial up must not exceed 5 secs., and between pendant up and any other position, 10 secs. ; change of temperature must affect the daily rate by less than one-third of a second per degree Fahrenheit. The behaviour of a watch keeping *just within* the limits throughout the trial is appended.

Mean daily rate	...	...	...	...	...	gaining 9·9 secs.
,, variation of daily rate	...	...	...	...		1·9 ,,
,, ,, ,, for 1° F....			...	...		0·3 ,,
Difference of mean rate between pendant up and dial up						4·9 ,,
,, ,, ,, ,, pendant right						9·9 ,,
,, ,, ,, ,, pendant left						9·9 ,,
,, ,, ,, ,, dial up and dial down						9·9 ,,

It will be observed that the MEAN variation is taken into account. For instance, between pendant up and dial up the pieces may vary more than 5 secs. between one day and the next, provided the MEAN daily variation of the pendant up period does not differ from the MEAN daily rate of the dial up period by more than 5 secs. Marks, showing the excellence of the watch, are awarded to A certificates : 100 marks representing absolute perfection, that is, 40 for no variation of rate, 40 for no change of rate with change of position, and 20 for no temperature error. A watch just touching the limit of variation allowed would get no marks, and for every mark obtained the 2 secs. limit would have to be reduced ·05, the 10 secs. limit ·25, and the temperature error ·015. The words " *especially good* " are added to an A certificate when the watch obtains not less than 80 marks.

For B certificates the watches are tried 14 days pendant up, 14 days dial up, one day in the oven, one day at temperature of the room, and one day in the refrigerator ; variation limit in each position 2 secs., and between hanging and lying 10 secs. " *Especially good* " when the 2 secs. limit is reduced to 0·75 sec.,

the 10 secs. to 5 secs., and the temperature limit 0·2 per sec.
Fahr.

Keyless.—[*Sans clef.—Uhr mit Bügelaufzug. Die Remontoiruhr.*]—Capable of being wound without a key.

Rocking Bar Keyless Mechanism.—The action of the keyless
mechanism most generally adopted in English going-barrel
watches will be understood from Fig. 1, which is an arrangement
of a simple character, well suited for low priced watches. For
winding the watch, connection has to be made between the serrated
button projecting above the pendant, and the wheel to the left
hand of the figure which is attached to the barrel arbor. For
setting the hands the winding connection must be broken, and
connection made with an intermediate wheel gearing with the
minute wheel on the right hand of the figure, so that it may be

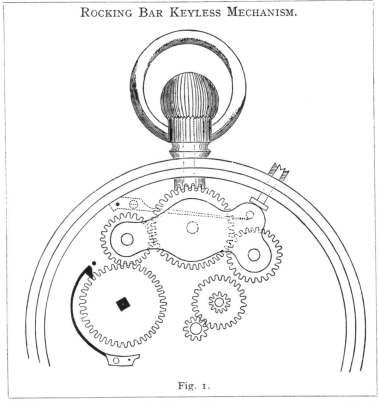

ROCKING BAR KEYLESS MECHANISM.

Fig. 1.

actuated in either direction by turning the button. Three wheels gearing together are planted on the rocking bar. The middle one rides freely on a stud which projects from the rocking bar. This stud forms the centre of motion of the rocking bar, which is free to move up or down so as to engage with either the barrel wheel or the minute wheel. In its normal position the connection is with the barrel wheel. A spring fixed at one end to the pillar plate presses against a small stud on the rocking bar just sufficiently to keep the winding wheels in gear. A contrate wheel squared on to the stem of the winding button gears with the middle wheel on the rocking bar. As the button is turned for winding, the resistance of the barrel wheel ensures the safety of its depth with the wheel on the rocking bar. When the knob is turned the reverse way, the teeth of this latter wheel slip over the teeth of the barrel wheel. To prevent the barrel wheel running back, a spring click is shown in the drawing. When any strain is thrown on the click, the end of it butts against a pin screwed into the plate, but during winding there is a space between this pin and the end of the click, so that if the mainspring

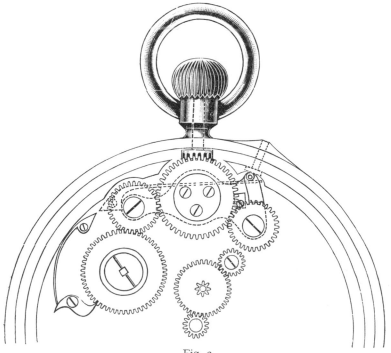

Fig. 2.

is wound tightly the wheel is allowed to recoil a little till the end of the click touches the pin. Undue strain is thus taken off the mainspring, and there is no fear of the escapement banking, which is often observed after careless winding where no such provision is made. (For another form of Recoiling Click see Barrel Ratchet.")

The arrangement more usual in watches of the better class is shown in Fig. 2. Here a solid click is used. A line drawn from the centre of the wheel through the point of the click will be a guide for the position of the click hole, which should form within an angle of about 100°. Sometimes the hole is much farther out, and then with careless winding the click may be jammed or forced from the wheel altogether. It is most important that the click should be planted so that the points of two teeth of the barrel wheel are never presented to the wheel on the rocking bar with which it gears. If this precaution is disregarded, these wheels will butt on going into action, the winding will be unsatisfactory, and there will be danger of stripping the teeth of the wheels. The hole should be slightly oval, the nose of click is then certain to reach the bottom of the tooth and the click allows of a little recoil after the winding is completed. Sometimes a banking pin is placed a little way from the back of the click to prevent it from being jerked out to the other side of the spring, which may be possible with careless winding, and occasionally there are two clicks.

A slotted hole may be noticed in the rocking bar on the right-hand side. Into this projects the head of a screw which is screwed into the plate ; this acts as a banking and limits the motion of the rocking bar in each direction so that the wheels can never be thrust together too deeply.

The balance of opinion is that a bevelled pinion working into a bevelled wheel does not answer so well as the contrate wheel arrangement shown in the drawing. With the bevelled pinion the strain in working tends to tilt the rocking bar. When the contrate wheel is used, the teeth on the underside of the wheel with which it gears should be rounded. To ensure smooth working this wheel should not be exactly opposite to the pendant, but a trifle to the left towards the barrel wheel. The smaller wheel running between this and the barrel wheel should be pitched rather shallow into both of the wheels with which it gears. This intermediate wheel is sure to get deeper with wear, and if full at first the winding is likely to become rough in consequence.

For setting the hands a push piece, projecting through the band of the case, is pressed with the thumb-nail, so as to depress the right hand side of the rocking bar till the wheel on that side engages with the intermediate wheel gearing with the minute wheel. The thumb-nail presses on the push piece till the operation of setting the hands is completed, and directly the push piece is released the winding wheels engage again. In many instances the wheel placed between the rocking bar wheel and the minute wheel is omitted. At first sight it may appear to be redundant but it certainly ensures a better action when the setting connection is made. Whether an intermediate wheel is used or not, a curve struck from the centre of motion of the rocking bar through the centre of the right hand wheel of the bar should pass through the centre of the wheel with which it engages. OTHERWISE THE HAND IS LIABLE TO JUMP when the setting wheels enter or leave contact.

Pendant Set Hands.—In most of the latest designs the set hands wheels fall into gear automatically with the pulling out of the pendant. The push piece in the edge of the case for setting hands is at the best an eyesore, and affords an inlet for dirt. The keyless work may be of the usual except that a light spring keeps the setting wheel in action. Fig. 3 is a section of a pendant showing a steel spring sleeve, split into four from the lower end for nearly its whole length. This sleeve is kept into place by a screwed brass plug. Normally, the spring that would keep the setting wheels into gear is overcome by the pressure of the stem, and

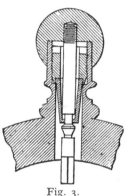

Fig. 3.

the winding wheels are kept into gear by the lower end of the sleeve pressing on a collar on the winding arbor. When it is desired to set hands, the button is pulled out, the lower end of the sleeve is sprung open by the pressure of the bevelled face of the collar, and the sleeve closes on to the arbor again below the collar, allowing the set hand wheels to be in contact. When the setting is completed, the button is pushed down again, and the arbor returns to the position shown in the sketch.

Shifting Sleeve Keyless Mechanism.—A form of keyless work often used in Swiss watches is shown on page 86. A bevelled pinion with clutch teeth underneath rides loose on the stem of the winding button, and gears with bevelled teeth on the face of the large wheel which is just below the pendant. The part of the winding stem below the bevelled pinion is square, and upon this part is fitted a sleeve with clutch teeth corresponding with those on the bevelled pinion at its upper extremity, and a contrate pinion at its lower extremity. A groove is formed around the sleeve, in which is a spring pressing the sleeve upwards so as to keep the clutch teeth engaged. While the clutch teeth are so engaged the winding may be proceeded with. To set hands a push piece projecting through the band of the case acts on a knuckle of the spring just mentioned, so that as the push piece is pressed in the spring draws the sleeve away from the clutch teeth of the bevelled pinion and brings the contrate pinion into gear with a small wheel, which latter gears with the minute wheel.

The remarks on the depths, click, &c., in the description of the Rocking Bar Mechanism are equally applicable to this form. If a stiff click is used, as shown on page 86, it should be short. A short click planted in this way permits a little recoil after the watch is tightly wound.

Fusee Keyless Work.—The simple keyless mechanism used for going barrels is not suitable for the fusee, because in the latter the mainspring is wound by turning the fusee, and accidental pressure on the button would most likely stop the watch if the winding wheels were left in action after winding.

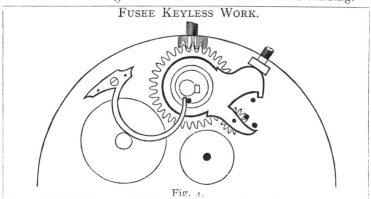

FUSEE KEYLESS WORK.

Fig. 4.

Fig. 4 shows a clever arrangement of fusee keyless work by W. Chalfont. The large circle on the left represents the winding wheel on the fusee, and a smaller circle to the right of it the minute wheel. A wheel and pinion gearing together are mounted on a platform as shown. The platform is screwed to the plate, but the hole through which the screw passes is slotted, so that the platform is free to move horizontally for a limited distance. A spring fixed to the plate pressing on a pin in the platform pushes it to the right till it is stopped by a space between two teeth of the pinion coming into contact with a pin in the plate. This is the position shown in the drawing. If now, by means of the winding button, the contrate wheel is turned to the right as

o

in the act of winding, the wheel on the platform cannot turn because of the pin which is between two teeth of the pinion. The platform is therefore carried to the left, and the wheel on the platform engages with the winding wheel on the fusee. By this time the pinion is drawn free of the pin, a projecting tail of the platform is caught by another pin in the plate, and the winding may be proceeded with. As the button is released, the mechanism returns to the position shown in the drawing, and when the button is turned the reverse way the pinion trips over the little pin in the plate. This pin should be placed so that as the winding wheel on the platform is carried forward it presents the point of but one tooth to the fusee wheel. By this means butting as these two wheels go into gear is avoided. To set hands, the push piece which passes through the band of the case presses the platform down, the pinion is carried free of the pin and into gear with the minute wheel. The wheel on the platform is kept free of the fusee wheel by another pin in the plate catching a hooked projection on the platform.

For Hermann's Fusee Keyless Work, see Maintaining Power.

Kendal's Winding Work for Marine Chronometers.—Though marine chronometers are still generally wound by the clumsy plan of turning the instrument upside down and attaching a key to the fusee square, several arrangements of keyless work have been devised, of which this is a good example. In steadying the chronometer to wind it, the lever l is pressed in by the thumb

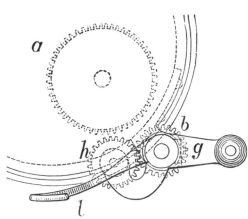

of the operator, and the wheel h is thus brought into gear with the wheel a, which latter is fixed to the lower extremity of the fusee arbor. The winding handle g is just above the dial and is fixed to a spindle carried in two bearings attached to the brass box; at the lower end of the spindle is the wheel b gearing with h. The bracket carring h and the lever l are both fixed to a pipe, fitting the winding spindle freely. When the winding is completed, and the lever is released, a light spring immediately disengages the wheels h and a.

Adjustable Spanner for Caps of Keyless Wheels.—This will  also be found useful for setting up or letting down mainsprings, removing barrel arbor collets and the like.

Kidney Piece.—[*Excentrique pour équation.—Die excentrische Scheibe (Equationsuhren).*]—A cam shaped something like a kidney which is used in clocks that denote the difference between true and mean solar time. (See Calendar Clock.)

Kinetic Energy.—The amount of work stored in a moving body. Half the mass of a moving body multiplied by the square of its velocity gives the measure of its kinetic energy : formulated thus—$KE = \frac{1}{2} mv^2$.

Knee Punch.—[*Poinçon à coude.—Der Tamponbunzen.*]—A cranked punch for removing plugs from cylinders. (See Cylinder Escapement.)

Knuckles.—[*Charnons.—Die Scharnierteile.*]—The rounded portions of a watch case that form the hinges or joints. There are generally two knuckles on the cover, and one on the middle for each joint.

Kullberg, Victor, born at Gothland, Sweden, in 1824, was one of the most brilliant and successful horologists of the nineteenth century. In 1851 he came to London, where he died, in 1890. The close going of his chronometers at the Greenwich Observatory during many years was remarkable. Among the many elaborations of Hardy's principle, Kullberg's flat rim balance, shown in the engraving, is one of the most successful. Here the central arm (A) and the rim (B) composed of brass and steel are in one piece, but in the arm the brass is on top and in the rim underneath, so that with a rise of temperature the ends of the arm bend down and the free ends of the rim are lifted upwards and inwards. The weights (C) are carried on stalks, which also serve as supports for the timing nuts (D).

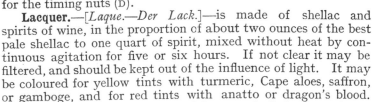

Lacquer.—[*Laque.—Der Lack.*]—is made of shellac and spirits of wine, in the proportion of about two ounces of the best pale shellac to one quart of spirit, mixed without heat by continuous agitation for five or six hours. If not clear it may be filtered, and should be kept out of the influence of light. It may be coloured for yellow tints with turmeric, Cape aloes, saffron, or gamboge, and for red tints with anatto or dragon's blood.

Lacquering Brass..—The articles are boiled in a strong solution of equal parts of pearl ash and slaked lime, to thoroughly remove all old lacquer and greasy matter ; then rinsed in water and scoured with sand, or scratched-brushed, rinsed again and dried. They are then immersed for a moment in a solution of 1 part sulphuric acid, 1 part nitric acid, two parts water, and a very little hydrochloric acid, and then well rinsed in clean water. If the colour is not uniform, the dipping and rinsing are repeated. The articles are then dried in warm sawdust, and rubbed with a clean cotton cloth to remove any stain of finger-marks from handling. They are placed on a stove or heated iron plate until warm enough to hold in the hand ; the lacquer is then applied evenly by means of a piece of soft sponge, and the articles set aside in a dust-free place to dry.

Lantern Pinion.—[*Pignon à lanterne.—Das Hohltrieb.*]—A pinion formed of two shrouds or sides connected by cylindrical rods or trundles, as they are called. Lantern pinions answer admirably as followers, but are not suited for driving. To obtain good performance it is necessary that the spaces between the trundles should be equal. There will be no difficulty in drilling off the shrouds in a lathe with a division

Fig. 2.

plate, or in its absence with the large part of the lathe pulley carefully divided, as a guide, into the number of parts that the pinion is to have trundles. There is no benefit in allowing the trundles to

Lamp. Fig. 1.

run loose in the shrouds. They should be slightly shouldered and riveted tight.

Lamp.—[*Lampe* —*Die Lampe.*]— As an artificial illuminator the paraffin lamp shown in Fig. 1 fulfils nearly all the requirements of the watchmaker. It yields a strong white light, is cooler than gas at less cost, is readily adjustable for height as well as horizontally. The one drawback incidental to all oil lamps is the necessity of frequent cleaning. The Gas Lamp shown in Fig. 2 is furnished with a sub- sidiary shade, which is a protection to the eyes.

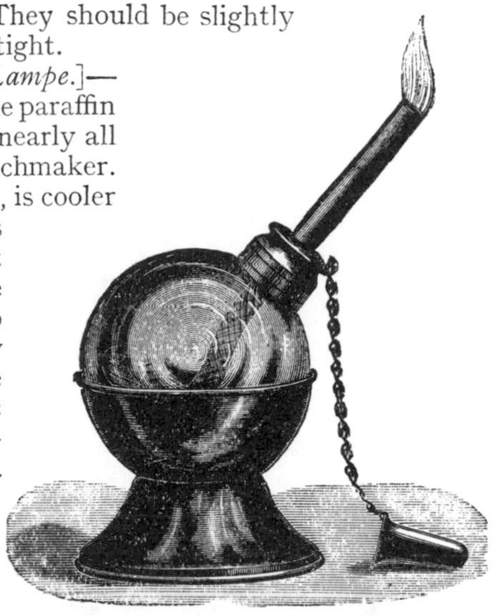

Spirit Lamp.—The ordinary spirit lamp used by watchmakers for har- dening, tempering, soften- ing shellac, and many other

Fig. 3.

purposes, is of glass, with an air-tight extinguisher. The form shown in Fig. 3 is handier, the spout for the flame may be upright or adjusted at any convenient angle. The wick should not be too tight, and the outside of the lamp should be kept clean.

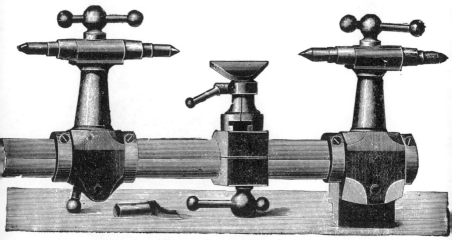

Fig. 1.—Sanspareil Lathe.

There is danger in allowing droppings of shellac and other inflammable material to accumulate.

Lantern.—[*Lanterne.—Die Arrondierscheibe (Zapfenrollierstuhl).*]—The perforated disc of a Jacot tool, used when burnishing the ends of pivots.

(2) The chuck of a screw-head tool used for the points of screws.

Lap.—[*Lapidaire.—Die Schleifscheibe.*]—See Mill.

Lathe.—[*Tour.—Die Drehbank.*]—Almost any tool devised to give circular motion to work while being shaped really comes under the denomination of lathe. "Turns," "Throw," "Mandril," "Jacot Tool," and others arranged for special purposes, are described under separate heads.

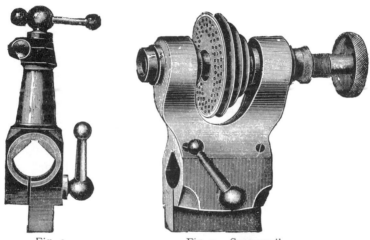

Fig. 2.

Fig. 3.—Sanspareil.

Fig. 4.—Sanspareil.

Boley's Sanspareil Lathe (Fig. 1, p. 213) consists of a pair of turnheads on a bar, one of which may be changed for a headstock, the headstock having a hole running through its base to suit the bar, which is of a rounded triangular section, as will be seen from the end view of the back centre head, Fig. 2. The headstock with double cone bearings gives more stability than is possible with a turn head, and also allows of an ample-sized hole for the reception of split chucks of adequate dimensions. There is a divided plate with three circles of divisions. The chucks for wire are pierced throughout their length, a convenience when cutting pieces from a long rod. The height of centres is 1·7 inch, and the bar may be 8·6, 10·4, or 12 inches long.

Each lathe is provided with a slide rest, chucks, runners, and a great many other useful accessories.

Fig. 5 is a larger tool by Boley, and is more suited for clockmakers. The centres

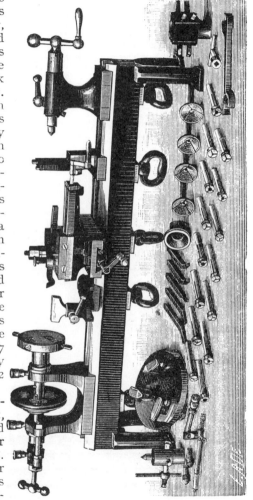

Fig. 5.

are three inches from the bed, which is made of three lengths : 20 inches, 30 inches, and 36 inches.

Fig. 6 is an excellent specimen of a cheap, plain lathe of London make, well adapted for ordinary turning of medium size.

The centres are $3\frac{1}{2}$ inches from the bed, which is of the V and flat kind, and three feet long.

Watchmakers from custom are apt to file male centres on the work when engaged in large turning. This is a mistake. When the size of the work allows, it should have sunk centres, and be mounted between the male centres of the lathe. After the centre of the work has been marked with a centring punch, it should be drilled up a little distance with a fine drill, and the conical mouth formed with a round milled countersink of the same angle as the lathe centres. A perfectly round mouth is essential, and this cannot be ensured if a flat countersink is issued. Both the back and front centres should be turned to the same angle ; they should be interchangeable and kept in good order by means of a small emery wheel held in a frame which may be attached to the slide rest and driven, either from a pulley fixed to the face plate or in any other convenient way.

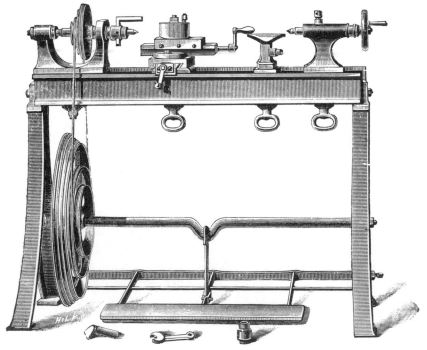

Fig. 6.

A Screw Cutting Lathe is provided with a leading screw, by means of which the saddle whereon the slide rest is planted is caused to traverse the bed with regular progression as the mandril rotates. The connection between the mandril and leading

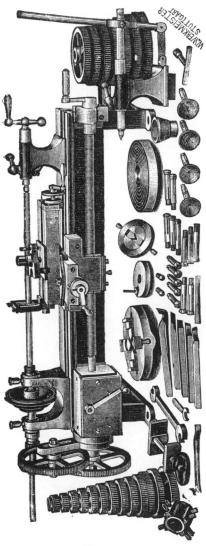

Fig. 7.

screw is by means of the train of wheels seen at the end of the lathe on the left hand, and there is a supply of wheels and pinions of different numbers, so that the train may be arranged to actuate the saddle slowly enough to give sufficient feed in ordinary turning, or quicker as required to suit any pitch of screw it may be desired to cut. Those familiar with the operation of calculating the train of a watch or clock would find no difficulty in deciding the numbers required for cutting a particular screw. For instance if the leading screw has 8 threads to the inch, it is clear that any two wheels of the same number on the mandril and leading screw connected by idle wheels would cause the saddle to traverse one inch for 8 turns of the mandril, and a tool cutting a screw between the centres would consequently cut it 8 threads to an inch. But if instead of the idle wheels, the wheel on the mandril drives a pinion of 15 fastened to a wheel of 60 (like a minute wheel and nut), and the wheel of 60 drives the one on the leading screw the saddle would be traversed 4 times slower, and would cut a screw of 32 threads to an inch. In cutting a screw the cutting sides of the tool must not only match and be the exact counterpart of the screw it is desired to cut, but when viewed from the top as the tool is fastened in the slide rest must incline to the right, if a right-handed screw is to be cut, and to the left if a left-handed one, just as the screw itself when cut would incline, or the tool will not clear in the thread. For a left-handed screw an idle wheel is usually introduced into the change train to reverse the motion of the leading screw, but in Boley's screw-cutting lathe shown Fig. 7 this is unnecessary, for it is furnished with a pair of reversing clutches between the change wheels and the screw. For a double screw the thread is merely cut one-half of the width and depth it otherwise would be, and the second thread then started in the middle of the space left between the threads. For any one with a love of mechanical pursuits a screw-cutting lathe is the most useful of tools, and one that affords unceasing pleasure by its capability of a wide range of work.

Back Gear. Double Gear.—For heavy work and where slow speed is required, this device, which is generally attached to headstocks of large lathes, is brought into operation. The speed pulleys are loose on the head spindle, and have a pinion fixed to their smaller end. Under ordinary working the speed pulleys are coupled to a toothed wheel adjacent to their larger end. This wheel is tight on the spindle, and therefore the speed pulleys, wheels and spindle, all go round together. At the back of the lathe head coupled together are a wheel and pinion mounted on a spindle, either with an eccentric bearing, or capable of a to and fro motion, so that they may readily be brought into gear with the wheel and pinion on the head spindle. When it is desired to use the double gear, the speed pulleys are uncoupled from the

Fig. 8.—Hopkins' Lathe.

wheel on the head spindle, and the double gear is brought into depth. Then the pinion on the speed pulley drives the wheel at the back, and the pinion at the back drives the wheel on the head spindle.

The American Lathes are admirable tools, but their greater cost limits their use. Fig. 8 shows the "Hopkins" combination Mandril and Lathe, with swinging tailstock (A). This tailstock is much used for drilling, jewel-setting, cutting sinks in watch plates and many other purposes for which it is more convenient than the slide rest. Fig. 9 is the slide rest. The upper slide is swivelled and may be fixed at any desired angle. The tool post may be angled as desired and by means of the nut G adjusted to any convenient height. The nut D locks it. The cutter may then be removed without disturbing the position of the tool post

Fig. 9.

The "Lorch" or "Triumph" Lathe.—This is an excellent form of watchmakers' lathe which has very rapidly come into favour. The bar is a turned one with a key bed running along

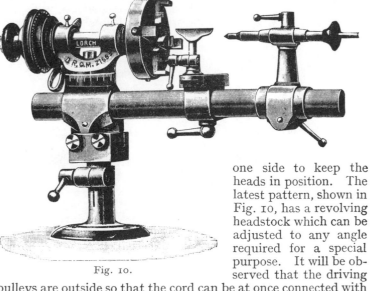

Fig. 10.

one side to keep the heads in position. The latest pattern, shown in Fig. 10, has a revolving headstock which can be adjusted to any angle required for a special purpose. It will be observed that the driving pulleys are outside so that the cord can be at once connected with hand or foot wheel without the trouble of unfastening it. A good slide rest (Fig. 11), mandril head and accessories for drilling and polishing accompany it. When required the lathe heads can be removed and the tool converted into turns.

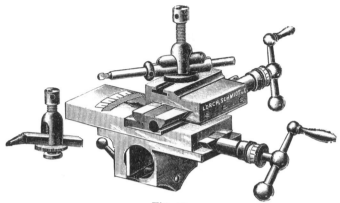

Fig. 11.

"*Go Ahead*" *Lathe.*—This lathe is on the principle of the "Lorch" or "Triumph." It is shown as turns in Fig. 12, and complete as a lathe in Fig. 13. Fig. 14 is its mandril head, and Fig. 15 its slide rest. By curving two of the slots in the face plate of the mandril head the dogs are enabled to take in a larger diameter than they could with radial slots. The slide rest

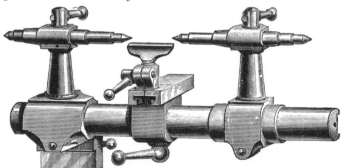

Fig. 12.

is very well planned. Both screws are covered by the slides so that dirt is excluded. The bearing surfaces are large, and the top slide may be angled. On the upper surface of the top slide is a boss to receive the fixing screw of the tool post. Around this boss a spherical hollow is turned in the thickness of the slide. The square tool post is turned at the bottom to fit the hollow in the top slide ; the plate which screws down on the tool has a spherical hollow turned in its upper surface, and a washer to fit, so that the tool post can turn in any direction on a kind of universal join and can also be tilted quite sufficient for any required adjustment of the tool. Provision is made for either

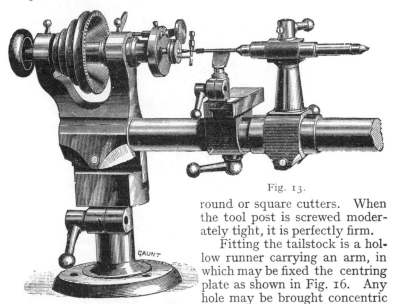

Fig. 13.

round or square cutters. When the tool post is screwed moderately tight, it is perfectly firm.

Fitting the tailstock is a hollow runner carrying an arm, in which may be fixed the centring plate as shown in Fig. 16. Any hole may be brought concentric with the lathe centres by pressing the centring runner into it The plate is then secured by tightening the screw at the top of the arm. If a centre has to be turned in the end of an arbor to drill a hole for a new pivot, the head-rest can be run up between the plate and the end of the runner. A drill of the proper size may then be fitted to the drilling spindle, which also fits the hole in the runner. Unless the object to be drilled is very long, the centring plate may with advantage be slackened from it before the drilling is proceeded with.

The pivoting bed (Fig. 17) also fits in the arm in place of the centring plate, and is sized for large watch pivots, and pivots of French and carriage clocks.

Fig. 14.

Driver Chuck.—This adjunct (Fig. 18) is useful for fine work supported on male centres. The mandril is fixed by inserting the index pin in one of the holes of the pulley, and a dead centre lathe is formed ; the driver chuck is shown in position in Fig. 13, together with a friction pulley for taking some of the strain of the cord off the work, which is, however, not often required.

The " International " is another lathe on the principle of the Triumph.

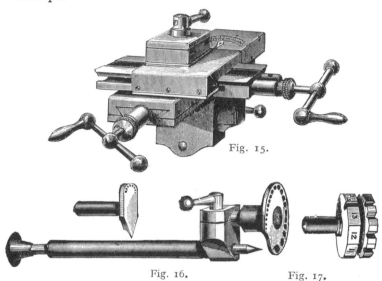

Fig. 15.

Fig. 16. Fig. 17.

The Martin Lathe, shown in Fig. 19, is constructed on the lines of the American lathe, but is much cheaper. The length of the bed is $9\frac{7}{8}$ inches, and the height of centres 2 inches. The slide rest, Fig. 20, is strong and convenient. The tool post is free

Fig. 18.

to slide in a grove so that it may be adjusted at either corner or at any other point of the rest, and may be swivelled to any desired angle. Fig. 21 shows an additional holder for *round* cutters, which are often preferred for light work. A steel washer grooved to suit the holder is first slipped over the tool post, and a tube with the hole slightly eccentric placed in position. The tube has a solid boss at one end, but the body of it is split on opposite sides so that it may grip the cutter. By turning the tube slightly round the cutter may be raised or lowered as required. There is a separate mandril head and a full line of accessories.

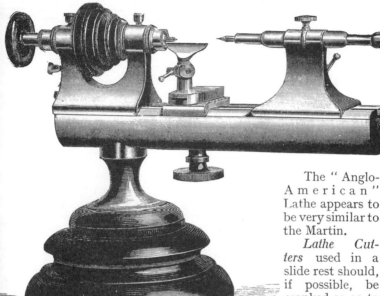

Fig. 19.—Martin Lathe.

The "Anglo-American" Lathe appears to be very similar to the Martin.

Lathe Cutters used in a slide rest should, if possible, be cranked so as to bring the cutting edge below the centre of the body of the tool, as shown in Fig. 22. Downward pressure on the tool then tends to relieve it from the cut, but if the cutting edge is above the centre of the body, pressure sends it deeper into the cut, and causes it to "chatter." Tools are often cranked evidently without any idea of the object to be gained, for the

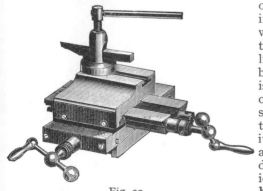

Fig. 20.

bend is so managed as to bring the cutting edge above the centre of the body. The front of the cutter for a little distance below the cutting edge should be all but upright. If ground back more than sufficient for clearance, it will not only be weaker, but will have a tendency to dig into the work. The top

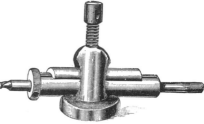

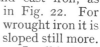

Fig. 21.

of the tool though, which is flat, or nearly so for brass, is sloped downwards from the cutting edge for steel and cast iron, as in Fig. 22. For wrought iron it is sloped still more.

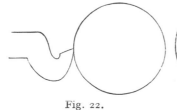

Fig. 22.   Fig. 23.

In all but the very smallest lathes, small cutters held in a bar are rapidly taking the place of large ones fixed in the slide rest. A good example of such a bar, with a swivelled holder to allow of instant adjustment of the cutter to any required angle, is shown in Fig. 24, together with a case of useful cutters.

Disc cutters like Fig. 23 are used in many of the American lathe-tools. They are fixed to the side of the rest by a bolt passing through their centre. The edge of the disc may be turned to any form desired, and if whetted on the face, the shape of the cut will not be altered. For roughing out pinions and the like, the disc is thick enough to embrace the whole surface to be turned, and is fed inwards, the feed motion parallel to the centre of the arbor being dispensed with. Discs are sometimes apt to

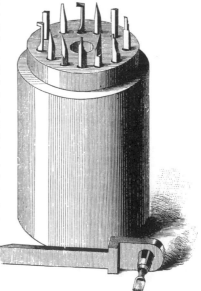

Fig. 24.

fly in hardening, but this danger may be obviated by drilling a ring of holes round the centre one.

Backstay.—This is a great assistance when turning long, slender objects ; though not often supplied with a lathe it may be easily made. The support cut from sheet brass may have two slots at the base, so that a couple of screws will attach it to the saddle of the lathe, as shown in Fig. 25. The V-shaped bearing is of course the same height as the lathe centres, and is made to adjust to work of different diameters by means of a slot running through its length. When using the backstay, a portion of the work is turned true as far from the back centre as can be without its springing from the graver, the stay is then brought to bear against this part, while a further portion is turned true, repeating the process as often as necessary.

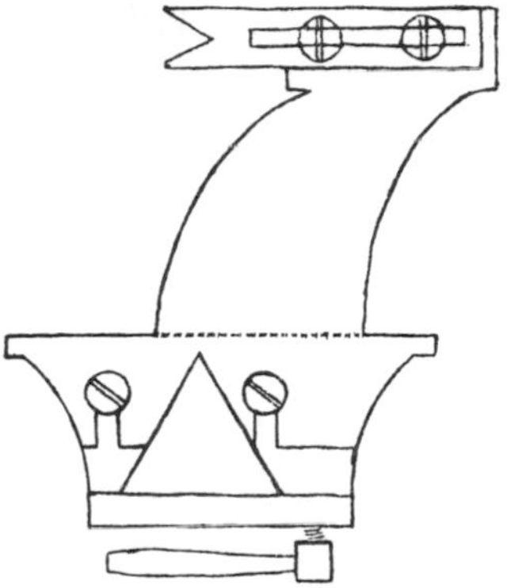

Polishing Mills or Laps.— Fig. 26 is an accessory to the Whitcombe lathe which is used for polishing pivots and shoulders. It consists of a carriage which may be attached to the slide rest as shown. At one extremity of a spindle is a lap, and at the other a handle for

Fig. 25.

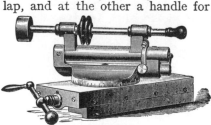

Fig. 26.

giving the spindle longitudinal motion, either parallel to the lathe spindle, or at an angle as may be desired. By means of a belt from the transmission pulleys at the back of the lathe, the lap spindle is rotated rapidly in the opposite direction to the work in the lathe, against which the lap is brought to bear. This device is applicable to conical as well as straight pivots, and for polishing or snailing wheels, etc. It may also be adapted for drilling or milling.

Fig. 27 is the sanspareil accessory for polishing or snailing. It consists of a slide block on which the mill-holder is mounted. The object to be operated on is held in the lathe-head, and may be rotated in the usual way through the intervention of the cord from the transmission pulleys. From the transmission pulleys also the mill-holder is rotated in the opposite direction to the motion of the lathe by means of a cord passing round two hori-

P

Fig. 27

zontal guide pulleys on the slide block to a pulley on the mill-holder. As in working the slide to and fro this cord would be tightened or slackened, an iron arm is provided, one end of which is hooked on to the transmission pulley holder, while a loose

pulley on the other extremity presses on the cord so that the weight of the arm keeps the cord uniformly taut. Mills of steel, copper, zinc, and wood accompany the accessory, and also an abrader to re-grain the faces of the mills so that they retain the polishing medium. This fits the nose of the lathe mandril, as do

Fig. 28.

other chucks for holding the work. (See Snailing.)

Fig. 28 is the milling fixture for the "Go-ahead" lathe which is to be attached to the hand-rest holder.

Filing Fixture.—This attachment is now supplied with most watchmakers' lathes. Fig. 29 shows the one used by Whitcombe. The lathe pulley is divided into 60 and furnished with an index pin, so that squares, hexagons, and other angular forms can be filed correctly to shape.

Fig. 29.

The " fixture " consists of an arm carrying two hard steel guide
rollers, with a flange at the left side of each to keep the file at right
angles to the work. The arm carrying the rollers may be raised
or lowered, as required for adjustment to work of various sizes.
The work is fixed in a chuck, and if necessary the back centre can
be used to steady it. A " filing fixture " will be found useful for
squaring barrel arbors, making punches for punching stop-fingers,
etc., as a perfectly true figure can be filed by this means very
quickly.

Step Spindle in Chuck.—This adjunct to some of the
American lathes is for opening or polishing jewel holes, or doing
any other light work that requires great speed. The chuck has
a conical toe running in the frame and carries a small driving
pulley as shown in Fig. 30.

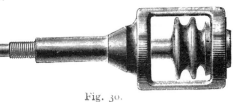

*Self Centring-
Chucks.*—Most mod-
ern lathes are supplied
with split chucks which
close centrically on the
work. Chucks of this
kind were made here
many years ago, though

Fig. 30.

the credit of a most important improvement in the method of
gripping the work, which is worth description, is, I believe, due to
the American Watch Tool Company ; at all events, it is taken
from one of their lathes. Fig. 31 is a longitudinal section of the
upper part of the lathe head. A is the chuck, which terminates
with a screw. The mandril, B, is hollow, and contains a barrel

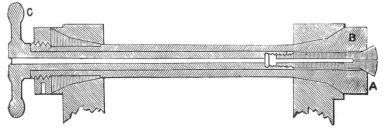

Fig. 31.

through which also a hole is pierced. At one end of the barrel
is a female screw to match the screw on the chuck, and at the
other end a small hand wheel C. By turning the hand wheel so
as to screw the end of the barrel on to the chuck, the chuck,
which is slit crosswise, is drawn further into the cone of the
mandril and contracts, thus gripping the work. There is a
feather in the straight part of the mandril hole to prevent the

chuck turning round. These self-centring chucks save a lot of time in fixing the work. If well made they are really excellent in the hands of a careful workman, but if strained by rough usage or by gripping work of a disproportionate size, they cease to centre accurately, and are then a great nuisance.

Figs. 32 and 33 show two of the chucks, one for wire and the other for wheels. I notice that in all but the smallest wire chucks of the " Go-ahead " lathe, there is a shallow square sink turned in the face, which is exceedingly handy for gripping jewel settings and the like. One of the neatest applications of the wheel chucks is for barrel work, which forms an item in a considerable majority of watch repairs, for a barrel can be set up

Fig. 32.

Fig. 33.

true, recess turned for cover, faced or otherwise operated on, quicker than by any other way of fixing. Latterly there are issued with the Triumph lathe, chucks with the steps outside. They open to grip pieces from the inside and would be a convenience in many instances.

The " Hopkins " American self-centring chuck on another plan seems to be a very clever idea. Here separate split plugs for different-sized objects are provided. One of these holding the

Fig. 34.

work is fastened by means of a screw to the plate, shown in Fig. 34, which is then slipped under the heads of two spring bolts (a a) on the chuck proper (Fig. 35), the back end of the work being centred by the countersink end of a pump centre kept forward by a spring. Then by slightly loosening the nut k the chuck is free to

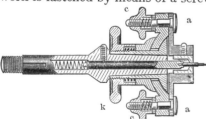

Fig. 35.

move in a basin-shaped joint, and the front end of the work may be trued with as much facility as in a wax chuck. The nuts c c and k are then tightened.

Fig. 37 is a dog chuck, with stepped dogs in radial slits on a face plate. At the back of the face plate is a disc with a spiral groove cut thereon, and each of the dogs has tongues inserted in this groove. As the grooved disc is turned the dogs move equally to or from the centre and so grip the work centrally, either from the inside or the outside. Spare dogs are provided for gripping small objects from the inside. One of these is shown separately.

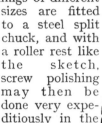

Screw Polishing.—For holding screws split brass linings of different

Fig. 36.

sizes are fitted to a steel split chuck, and with a roller rest like the sketch, screw polishing may then be done very expeditiously in the lathe, though many still declare the to and fro motion to be preferable.

Saw Arbor. —This, supplied

Fig. 38.

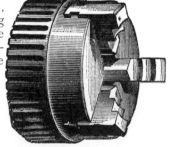

Fig. 37.

with many lathes, receives small circular saws and cutters, and is useful for cutting a groove out of the band of a case to free the barrel or chain, and many other purposes.

Wax Chucks.—For many operations required in watch jobbing Mr. Ganney recommends fixing the work to the chuck with wax or cement, and no doubt workmen used to this plan get through their work succesfully and expeditiously. A number of brass chucks for shellac usually accompany the modern lathe. They are generally screwed into a holder fixed in the lathe spindle. The chuck then projects far enough to avoid heating the lathe when the shellac is melted. Fig. 39 shows the holder with a form of chuck attached which is used for receiving wheels and other flat pieces. It has concentric circles on the face for

Fig. 39.

facility of centring the work. A cylinder, verge, or other fragile piece of work would be almost covered up with shellac ; the truth of the back end being ensured by the hollow in the brass chuck (*d* Fig. 40), which should be carefully turned to a point without leaving a tit in the centre ; the front end of the work is supported and centred with a peg as shown, while the shellac is softened by the flame of a lamp placed underneath. Fig. 43 is one of the brass chucks bored out, and having a cover soldered or screwed to it. The cover has a hole drilled through it, and is faced true so that a barrel arbor or other object with an enlarged part and long projecting end laid on it will be flat and steady, and require but a little shellac to hold it. Fig. 44 is also a hollow chuck

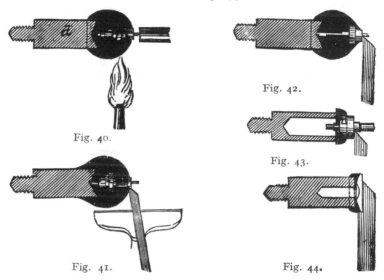

Fig. 40.

Fig. 42.

Fig. 43.

Fig. 41.

Fig. 44.

used generally for jewel holes which require opening or polishing. A peg notched away as shown serves to get the jewel hole quickly true on face and edge.

After the work is finished it is usually placed in spirits of wine in a copper boiling-out pan, and held over the flame of the lamp until the boiling spirit dissolves any cement remaining on it. Mr. Gray recommends, instead of this open pan (which is constantly igniting), using ordinary chemical test tubes of different sizes for this purpose. As these tubes are transparent, it can be seen exactly how the work is progressing, and there is also a considerable saving in the amount of spirit used.

Treadles.—The " Webster " Swing Treadle, shown in Fig. 45, is the best I have seen for watchmakers' use. It does not require the swaying motion of the body incidental to the up and down treadle, works with little friction, and is well under control. The wooden treadle arranged as shown under the head of Bench does fairly well for the ordinary run of light turning, but there has been lately introduced an excellent and inexpensive driving wheel and treadle wherein the treadle patten, shown in Fig. 46, affords a better rest for the foot, which will be found to be conducive to the steadiness essential for fine work. The treadle or patten rests by means of pivots on a cast iron base plate screwed to the floor, and if anything is dropped the patten can be instantly lifted off for a search to be made. For heavier work or for occasionally actuating a lathe usually driven from shafting, the self-contained foot wheel and treadle shown in Fig. 47 are well adapted.

Counter shaft with Transmission Pulleys fixed at the back of the lathe offers advantages in driving small lathes. The

Fig. 45.   Fig. 46.

driving cord, instead of passing from the treadle wheel direct to the lathe head, drives the transmission pulleys. Where great speed is required, as in working jewels and stones, the driving cord is taken over a small pulley on the countershaft, and another cord from a large pulley on the same shaft passes round the pulley on the lathe head. By using a large pulley on the countershaft in connection with the treadle wheel, and a small one in connection with the lathe head, a slow motion is obtained for turning articles large in diameter. A countershaft is also used when a live spindle is worked on the slide rest, as in polishing mills and other accessories. The countershaft attached to American lathes is shown under the head of "Wheel Cutting."

Fig. 47 also shows Boley's transmission pulleys, which are very good. The shaft frame is mounted in a socket standard which allows it to be moved backward or forward, higher or lower, as may be required to adjust the cords. His arrangement of these pulleys for driving his watchmakers' lathe is shown under the head of Bench, page 57.

The driving cord of a lathe should be moderately large, so as to work without being unduly tight. A small, tight cord is a nuisance. It soon slips, wears away rapidly, and throws unnecessary friction on the lathe bearings. Twisted leather cord is much used lately in preference to gut, and being more elastic it throws less strain on the bearings, absorbs less power and works with a much smoother action. The ends are fastened by means of a plain figure of 8 hook, and the cord may be readily tightened by giving it an additional twist or two. Another kind of line, if it can be so called, which is in favour, is a steel wire in the form of an endless spiral spring; the ends are soldered together, and it is said to be the smoothest working and

Fig. 47.

most lasting of any.

Screw Stump for Wood Chucks.—This is intended to receive boxwood chucks, which are useful for turning out the groove of a bezel when it is out of shape, and many other purposes.

Lead.—[*Plomb.—Das Blei.*]

Leading off Rods.—[*Verges de connexion.—Die Zeigerverbindung (Turmuhren).*]- The rods in a turret clock that connect the movement with the hands.

Left-Handed Movement.—[*Rouage à gauche.—Das links angelegte Uhrwerk.*]—A watch movement in which the third wheel is on the left and the balance on the right of the fourth wheel looking from the top plate.

Lepaute, J. A.—Born 1709, died 1789. He was a French clockmaker, and the inventor of that variety of the pin wheel escapement with pins on both sides of the wheel. Lapaute constructed several fine turret clocks and clocks for the Louvre at Paris, wound by means of an air current and fan, a method reinvent⸱ ! re.ently. He made many curious timepieces (equation, one-wheel clocks, etc.), and was the author of an excellent *Traite d'Horlogerie*, revised and augmented, says Moinet, by the celebrated Lalande. In the second edition of this work appears Lalande's treatise on " perfect pitching."

Lépine Movement.—[*Mouvement Lépine.—Das Klobenwerk.*]—Originally a bar movement with a hanging barrel, now generally understood to be one in which the bar supporting the top pivot of the barrel arbor is straight.

Le Roy, Julien.—A scientific French watchmaker, born 1686, died 1759. He devised a form of repeating mechanism much used in French watches.

Le Roy, Pierre, son of Julien Le Roy (born 1717, died 1785), surpassed his father in inventive genius. Among his conceptions was a form of duplex escapement and an escapement on which the present chronometer escapement is founded.

Lever Escapement.—[*Echappement à ancre.—Die Anker-hemmung.*]—Invented about 1765 by Thomas Mudge. An escapement in which the communication between the pallets and the balance is made by means of two levers, one attached to the pallets, and the other, in the form of a roller with a pin projecting from its face, to the balance staff.

Although possibly inferior for timekeeping to the Chronometer the Lever Escapement, when made with ordinary care, is so certain in its action that it is generally preferred for pocket watches. Its weak point is the necessity of applying oil to the pallets. However close the rate of the watch at first, the thickening of the oil in the course of time will inevitably affect its going.

Action of the Escapement.—Fig. 1 (page 234) shows the most usual form of the lever escapement, in which the pallets " scape " over three teeth of the wheel. A tooth of the escape wheel is at rest upon the locking face of the entering left-hand pallet. The impulse pin has just entered the notch of the lever, and is about to unlock the pallet. The action of the escapement is as follows : —The balance, which is attached to the same staff as the roller, is travelling in the direction indicated by the arrow which is around the roller, with sufficient energy to cause the impulse pin to move the lever and pallets far enough to release the wheel tooth from the locking face, and allow it to enter on the impulse face of the pallet. Directly it is at liberty, the escape wheel, actuated by the mainspring of the watch, moves round the same way as the arrow and pushes the pallet out of its path. By the time the wheel tooth has got to the end of the impulse face of the pallet, its motion is arrested by the exit or right-hand pallet, the locking face of which has been brought into position to receive another tooth of the wheel. When the

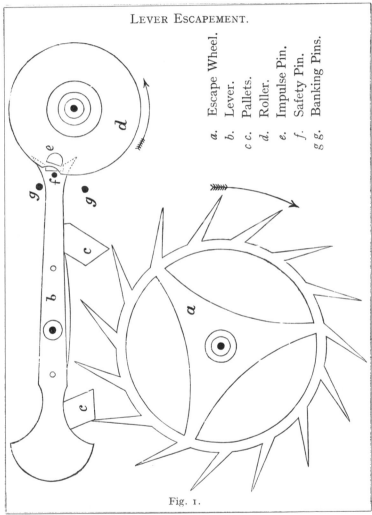

LEVER ESCAPEMENT.

a. Escape Wheel.
b. Lever.
c c. Pallets.
d. Roller.
e. Impulse Pin.
f. Safety Pin.
g g. Banking Pins.

Fig. 1.

pallet was pushed aside by the wheel tooth it carried with it the lever, which in its turn communicated a sufficient blow to the impulse pin to send the balance with renewed energy on its vibration. So that the impulse pin has the double office of unlocking the pallets, of giving a blow on one side of the notch of the lever, and of immediately receiving a blow from the opposite side of the notch. The balance proceeds on its excursion, winding up as it goes the balance spring, until its energy is expended. After it is brought to a state of rest its motion is reversed by the uncoiling of the balance spring, the impulse pin again enters the notch of the lever, but from the opposite direction, and the operation already described is repeated. The object of the safety pin is to prevent the wheel from being unlocked except when the impulse pin is in the notch of the lever. The banking pins keep the motion of the lever within the desired limits. They should be placed as shown, where any blow from the impulse pin on to the outside of the lever is received direct. They are sometimes placed at the tail of the lever, but in that position the banking pins receive the blow through the pallet staff pivots, which are liable to be broken in consequence.

Proportion of the Escapement.—The escape wheel has fifteen teeth, and the distance apart of the pallets, from centre to centre, is equal to 60° of the circumference of the wheel. The pallets are planted as close as possible to the wheel, so that the teeth of the wheel in passing just clear the belly of the pallets.* The width of each pallet is made as nearly as possible half the distance between one tooth of the escape wheel and the next. As the teeth of the wheel must be of an appreciable thickness, and the various pivots must have shake, it is not found practicable to get the pallets of greater width than 10° of the circumference of the wheel instead of 12°, which would be half the distance between one tooth and the next. This difference between the theoretical and actual width of the pallet is called the drop. The lever is pinned to the pallets, and has the same centre of motion. The distance between the centre of the lever and the centre of the roller is not absolute. The distance generally preferred is a chord of 96° of a circle representing the path of the tips of the escape wheel teeth, that is, the distance from the tip of the tooth to the tip of the fourth succeeding tooth. The proportion, as it is called, of the lever and roller is usually from 3 to 1 to $3\frac{1}{2}$

* When the tooth is pressing on the locking, the line of pressure should pass through the centre of the pallet staff. But as the locking faces of the two pallets are not equidistant from the centre of motion, a tangent drawn from the locking corner of one pallet would be wrong for the other, and, as a matter of fact, if a diagram is made, it will be found that even when the pallets are planted as close as possible they are hardly as close as they should be for the right-hand pallet. To plant as close as possible is, therefore, a very good rule, and is the one adopted by the best pallet makers : though in setting out the escapement a chord of the width of the pallet is produced to find the centre of the staff, as shown in Fig. 3.

to 1. In the former case the length of the lever (measured from the centre of pallet staff to centre of the mouth of the notch) is three times the distance of the centre of the impulse pin from the centre of the roller, and in the latter case $3\frac{1}{2}$ times. The portion of the lever to the left of the pallet staff hole acts as a counterpoise, and should really have the metal in it disposed at as nearly as possible the same distance from the centre as that in in the other end of the lever, though this is rarely the case.

In this form of the lever escapement the pallets have not less than 10° of motion. Of this amount, 2° are used for locking, and the remainder for impulse. The amount of locking is to some extent dependent on the size of the escapement. With a large escapement less than $1\frac{1}{2}$° would suffice, while a small one would require rather more than 2°. The quality of the work, too, is an element in deciding the amount of locking. The lighter the locking the better, but it must receive every tooth of the wheel safely, and where all the parts are made with care the escapement can be made with a very light locking.

Presuming that the staff hole is correctly drilled with relation to the planes, a rough rule used for testing 10° pallets is

that a straight edge laid on the plane of the entering pallet should point to the locking corner of the exit pallet, as indicated by the dotted line in Fig. 2. But this is clearly only an approximation, for any variation in the amount allowed for locking alters the direction of the planes.

Fig. 2.

When from setting the hands of a watch back, or from a sudden jerk, there is a tendency for the pallets to unlock, the safety pin butts againts the edge of the roller. It will be observed that when the impulse pin unlocks the pallets, the safety pin is allowed to pass the roller by means of the crescent which is cut out of the roller opposite the impulse pin. The teeth of the escape wheel make a considerable angle with a radial line (24°), so that their tips only touch the locking faces of the pallets. The locking faces of the pallets, instead of being curves struck from the centre of motion of the pallets, as would be otherwise the case, are cut back at an angle so as to interlock with the wheel teeth.* This is done so that the safety-pin shall not drag on the edge of the roller, but be drawn back till the lever touches the banking-pin. When the operation of setting the hands back is finished, or the other cause of disturbance removed, the pressure of the wheel tooth on the locking-face of the pallet

* The locking face forms an angle of 6° or 8° with a tangent to a circle representing the path of the locking corner.

draws the pallet into the wheel as far as the banking pin will allow. The amount of this " run" should not be more than sufficient to give proper clearance between the safety-pin and the roller, for the more the run, the greater is the resistance to unlocking. This rule is sometimes sadly transgressed, and occasionally the locking

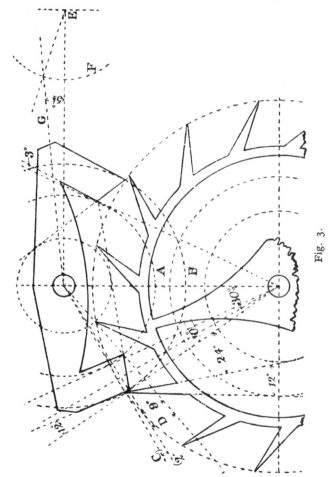

Fig. 3.

is found to be, from excessive run, almost equal in extent to the impulse. It will generally be found that in these cases the escapement is so badly proportioned that the extra run has had to be given to secure a sound safety action. In common watches the safety action is a frequent source of trouble. The more the path

of the safety-pin intersects the edge of the roller, the sounder is
the safety action, and if the intersection is small the safety-pin is
likely to jamb against the edge of the roller, or even to pass it
altogether. With an ordinary single roller escapement a sound
safety action cannot be obtained with a less balance arc than
33°, 10° pallets with one degree of movement added for run, and
with a lever and roller of 3 to 1, give a balance arc of 33°—that
is to say, the balance in its vibration is freed from the escapement except during 33°, when the impulse pin is in contact with
the lever. Even with a balance arc of 33°, the roller must be
kept small in the following way to ensure soundness of the
safety action. The hole for the impulse pin must not be left
round. After it is drilled, a punch of the same shape as the
impulse-pin—that is, with one-third of its diameter flattened off
—should be inserted, and the edge of the roller, where the
crescent is to be formed, beaten in. By this means the roller
can be turned down small enough to get a sufficient intersection
for the safety pin.

It is useful in estimating the balance arc of a watch, to re-
member if it has a three-armed balance that 30° is one-fourth of
the distance between two arms. With a compensation balance
a third of the distance between two of the quarter-screws is 30°.

A round impulse pin, although it is sometimes used in common watches, gives a bad action and necessitates a very large
balance arc.

Figs. 3 and 11 are appended as a guide to students in setting
out the escapement. Referring to Fig. 3, a circle representing
the extreme diameter of the escape wheel is taken as a basis,
and on the left of the centre line is set off, by means of a
protractor, the middle of one pallet (30°) and its width (10°).
The chord of this arc of 10° is then produced till it cuts the
centre line, and this intersection is taken as the centre of the
pallet staff. From the pallet staff centre curves, A and B,
representing the paths of the pallet-corners, are drawn. The
amount of locking C (say 2°) and impulse D (say 9°) are set off
from the chord of the left-hand pallet. The impulse plane is
traced through the intersection of the angular lines with the
curves A and B, and the line of the plane produced towards the
centre of the staff as shown. From the centre of the staff is
described a circle just touching the line so produced. The
impulse plane of the other pallet forms a tangent to this circle.
In this position of the pallets, a line drawn from the locking
corner of the left-hand pallet to form an angle of 12° with the
radial line from the centre of the wheel, will be required to show
the locking face of the pallet, and a similar line forming 3° will

answer for the locking-face of the right-hand pallet. Mark off the centre of the roller (E), and take, say, one-fourth of the distance between this centre and the centre of the pallet staff for the position of the centre of the impulse pin, and describe the arc F to represent its path. The line G, forming with the centre line running through the roller an angle equal to half the total angle of the motion of the pallets, or $5\frac{1}{2}°$, will represent the centre of the lever. The wheel teeth are set back about 24° from a radial line, so as to bear on their points only. The tip of each tooth may have a width equal to about half a degree, and if the drawing is to a considerable scale this amount may be set off. The rim of the wheel extends to about three-fourths of the whole radius. The remaining parts may be readily filled in from the foregoing remarks on the proportion of the escapement, and a study of Figs. 1 and 5.

Double Roller Escapement.—The Horn of the Lever.— Low angled pallets (*i.e*, pallets having but little motion) and small balance arcs are preferred for fine watches ; the low-angled pallets as being less affected by changes in the condition of the oil which is used to lubricate the faces of the pallets than when the motion is greater, and the small balance arc because it allows the balance to be more perfectly detached from the escapement. With a double roller escapement, pallets with from 8° to 9° of motion are generally used, with a lever and roller to give a balance arc of from 28° to 32°. With low-angled pallets, and less than 30° of balance arc, a different arrangement than the usual upright pin in the lever must be made for the safety action. A second roller, not much more than one-half the diameter of the one in which the impulse-pin is fixed, is mounted on the balance staff for the pupose, and a

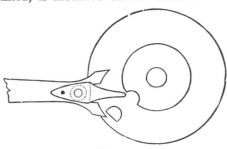

small gold finger, projecting far enough to reach the edge of the smaller roller, is screwed to the lever. The safety roller should not be less than half the diameter of the impulse roller, for the smaller the safety roller, the farther the safety finger enters the crescent before the im-

Fig. 4.

pulse pin enters the notch of the lever ; and, as directly the safety finger enters the crescent, the impulse pin must be within the horn of the lever, the smaller the safety roller, the longer must be the horn. Then, if the horns are excessively long, the extent of the free vibration of the balance is curtailed, because the impulse pin touches the *outside* of the lever sooner. It will be

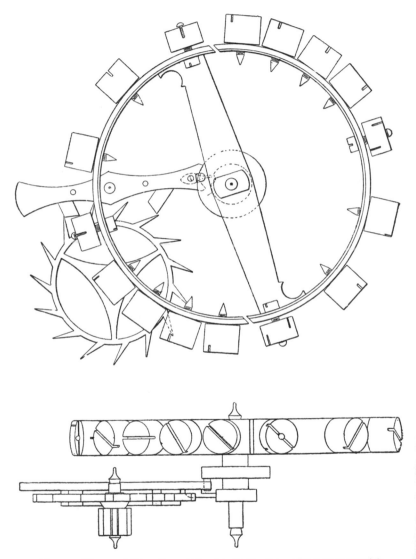

Fig. 5.—Plan and Elevation of Double Roller Lever Escapement with Compensation Balance.

seen that in the single roller escapement (Fig. 1) the safety-pin does not enter the crescent before the impulse pin enters the notch, and, therefore, in the single roller escapement the lever really requires but the smallest possible amount of horn. Fig. 4 shows the double roller arrangement. Here it will be seen that the safety finger enters the crescent some time before the impulse pin gets to the notch. During this interval, should the hands of the watch be set back, the pallets could not trip, for the horn of the lever would be caught on the impulse pin. I have tried to explain this fully, because the Double Roller Escapement occasionally fails to give satisfaction owing to the lever having insufficient horn. On the other hand, the levers of single roller escapements, where scarcely any horn is required, are often made with very long ones.

Besides getting a sound safety action with small balance arc the double roller has three other advantages. (1) The impulse is given more nearly on the line of centres, and consequently with less engaging friction. (2) The safety roller being of a lesser diameter, the safety finger when in contact with it offers less resistance to the motion of the balance ; and (3) the requisite amount of shake between the safety roller and banking pins is obtained with less run on the pallets. Double roller escapements are sometimes seen with pallets having 10° of motion, and even more, and with the safety roller nearly as large as the impulse one. An escapement made in this way really appears to lose most of the advantages of the extra roller. On the other hand, low angle pallets are sometimes used with a long lever to get increased balance arc. This also is objectionable, for the pallets must have more draw to pull the longer lever up to the banking, and more draw means harder unlocking. It is really only to watches of a high character throughout that double roller escapements with low angle pallets and small balance arcs should be applied. For the ordinary run of work, the single roller escapement with 11° pallets and a balance arc of from 36° to 40° is well suited.

Fig. 5 shows the escapement in plan and elevation complete with compensation balance.

Size of the Lever Escapement.—Lever escapements are classed by the trade into the following sizes :—

No. 1 is the smallest and No. 10 the largest size used in the ordinary run of work. The practice of J. F. Cole was to have the escape-wheel three-sevenths of the diameter of the balance, but there is no strict rule for the size of an escapement to a watch, though there has been a disposition of late years to use smaller escapements than formerly, as they are found to yield

Q

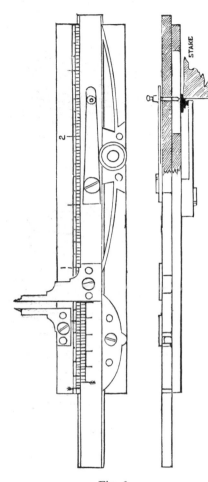

Fig. 6.

better results. In course of time a ridge is formed at the beginning of the impulse planes of the pallets, where the wheel teeth fall. This ridge is more marked and farther along the impulse plane when there is much drop and the escape wheel is large and heavy, because the inertia of the wheel, which increases in proportion to its weight and the square of its diameter, is so great that the balance after unlocking the pallets carries them farther before the wheel acquires sufficient velocity to overtake them. Undue shake of the impulse pin in the notch will also cause this ridge to be accentuated. The practice of some of the best London makers is, for 6 and 8 sized movements, No. 2 escapement; for 10 and 12 sized movements, No. 4 escapements; for 14 and 16 sized movements, No. 6 escapement; and for 18 and 20 sized movements, No. 8 escapement. Many manufacturers confine themselves to two sizes, "two's" for repeaters and ladies', and "sixes" for gentlemen's watches.

No. 0 in which the Escape Wheel is ·185 of an inch in diameter.

1	,,	,,	,,	·205	,,	,,
2	,,	,,	,,	·225	,,	,,
4	,,	,,	,,	·245	,,	,,
6	,,	,,	,,	·265	,,	,,
8	,,	,,	,,	·285	,,	,,
10	,,	,,	,,	·295	,,	,,
12	,,	,,	,,	·305	,,	,,

The escape wheel is of hard, well-hammered brass ;] the pallets are of steel* wider than the wheel, with the acting parts of ruby in the best, and garnet in the commoner escapements. The pallets are slit longitudinally, and the stones fixed in with shellac. The Swiss generally insert the stones across the pallets, so that they are visible. The impulse planes are curved so as to present a smaller surface to the wheel. The impulse pin is of ruby, fixed in the roller with shellac ; the safety-pin of gold, and the banking pins of brass. Non-magnetizable watches have the lever and pallets of some other metal than steel, generally aluminium bronze.

In a good lever escapement all the moving parts are extremely light.

C. Curzon introduced the ingenious arrangement shown in Fig. 6 for marking off the distances on the lever and roller. To the bottom of the stock of a Vernier slide gauge he attached a spring with a conical tit at its free extremity. To the top of the slide another spring is fixed, carrying a fine centring punch. When the gauge is closed, the punch and the centre of the conical tit are coincident. The gauge being opened so that its points mark the whole distance from the centre of the pallet staff to the centre of the balance staff, the measurement on the scale of the gauge is noted. Say it is ·28 inch, and that it is desired the lever and roller should have a proportion of 3 to 1. Three parts of this ·28 inch, or ·21 inch, would be the length of the lever, and the remaining ·07 the distance of the centre of the ruby pin from the centre of the roller. The roller would be placed on the conical tit as shown in the engraving, Fig. 6, the slide closed till the scale marked ·07, and then a tap on the punch would accurately mark the centre desired. The pallet staff hole of the lever would then be placed on the tit, the gauge opened till it marked ·21, and a tap on the punch would give the position of the mouth of the notch.

In making a new lever it is well to start with it full long, because a deep notch is much easier to polish than a shallow one. When the notch is finished the horns can be filed off as required.

Savage's Two-Pin Escapement—With a view to avoid the somewhat oblique action of the impulse pin, Savage introduced the two-pin escapement (Fig. 7). In place of the ordinary impulse pin, two very small pins are placed in the roller so that one of them begins to unlock just before crossing the line of centres. The passing space for the safety pin, instead of being formed like a crescent, is a notch into which the safety-pin fits, and by the time the unlocking is finished, the safety-pin has been drawn into the notch and gives the first portion of the impulse.

* The practice of rolling the pallet steel to harden it is not a good one, as there is danger in magnetizing it in the operation.

244

After it has left the notch, the impulse is completed by the notch of the lever striking the second small pin in the roller, which

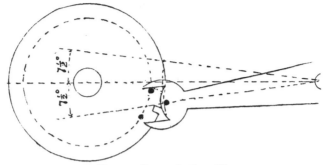

Fig. 7.—Savage's Two Pin.

has by that time reached the line of centres or nearly so. In order to get the safety pin well into the notch, this escapement requires pallets having 12° to 15° of motion, which is objectionable, and the lever and roller action is besides a very delicate job, and fails if not thoroughly well done ; so that, although the idea is taking, this form of the escapement has never come much into use, and when it is made one wide stone is generally substituted for the two pins in the roller.

The unlocking nearer the line of centres is also accomplished in what is called the anchor or dovetailed escapement, in which the impulse-pin is wider than usual, and of a dovetail form. It is open to the objection that, on account of the increased width of the impulse stone and of the lever, banking will occur with a smaller vibration of the balance than with the usual form.

Resilient Escapements.—A watch balance in general use rarely vibrates more than a turn and a half, that is, three-quarters of a turn each way ; yet occasionally, from pressing on the key after the watch is wound in going-barrel work, sudden movements of the wearer, or other cause of disturbance, the balance will swing round till the impulse pin knocks the *outside* of the lever. If this banking is violent,

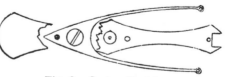

Fig. 8.—Spring Bankings.

the time-keeping of the watch is deranged, and a broken pivot may also result if the pivots are small. To obviate the evil of such banking, various plans have been tried. The most usual is to make the banking pins yield to undue pressure, and allow the impulse pin to pass the lever, the wings of which are omitted, as

shown in Fig. 8.   J. F. Cole devised a resilient escapement without
any banking-pins, in which the teeth of the escape wheel were so
formed as to resist the entrance of the pallet into the wheel more
than was required for ordinary locking.   (See Fig. 9.)   In the event
of over-banking, the pallet compelled the escape wheel to recoil, so
that the mainspring was really utilized as a spring banking.
But in the use of any of these resilient arrangements there is a

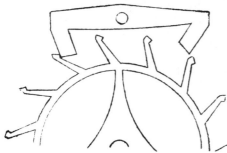

danger of "setting."
When the banking is
so violent that the im-
pulse pin drives the
lever before it, all is
well, but it is sure to
happen sometimes that
just as the impulse pin
is passing the lever its
motion is exhausted,
and it jambs against the
point of the lever and
stops the watch.   In a

Fig. 9.—Cole's Resilient.

recent arrangement, W. G. Schoof claimed to overcome this
tendency to set by using *very weak* spring bankings.   Another
objection to spring bankings is that in their recoil they are likely
to drive the safety pin against the edge of the roller.

Taylor's Resilient.—This, shown in Fig. 10, is one of the
latest forms of resilient action.   On the lever A are springs with
horse-shoe curves B and points C, which form the notch ; E, the

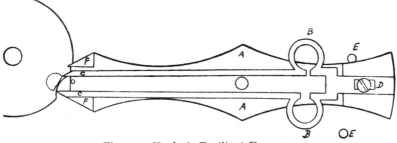

Fig. 10.—Taylor's Resilient Escapement.

usual solid banking pins for lever ; F, banking for the springs C.
It is claimed that when the impulse-pin, after a vibration of
excessive amplitude, strikes the notch formed by the springs, the
spring, yielding backward and inward, allows it to pass freely,
and there is no possibility of the impulse pin setting against the
springs.

Varieties of the Lever Escapement.—Swiss watchmakers were the first to form the tips of the teeth of the escape wheel into inclined planes, so as to divide the impulse between the wheel teeth and the faces of the pallets, like Fig. 11. It is urged that the wheel is not so fragile when made in this way, that less drop is required, and that the oil is not drawn away from the tip of the tooth by capillary attraction. On the other hand, English watchmakers maintain that as at some time during each impulse the planes of the wheel and pallet nearly coincide, the increased surface then presented to the varying influence of the adhesion of the oil is a serious evil. Then with " club teeth," as they are called, there is more difficulty in satisfactorily replacing a wheel than with ratchet teeth, for in the former case the planes must be of exactly the same angle and of the same length in the new wheel as in the old one. With brass wheels the impulse faces on the wheel get cut into ruts, but the Swiss avoided this by using steel wheels, and also much reduced the extra adhesion due to increased surface by thinning the impulse planes of the teeth. That the balance of advantages lies with the club tooth may be assumed by the preference given to it lately. Students will have but little difficulty in setting out this form of the escapement after a study of Fig. 3 and Fig. 11.

In another form of the escapement the whole of the impulse plane is on the wheel teeth, the pallets being small round pins, as in Fig. 12. The draw on the pallets is necessarily unequal and there is consequently a tendency for the exit pallet to set.

In most foreign lever escapements and occasionally in English, the roller is planted in a line with the escape wheel and pallet staff holes, as in Fig. 11 instead of as shown in Fig. 1. This alteration of the position of the lever with relation to the pallets has often provoked controversy, but there is practically no advantage either way except as a matter of convenience in arranging the caliper of the watch or in manufacturing the parts, though the straight line escapement certainly allows of the poising of the lever and pallets with less redundant metal.

Number of Teeth in Escape Wheel.—Schoof strongly advocated the use of escape wheels with but few teeth. There is no virtue in the particular number of fifteen, and with a very large number the planes of the pallets may become so oblique with relation to the path of the teeth that the wheel will start them with difficulty. There is no doubt also that with fewer teeth the proportionate loss from drop is lessened, but on the other hand, using for a watch an escape wheel with less than the usual number of teeth necessitates an extra wheel and pinion in the train or an increased number of teeth in the fourth wheel with, of course, proportionately increased wear of the escape

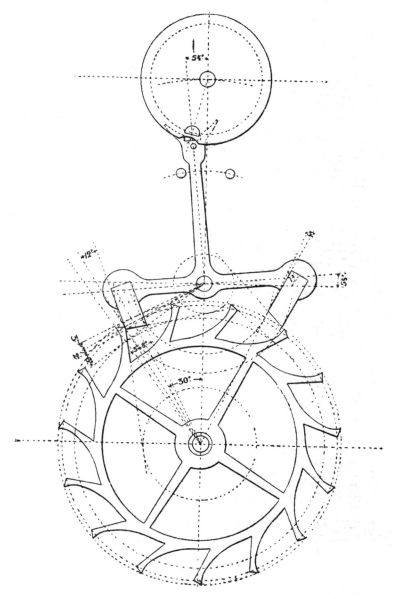

Fig. 11.—Club Tooth Lever Escapement.

pinion, unless there is a reduction in the number of vibrations of the balance, which again is open to objection. Mr. Alfred Curzon points out that for a carriage clock, where the number of vibrations is materially less, an escape wheel of twelve teeth answers much better than one of fifteen, because there is less danger of the teeth setting on the impulse planes ; and those who have to do with carriage clocks with the fifteen-toothed wheel know that if allowed to run down they do sometimes set so hard that, after winding, inexpert owners are unable to start them again.

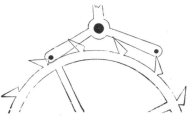

Fig. 12.

Pallets with Equidistant Lockings.—The drawing, Fig. 1, shows the pallets at an equal distance from their centre of motion. But then, although the impulse planes are equal, the locking faces are not the same distance from the centre, and the locking resistance is therefore unequal. Pallets having the lockings equidistant have in consequence been advocated by Grossman and other authorities, but such a construction is not without its disadvantages. The action of the wheel tooth on the impulse plane of the entering pallet before the line of centres is an engaging action, and on the exit pallet after the line of centres is a disengaging action. The friction is therefore greater on the entering pallet, and when an escapement sets on one impulse face, it is in nine cases out of ten the impulse face of the entering pallet. From this it is argued by some that if either pallet should be placed further from the centre of motion it should not be the exit, but the entering pallet, so as to give it a more favourable leverage wherewith to encounter the greater friction which undoubtedly exists. But there is really no advantage in the longer arm, for it has to be pushed through a greater distance by the wheel tooth than the shorter one. Arrange the length of the pallet arms how you will, you get but the force of the wheel passing through half the distance between two teeth. As far as the relative adhesion of the oil goes, the advantage is with a shorter arm. But the chief objection to the equidistant lockings is that with them the leaving corner of the exit pallet dips further into the wheel than with circular pallets, thereby requiring more drop to give the requisite freedom. However, pallets with equidistant or what is called " half-equidistant " lockings are now often used with the club-toothed escapement.

Examination of the Lever Escapement.—See that the

balance staff is perfectly upright. See that the wheel is perfectly true on edge and on face, and that the teeth are equally divided and smooth ; also by gently turning the wheel backwards see that the pallets free the backs of the teeth. If the wheel is out of truth it must be set up in the lathe and re-bored. It can be fixed either with shellac or in a brass sink bored out the exact size to receive it. If the divisions are unequal, or the wheel has some thick teeth, it should be discarded. It is useless to attempt to make the wheel right, and to reduce the corners of the pallet to free the wheel is simply to spoil the escapement for the sake of the wheel. At the same time, it must be left to the operator to judge whether the amount of the inaccuracy is serious. The whole affair is so minute that no rule can be given.

Is the wheel the right size ? If the lockings are too light and the greater part of the shake INSIDE, the wheel is too small, and should be replaced by one larger. Before removing the wheel gently draw the balance round till the point of the tooth is exactly on the locking corner, and see if there is sufficient shake. If not, it will be prudent to have the new wheel with the teeth a little straighter than the old ones. If the lockings are too deep and most of the drop OUTSIDE, the wheel is too large and should be topped.*

The wheel is so fragile that care is required in topping, which is done by revolving it in the turns against a diamond or sapphire file. A brass collet is broached to fit friction-tight on one of the runners of a depth tool : one side of this collet is then filed away, leaving sufficient substance to avoid bursting into the hole. On this flat a small piece of sapphire file is attached with shellac, taking care that the *face of the file is parallel to the centre of the runner.* The escape wheel on its pinion with the ferrule attached is placed in the centres of the depth tool *farthest from the adjusting screw*, and the collet and file on one of the opposite centres, and that centre fixed firmly by its clamping screw. A very light hair bow is used to rotate the pinion, and the depth tool laid on its side on the work board— the tool being closed by its screw until the teeth of the wheel *nearly* touch the surface of the file ; now if a slight pressure is made by the fingers on the uppermost limb of the tool, at the same time rotating the wheel by the bow, the *spring* of the tool will allow the teeth to be brought into contact very slightly and without fear of bending the teeth ; the wheel can be reduced as much as is necessary.

If the wheel is the right size and there is no shake (which

* In planting the wheel and pallets it is always best to err, if at all, by making them too deep rather than too light. If they are a shade deep, topping the wheel soon puts matters right.

try as before directed), the discharging corner of the pallets may be rounded off by means of a diamond file if they are of garnet. If they are of ruby, they may be held against an ivory mill charged with diamond powder. If the lockings are too light and there is but little shake they may be made safe by polishing away the locking face as before, or the pallet may be warmed and the stone brought out a bit. The locking faces of the pallet should be sufficiently undercut to draw the lever to the banking pins without hesitation. If they require alteration in this respect, polish away the upper part of the locking faces so as to give more draw, leaving the locking corner quite untouched. But proceed with great care, lest in curing this fault the watch sets on the lockings, as small watches with light balances are very liable to do. If a watch sets on the lockings, or on one of them, the locking face or faces may be polished away so as to give less draw—*i.e.*, have most taken off the CORNER of the locking. If the watch sets on the impulse, the impulse face may be polished to a less angle if the locking is sufficiently deep to allow it. For it must be remembered that in reducing the impulse the locking of the opposite pallet will also be reduced. In fact, the greatest caution should be exercised in making any alteration in the pallets.

Sometimes, in new escapements, the oil at the escape wheel teeth will be found to thicken rapidly through the pallet cutting the wheel, showing that one or both corners of the pallet are too sharp. If ruby, the corner may be polished off with a peg cut to the shape of a pivot polisher, and with a little of the finest diamond powder in oil ; if garnet, diamantine on a peg will do it very well. Great care should be taken to remove every trace of the polishing material, or the wheel may be charged with it.

See that the pivots are well polished, of proper length to come through the holes, and neither bull-headed nor taper. A conical pivot should be conical only as far as the shoulder : the part that runs in the hole must be perfectly cylindrical. They must have perceptible and equal side shake, or if any difference be made the pallet pivots should fit the closest. Both balance staff pivots should be of exactly the same size. The end shakes should all be equal. Bad pivots, bad uprighting, excessive and unequal shake in the pivots are responsible for much of the trouble experienced in position timing. With unequal end shakes the pallet depth is liable to be altered owing to the curved form of the pallet faces. The action of the escapement will also be affected if the end shakes are not equal, by a banking-pin slightly bent, a slight inaccuracy in uprighting, and other minute faults. The infinitesimal quantity necessary to derange the wheel and pallet action may be gathered from the fact that a difference of ·002 of an inch is quite enough to make a tripping pallet depth safe or a correct depth quite unsound.

When the wheel and pallets are right see that the impulse pin is in a line with an arm of the balance, and proceed to try if the lever is fixed in the correct position with relation to the pallets. Gently move the balance round till the tooth drops off the pallet. Observe the position of the balance arm, and see if it comes the same distance on the other side of the pallet hole when the other pallet falls off. If not, the pins connecting pallet and lever are generally light enough to allow of the lever being twisted. To do this successfully a clamp to grasp the back and

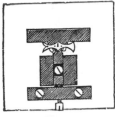

Fig. 13.

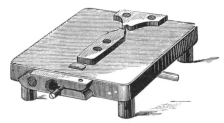

Fig. 14.

belly of the pallets, as shown in Fig. 13, or some similar tool, is necessary. R. Bridgman devised a very superior clamping tool for pallets, which with some modification by C. Curzon, is shown in Fig. 14. There is a spring for keeping the slide in contact with the pallets when they are placed in position, and until they are gripped tight by the screw, and a lever under the body of the tool by means of which the spring is overcome and the pallets released when the operation is completed. It will be seen that the screw passes through a split nut, which may be set up as it wears. The jaws of the tool should be faced with tin to avoid marking the pallets. When the lever is right with relation to the pallets, see that the pallets are quite firmly fixed to the lever, and that the lever and pallets are perfectly in poise. This latter is an essential point in a fine watch to be timed in positions, but it is often neglected.

See that the escapement is in beat. When the balance spring is at rest, the impulse pin should be on the line of centres, that is in the middle of its motion. If this is not so, the spring should be drawn through or let out from the stud if the position of the index allows ; if it does not, the roller may be twisted round on the staff in the direction required.

Is the roller depth right ? If the safety pin has insufficient freedom while there is enough run, the roller is probably planted too deep. On the other hand, if it is found that while the safety pin has plenty of freedom there is no shake between the bankings the roller depth is probably too shallow. When the impulse pin is led round there should be an equal clearance all round the inside of the horn, and the pin must fall safely into the notch. If it binds in the horns and bottoms in the notch it is too deep, and, on the other hand, if with excessive clearance in the horn the pin when it falls does not pass well into the notch, it is too shallow. The readiest method of altering is to warm the roller, remove the impulse pin, and using a to-and-fro motion with a wire and oil-stone dust, draw the hole in the required direction. If a pin is deep in the notch and too tight in the roller to give a little, it should be removed and flattened off a trifle more. If too shallow a triangular pin, or one of some other shape with the point of contact more forward, can generally be substituted by polishing out the hole towards the crescent. If not, the staff hole in the lever may be drawn to allow of shifting the lever sufficiently ; or the recesses for the jewel settings of the balance staff pivots may be scraped away on one side and rubbed over on the other to suit. See as it passes round that the impulse pin is free when in the notch.

Just as the safety pin is about to enter the crescent, the impulse pin must be well inside of the horn. In the single roller escapement a very little horn is required, unless the crescent has been made of an unnecessary width. In very common work one occasionally sees a flat filed on the edge of the roller instead of a crescent. There is no excuse for such a piece of bungling.

A fault occasionally met with is that the impulse pin after leaving the notch just touches on some part of the inside of the horn in passing out. If a wedge of cork is placed under the lever, so that the lever moves stiffly, it can be readily seen whether or not the impulse pin is free to leave the notch, and is free all round the horn when the wheel tooth drops on the locking.

See to the safety action. When the tooth drops on to the locking, the safety pin should be just clear of the roller. If it is not clear, the edge of the roller should be polished down till it is right. If there is more than clearance, the safety pin must be brought closer to the roller. See upon pressing the safety pin against the roller that the tooth does not leave the locking, and that the impulse pin is free to enter the notch without butting on the horn of the lever : also that the safety action is sound, so that the pin is in no danger in passing the roller. If the action is not sound the diameter of the roller should be reduced and the safety pin brought towards it sufficiently to get a sound action if it can be done, but if the escapement has been so badly proportioned as not to allow of a sound action being obtained in this way, the pin must be shifted forward and the bankings opened to allow more run.

See if the banking pins are so placed as to allow of an equal run on each side. If not, they should not be bent, for with bent banking pins a difference in the end shakes of the pivots will cause a difference in the run. The banking pin allowing of the most run should be removed, and the hole broached out to receive a larger pin.

A fault rather difficult to detect, which is sometimes met with in the double roller escapement, is the end of the impulse pin slightly touching the safety finger, caused by excessive endshake, or from a longer impulse pin than was orginally intended having been put in.

Lifting Piece.—[*Détentillon.—Der Auslösungshebel.*]—In a striking clock, a two-armed lever that lifts the rack hook. (See the Striking Work.)

Line of Centres.—[*Ligne des centres.—Die Mittelpunktslinie.*] —A line passing through centres. In a lever escapement, the line passing through the centre of the pallet staff and the centre of the balance staff.

Lips. — [*Lèveres. — Die Hebungsflächen (Zylinder).*] — The rounded edges of the cylinder in a cylinder escapement, on which the escape wheel teeth press to give impulse to the balance.

Litherland, Peter.—Patentee of the Rack Lever Escapement (No. 1830, Oct., 1791).

Live Spindle. — [*Arbre tournant. — Die rotirende Welle (Drehbank).*]—A rotating spindle ; applied generally to the rotating mandril of a lathe.

Locking.—[*Repos.—Die Ruhe.*]—The act of stopping the escape wheel of a watch or clock. The portion of the pallet on which the escape wheel teeth drop.

Locking Plate. Count Wheel.—[*Roue de compte.—Die Schlosscheibe.*]—A circular plate with notches in the edge at distances corresponding to the hours struck, used in striking work of a kind rarely made except for turret clocks. The primitive form of this striking work is shown in the engravings of " De Vick's clock," and a modern arrangement is given under the head of " Striking Work." Though simple, it has the defect that the hours are struck in regular progression without reference to the position of the hands.

Locking Spring.—[*Ressort de fermeture.—Die Schliessfeder.*]— The spring of a watch case that keeps the cover closed against the force of the fly springs. (See Case Springs.)

Loseby, Edward Thomas.—Inventor of a compensation balance (Patent 1011, December, 1852). The rims are bimetallic of brass and steel, shorter than usual ; at the end of each segment of the rim is a cup-joint, in which is placed a glass vessel consisting of a curved arm and a bulb, which contains mercury. The curved arm is sealed with a little air in it to insure the continuity of the thread of mercury when it contracts. It is apparent that by bending the cup-joint, the direction of the glass

arms may be altered, and in this way a very exact temperature adjustment obtained. Chronometers with this balance were remarkably successful at the Greenwich trials from 1846 to 1853.

Lunette.—[*Verre de montre ordinaire.—Die gewöhnliche Uhrglasform.*]—The usual form of rounded watch glass.

Magnetism.—[*Magnétisme.—Der Magnetismus.*]—A magnetized watch may be demagnetized by placing it for a few seconds in the solenoid of a dynamo or other magnetic coil, through which

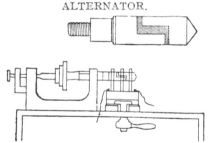

ALTERNATOR.

an *alternating current of high intensity* is passing, and then gradually withdrawing it while the current is flowing.

Where access cannot be had to such a coil, the following description of a more modest demagnetizing apparatus attached to a lathe may be useful. Here the current is obtained from a series of six or other number of Leclanche cells. A coil large enough to receive a watch is provided and an alternator fixed to the mandril of the lathe which is then rotated. The alternator consists of a core of box-wood with brass tubing over it, the tubing being cut as shown in the enlarged view. The cuts are filled up with sealing wax to prevent sparking. On a block of wood are four contact pieces, two for the coil on one side of the alternator and two to the battery on the other. The contact pieces are of brass with small dabs of silver where they rub on the alternator.

This apparatus will be found effective for demagnetizing small pieces even if the current is not sufficiently strong to treat the watch as a whole.

A way of demagnetizing without a battery is to secure to the chuck of a lathe with the points outwards either a strong horseshoe magnet, or if possible two or three clamped together in a block of wood so that the wood is between the bend of the magnets and the lathe chuck. While the magnets are rapidly rotated, the steel to be operated on is to be held as close to their points without touching as possible and then *very* slowly withdrawn till quite out of the range of the magnets. The

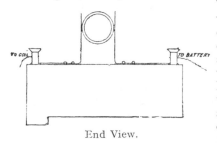

End View.

work may be held in ivory, wood, or all *brass* pliers, but not, of course, in a steel holder of any kind.

Under the title of "Matador" has been introduced an arrangement for demagnetizing as shown in the adjacent illustration. A magnet contained in the case is actuated by a spring which when wound would cause the magnet to make, it is said, nearly 4,000 revolutions a minute. After being subjected to this influence for about a minute the watch is slowly withdrawn while the magnet is still revolving.

To ascertain if any and what parts of a watch are magnetized without taking it down, Dr. Waldo recommends suspending a minute fragment of *iron* wire, previously annealed by subjecting it to a red heat, from a slit in the end of a peg, by means of a strand of the finest silk. This forms a kind of small fishing-rod, which may be approached to the suspected parts in succession. If the soft iron is attracted the presence of magnetism is assured. The parts most likely to derange the going of a watch are the balance, balance spring and case springs, and next to them the lever. Pieces proved to be magnetic should be removed and demagnetized. Case springs may be heated to redness to get rid of magnetism. Most parts of watches may be advantageously held during the operation by waxing them to a cork.

"Matador" Demagnetizer.

To divest steel work of magnetism with bar magnets is slow work, but is considered by many the most permanently effectual course. It is necessary to have three or four magnets of different sizes, also a good horseshoe-magnet for recharging, for straight magnets soon lose strength. A piece of bar-steel of the required size, hardened first and then charged by the horse-shoe magnet, answers the purpose, or an old worn-out round or square file, or stump of an old graver, will do equally well, and save the trouble of hardening. The size of the magnet used must be determined by the size of the article operated on. A watch balance, for instance, is one of the most troublesome things to treat. Take a magnet about 3 inches long and $\frac{1}{4}$ inch square. It will be found that polarity is situated principally in the neighbourhood of the arms of the balance, and these are the points to be first attacked.

Hang the balance by its rim on a piece of brass wire, and approach the magnet towards the rim in the direction of one of the bars. If it should be attracted towards the magnet, try the other pole, and it will be found to repel. Now take the balance in your hand and bring the *repelling* pole of the magnet in momentary contact with the balance at the point tried, then test it with a minute fragment of small iron binding wire ; if still magnetic bring the magnet in contact again, and so on—trying after each contact—till the magnetism is entirely out at that point. Suspend the balance on the brass wire as before, and proceed to try the rim at the point where the second arm comes, and the same with the third. Having got the magnetism out of these three points there will be but little remaining in the balance. However, try it carefully all round, when several places will probably be found containing sufficient magnetism to pick up a small fragment of iron. These must all be treated in the manner before described ; but when the magnetism is very feeble a smaller magnet must be used, for if the magnet is too powerful the article operated upon discharges what little remains, and before contact can be broken, begins to pick up again off the reverse pole. The balance spring may be successfully treated, though so strongly charged as to be "feathered" in iron filings after being immersed in them. A good way to try the polarity of many pieces is to suspend the article by means of a particle of wax to a piece of the finest silk. Steel filings, or fragments of chain wire, should on no account be used for testing ; for if not magnetic to begin with they speedily become so by contact with the article under treatment. Even with soft iron it is well to change the fragment you are testing with occasionally. (See also Non-Magnetizable.)

Mainspring.—[*Ressort moteur.*—*Die Zugfeder.*]—The long ribbon of steel used for driving a clock or watch. The mainspring is usually coiled into a circular metal box called the barrel, the outer end of the spring being attached to the rim of the box. The box is mounted on an arbor to which the inner end of the mainspring is fastened. By the act of winding a watch or clock with a going barrel, the barrel arbor is turned round and round till nearly the whole of the spring is drawn away from the rim of the box, and coiled round the arbor. The escapement of the watch or clock stops the barrel from turning with the arbor and a click on the plate taking into a ratchet on the arbor prevents the arbor recoiling when the winding is completed. The tension of the spring strives to turn the barrel, and so the force exerted in winding is utilized to drive the train of wheels until the barrel has made as many turns as were given to the arbor in winding.

In order that the mainspring may not be injuriously contracted, the part of the barrel arbor to which it is attached is enlarged or colleted. For watches and 2-day marine chronometers, the collet is one-third the diameter of the inside of the barrel. The barrel arbor gauge, shown in Fig. 1, is a pair of double-ended or proportional callipers; the long ends for the inside of the barrel are three times the distance from the centre of motion, and always open to three times the size of the shorter

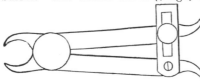

Fig. 1.

ends, which are for gauging the barrel arbor collet.

When a suitable spring and arbor are in the barrel, there should be as much unoccupied space as is equal to the bulk of the spring. Then if we suppose the inside diameter of the barrel to be divided into 100 equal parts, the spring will occupy, when nearest the circumference of the barrel, 26 of those parts (13 on each side); when wound close round the arbor, 41 parts; or, together with the arbor, 74 parts. These proportions may be readily measured with the sector.

In 8-day marine chronometers, taking the inner diameter of the barrel as 100, the arbor is generally made = 43, and then arbor and mainspring wound should measure 77·4, and the coil of mainspring when unwound 22·6, that is 11·3 on each side of the barrel.

Whatever the size of the arbor, IF THE OUTER DIAMETER OF THE SPRING WHEN IT IS WOUND IS THE SAME AS ITS INNER DIAMETER WHEN IT IS UNWOUND, IT WILL BE RIGHT, and the difference between the number of coils when the spring is wound and when it is unwound will be the number of turns the barrel will make. In a great many instances, too much spring is crowded into the barrel.

R

It may be required to find the proper thickness of the coiled ring of mainspring in other cases when the diameter of the arbor bears a different proportion to the size of the barrel.

Let $B =$ inner diameter of the barrel.

,, $A =$ diameter of arbor.

,, $T =$ thickness of the rim of mainspring when lying close to the rim of the barrel.

Then $T = B - \sqrt{\left(\dfrac{B^2 + A^2}{2} \right)}$

In order to get a given number of turns to find the number of coils there should be in the barrel when the mainspring is unwound :—

Let $M =$ the mean diameter of a spring when unwound.

,, $m =$ the mean diameter of spring when wound.

,, $P =$ the proportion between the mean diameter of spring when unwound and its mean diameter when tightly coiled about the arbor.

,, $t =$ the given number of turns.

,, $C =$ the number of coils when the spring is unwound.

Then $P = M \div m$.

And $C = (P \times t) + 1 \cdot 5$.

This gives an extra coil and a half on account of the ends of the spring not coming into action, and will be found to be an ample allowance.

Example.—The diameter of a barrel being ·7 and the diameter of arbor ·233, it is required to get 6·5 turns. The diameter of the circle, which represents the inner diameter of spring when unwound and the outer diameter of it when wound is ·52. The mean diameter of the unwound spring is ·61, and the mean diameter of it when wound is ·376.

Then ·61 ÷ ·376 = 1·62. And $\overline{1 \cdot 62 \times 6 \cdot 5} + 1 \cdot 5 = 12 \cdot 03$, or say 12 coils.

Twelve is a suitable number of coils for a going barrel, when the spring is unwound. As we see from the above example, it allows two turns for setting up, and half a turn to spare.

For a fusee watch five turns and a quarter, giving three-quarters of a turn for setting up and half a turn to spare, would be sufficient for the most extreme case. Taking the preceding proportions we have $1 \cdot 62 \times 5 \cdot 25 + 1 \cdot 5 = 10$, which is the number of coils required for 5·25 turns. Generally but three and a half turns are required for use, and unless an unusual length of the end of the spring is left soft, less than three-quarters of a turn will suffice for setting up. The ordinary custom of packing 13 or 14 turns into the barrel, filling it unnecessarily, and leaving room for but 4·5 or 5 turns of action, and little or nothing to set up involves the use of the weak end of the spring only. An equal adjustment with greater power and freer action may be obtained with fewer coils, as shown by the above examples.

If the vibration of a watch is too small, it may often be sufficiently increased by breaking off one or two excessive coils of mainspring.

To find the thickness of mainspring necessary, divide the thickness of the coiled ring of mainspring when unwound by the number of coils. Taking the thickness of ring and number of coils, as in the last example we have—

$$\cdot 09 \div 10 = \cdot 009 \text{ the thickness required.}$$

Practically the thickness would be a little less, for the coils do not lie absolutely close together.

Tapering Mainsprings.—It was at one time thought that the force of the spring in going barrels might be equalized by thickening the inner coils, but all such attempts have ended in failure, because of the extra coil friction thereby induced, and most springs for use with the fusee are thinned at the inner turns, to lessen the coil friction and induce a better action of the spring.

At page 211 is shown an adjustable spanner, suitable for setting up or letting down watch mainsprings, by engaging in the teeth of the ratchet.

Hooking the Mainspring.—The attachment of the outer end of the mainspring to the barrel is usually made either rigid by means of a square hook riveted to the mainspring, or free to

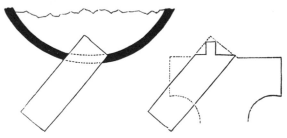

Fig. 2.   Fig. 3.

adjust itself by being hooked to a stud fixed in the barrel. A ready way of making a square hook is to fix a piece of rectangular steel to the hole in the barrel, as shown at Fig. 2, first seeing that the hole in the barrel is not at less angle than about 45°, or the hook will be apt to draw out when in use. When the steel is properly fitted, mark on it with a fine point the curve of the inside and the outside of the barrel, as shown by dotted lines in the figure, leaving enough of the steel inside the barrel to form the pivot. Then catch the steel in the vice at the same or a slightly less angle than it occupied in the barrel, and form the pivot as close to the slope on which the strain comes as possible. Fig. 3 shows clearly what is meant ; the jaw of the vice is there indicated by dotted lines. The object of placing the steel in the vice at a slightly less angle than it occupied in the barrel is so that when the hook is in action the strain shall be taken at the ROOT of the slope. Some watchmakers use a cutter for finishing the pivot and shoulder of the hook like a rose cutter. It is formed of a piece of round steel, up to the middle of which a hole is drilled of a size to just admit the nose of the pivot. At the end of the piece of steel around the hole serrations are formed (see Rose Cutter). A few turns of this tool quickly finish the pivot and shoulder. The pivot hole in the spring may now be made. If the spring is softened sufficiently to enable it to be drilled, it will be right. The spring should not be tempered, but just rounded at the end, and bent to the circle of the inside of the barrel. In large hooks the pivot should be annealed to a red or the rivet may crack.

To prevent the spring drawing away from the hook, Mr. Arthur Webb recommends the insertion between the mainspring and the rivet of a piece of mainspring about half the thickness of the spring which is being operated upon, and a sixth of the circumference of the barrel in length.

Mr. Bickley has favoured me with another method of procedure. Having selected a suitable spring, which, to ensure freedom in the barrel, must lie well under the groove of the barrel cover, temper the end, taking care not to soften it too much or too far up the spring. Drill the pivot hole (supporting the spring against a piece of brass while doing so) pretty near to the hard part of the spring, and at a point corresponding in height to the centre of the barrel hole. If the pivot hole is made too much into the soft, it will cause the spring to buckle or bend in riveting the hook. Proceed to make the hook by turning or filing the pivot on a piece of narrow, flat steel (Fig. 4). The pivot must be straight, or very nearly so, and fit the spring hole tightly ; the shoulder must be clean and square and lie close up to the spring. File the sides of the steel, keeping them parallel, and with the pivot in the centre, until the end of the steel on either side of the pivot will pass freely through the barrel hole. Now make the back slope of the hook (Fig. 5) at an angle of about 45° ; hold the steel

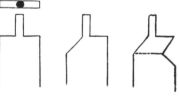

Fig. 4. Fig. 5. Fig. 6.

in the vice while doing so, and continue the slope as in Fig. 6 up to the pivot, taking care not to nick the latter at the root. Reverse the steel in the vice, and file the front slope of the same angle as the back (Fig. 6), shortening the base of the hook in the direction of the dotted line until it is right length. This will be determined by placing the back slope of the book on the outer slope of the barrel hole, and trying it carefully and frequently until the front slope will pass through. The hook thus formed should pass freely through the hole, fitting it closely at the sides and ends, for more than the height eventually required. The pivot is now to be shortened to about twice the thickness of the spring and riveted slightly, the inner side of the hole having previously been chamfered to receive the rivet, which must project as little as possible above the spring. The hook is now to be cut off the steel as indicated by the dotted line in Fig. 6, and filed to the right height. To ascertain the right height pass the hook through the hole in the reverse direction to that above mentioned, *i.e.* with the front slope of the hook towards the outer slope of the hole, and lower it until its height barely equals the thickness of the barrel rim. Polish the hook and before winding in see that the spring is perfectly flat, and that its circular form is not disturbed at the hook. If properly made the hook will slip freely into the hole, and the outer coil of the spring will lie close to the inner circumference of the barrel. The outside of the barrel should never be filed, neither should the hook be tampered with after it is riveted on the spring. If these directions are carefully followed, but little practice will be needed to ensure good fits, and to make the operation of hooking the mainspring a simple one indeed.

The advantage generally claimed for the hook riveted to the spring over the eye attached to a stud in the barrel, is that the former being a rigid attachment keeps the coils of the spring equally diffused when under tension, instead of allowing the turns to go over in a mass to the point of least resistance. But the spring must not be left perfectly hard at the rivet or it will break, and in many instances it is found that when the spring is wound it is bent at the rivet, forming an elbow. The advantage of the rigid attachment is then lost, and the spring goes over at once to the point of least resistance. Sometimes the end of the spring beyond the hook is thinned, and also filed to a point, but the chance of the spring breaking across the rivet is thereby much increased without serving any good purpose. This weakening of the projecting end appears to have been done at first to enable the watchmaker to tuck the end of the spring easily into the barrel when the spring is wound up on the spring tool ; but the same convenience may be obtained when the end is left as strong as possible by bending it to the shape of the inside of the barrel. It will then slip in without trouble. If a mainspring with a hook attachment keeps its shape without bending in to an elbow at the rivet, it will invariably be found upon examination that the barrel has been so full of spring that the angle of inflection when the spring is wound is very small. The spring is consequently but little strained at the hook. But this excess of spring is clearly not economy of room or power, and does not permit of any better adjustment.

Fig. 7 shows a fixed stud in the barrel. The round stud is a good form and is used generally for marine chronometers. The head of the stud should be in diameter about one-third of the width of the spring made as thin as possible, so as to project but little beyond the first coil of the spring.

Fig. 7.   Fig. 8.   Fig. 9.

After the stud is screwed home it should be riveted on the outside. The eye should be made close to the end of the spring, which should be rounded as shown in Fig. 9. It will then allow an amount of play on the detaining stud that will preclude all chance of breaking, no matter to what angle the spring is drawn. The spring may also be wound up quite tight as often as may be without any bend or kink in the attachment. It may, therefore, be assumed that the eye is the best attachment, at all events for the going barrel, in which the spring is required to be set up as much as possible in order to strengthen the lower coils and get a good adjustment.

Instead of a stud screwed straight into the barrel as shown in Fig. 7, a square hook is usually screwed in. The hook has more holding power if the hole is drilled at an angle as in Fig. 8. The hole is often too large. For the ordinary run of watch work if it is made to suit No. 13 or No. 14 tap it will be right. The hook should be screwed in from the inside of the barrel, the end being finished to proper length and angle before the pin is removed from the screw plate, and must not be left too long. A little more than the thickness of the first coil of spring is sufficient. More length, besides taking up room in the barrel, generally causes the barrel to bulge in the event of the spring breaking. Where the barrel is thin, care should be taken to have the thread of the screw sufficiently fine.

The advocates of the hook in the spring assert that a better adjustment is obtained with a rigid attachment than with a yielding one but I am told that the eye was used in preference for marine chronometers by the late John Poole, and most chronometer makers of the present day find there is no difference in the adjustment, and also adopt the same kind of attachment.

Pivoted Post.—To assist a free development of the spring, many French and Swiss watches with the eye attachment have a post interposed between the outer and next coil. It is a piece of steel the width of the mainspring and a little thicker, with rectangular ends fitting holes in the barrel and cover.

Other Methods of attaching the Mainspring to the Barrel.— Among other methods of attaching the outer end of the mainspring to the barrel, one of the most simple and effectual, by M. Philippe, is to coil inside the barrel a piece of thicker mainspring of a little more than one complete turn in length, so that the ends just overlap. The mainspring of the watch is riveted to this loose piece, the adhesion of which against the barrel is sufficient to drive the watch. Three or four half-round grooves are cut inside the rim of the barrel, and a corresponding projection riveted to the outside of the loose piece, and the clicking of this projection as it enters the grooves indicates when the spring is fully wound.

Pivoted Brace.—Another plan is to rivet to the end of the mainspring a tee piece forming two pivots, one of which passes through a hole in the bottom of the barrel, and the other through a hole in the barrel cover. In an arrangement by Mr. Glasgow (Fig. 10) these pivots instead of being retained in round holes, are free to move in slots forming tangents to the spring when it is wound, in order to obtain a better adjustment.

Fig. 10.

A device often adopted when the stop work is missing from a watch, is to connect the end of the mainspring with the barrel by means of a short

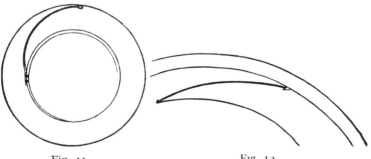

Fig. 11. Fig. 12.

piece of spring as shown in Fig. 11. This is sometimes riveted to the mainspring, but Mr. G. Kirton says that a better result is obtained by a loose piece rather longer than a third of the diameter of the barrel. A hook is formed on the end of the mainspring with pliers while it is held in the flame of a lamp, to catch over the end of the loose piece, as shown in Fig. 12. The loose piece may be made from the old mainspring where it nearly corresponds with the curve of the barrel, and should have nicely rounded or sharpened ends.

Attachment of the Mainspring to the Barrel Arbor. —The attachment of the inner end of the spring is universally made by a hole in the spring hooking on to a slight projection from the arbor collet. The collet should be snailed or formed into a volute, so that the part of the collet for the reception of the end of the spring is sunk equal to the thickness of the spring, thus giving an even surface for the next turn to pass over. The collet is usually of nearly the same width as the spring, and the hook is driven into it. In watches the spring is made as wide as the thickness of the watch will allow, and consequently the holes in the barrel and cover are often too thin to form a sound and lasting bearing for the arbor. Mr. Plose suggests the improvement depicted in Fig. 13. He makes the collet but one-third the

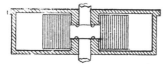

width of the spring, and thus obtains very long holes for the arbor. The oil sinks are cut in the collet, and the hook instead of being driven in, is formed out of the solid.

Fig. 13.

Donne's Resting Barrel.—Fig. 14 is a section of an arrangement invented by Mr. Lewis Donne. A is the upper and B the pillar plate of a keyless watch. The barrel is of steel, thin, and with a central projecting boss on the lower side, fitting a hole in the pillar plate. The ratchet wheel D is fixed to this central boss by three screws, and sunk into the under side of the ratchet wheel is the stop work. The great wheel (E) is tight on the arbor, and has a long boss to which the inner end of the

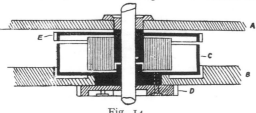

mainspring is hooked. Unless the watch is being wound the barrel is stationary. As the ratchet wheel is turned in winding the barrel moves with it.

Fig. 14.

In the going of the watch, the mainspring begins to unwind

from the centre. A very much wider mainspring is obtained than would be possible in the ordinary way. There are long holes for both pivots of the barrel arbor, and the steel ratchet wheel affords a sounder attachment for the Maltese cross than the brass barrel or cover would do.

Mainspring Hook.—[*Crochet de ressort.—Der Federhaken.*]— A little steel projection at the outer end of a mainspring, by means of which the mainspring is attached to the barrel.

Mainspring Punch.—[*Pince à percer les ressorts.—Die Federlochzange (Zugfeder).*]—Tongs with movable punches and dies for perforating mainsprings. An improved pattern shown in the engraving besides four punches for the spring carries also

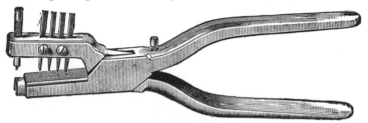

a barrel hook punch. The inside of the barrel is placed on the projecting circular bolster which contains a hole to admit the hook. The punch is then firmly closed and a tongue-like projection formed inside the barrel which answers as a hook for the spring. Punches for the barrel hook alone are also made.

Mainspring Winder.—[*Estrapade.—Der Federwinder.*]—An arbor on the end of which a mainspring is tightly wound in order to get it into the barrel. Usually the arbor, mounted in a standard, is furnished with a handle for turning, and a ratchet

to prevent it running back. The sketch appended shows a form of mainspring winder of American origin which is certainly much preferable to the old pattern in which the spring is coiled up between the fingers, whereby the fingers are often injured and the spring rusted, to say nothing of the chance of damaging the spring by kinking it when the operation of winding in is performed by unskilled hands. The new winder consists of a barrel with a slit in the side as shown, and a false bottom which may be pushed forward to eject the spring from the winder into the barrel. There are three winding barrels of different sizes to cover the whole range of watch barrels usually met with.

Maintaining Power. — [*Entretien.—Das Gegengesperr.*] — A device for driving a watch or clock during the operation of winding.

Harrison's Maintaining Spring.—To obviate the danger of a watch or clock stopping while being wound, Harrison introduced the contrivance now used in all fusee watches, as well as in regulators, and the better class of house clocks.

Fig. 1 is a view of the larger end of a watch fusee, which is fixed tight to the winding arbor. The great wheel rides loose

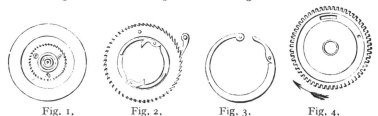

Fig. 1.   Fig. 2.   Fig. 3.   Fig. 4.

on the arbor, as does also a thin steel ratchet wheel as large as the fusee, which is placed between the fusee and great wheel. There is a smaller ratchet wheel whose teeth are cut the reverse way, let into and screwed to the fusee, as seen in Fig. 1. Fig. 2 shows that side of the larger ratchet wheel which is placed next to the fusee. The two clicks thereon take into the ratchet on the fusee and thus establish connection between the two pieces. A pin passes through the ball end of the spring (Fig. 3), and enters a hole in the larger ratchet wheel. Fig. 4 shows the great wheel round the inner face of which a recess is turned to cover the spring, so that the great wheel can be brought close to the large steel ratchet wheel. Near the other end of the spring it is fixed to the great wheel by means of the pin shown in the right-hand side of Figs. 3 and 4. It will thus be seen that while the ball end of the spring is fixed to the large ratchet wheel, the other end is fixed to the great wheel. The spring, being made rather weaker than the force of the mainspring of the watch exerted at the radius of the pin, is bent up till the tail touches the ball as the watch is going, when the great wheel rotates in the direction indicated by the arrow, and the teeth of the larger ratchet pass under the click, or detent, as it is called. In winding, the fusee is turned the reverse way, and the teeth of the smaller ratchet slip under the two clicks, which are pivoted on the larger ratchet. The spring connecting the larger ratchet to the great wheel then, in striving to unbend, drives the watch, the larger ratchet forming a resisting base ; for it cannot go back with the fusee because the click which takes into it is pivoted to the watch plate. At the top of Fig. 4 there is a concentric slot which serves as a sight hole to observe the bending of the spring. For it is important to see that on the one hand it is not stronger than the mainspring, yet on the other hand it is sufficiently strong to drive the watch. The pin through the ball of the spring which attaches it to the larger ratchet wheel is generally projected right through the spring, so that its other end traverses the slot. There appears to be no advantage whatever in this, and it often renders the mainspring power inoperative by binding in the slot, which it is sure to do if the slot is not exactly concentric. When this is found to be the case, the end of the pin may be safely filed off.

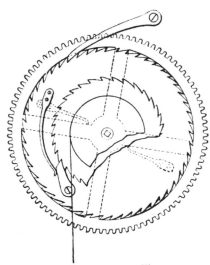

Click Pivoted to Clock Plate.

Cord for Weight.   Fig. 5.

Mr. Ganney points out that this slot is unnecessary, for the amount the spring moves may be ascertained by observing how many teeth of the great wheel pass a mark placed on the ratchet wheel. In many watches the detent spring is much too strong, thus absorbing the power of the watch, and in a few months the detent will probably be so worn as to slip over the teeth instead of locking the wheel.

The application of Harrison's maintainer to weight clocks is shown in Fig. 5. Here the smaller ratchet is attached to one end of the barrel, which is tightly fixed to the arbor. Next to the smaller ratchet is the larger one, and at the back of that is the great wheel, the larger ratchet and the great wheel being both free on the arbor. The click for the smaller ratchet is pivoted to the nearest face of the larger ratchet, and to the outer face of the larger ratchet are screwed two springs, whose free extremities bear on opposite arms of the great wheel. In the going of the clock the pressure of the weight acting on the click pivoted to the larger ratchet, bends the springs. There are two pins in the larger ratchet, one on each side of one of the arms of the great wheel, to circumscribe the action of the springs, which may be limited to three teeth of the great wheel. When winding the clock, the teeth of the smaller ratchet run under their click, but the teeth in the click of the larger ratchet, which is pivoted to the clock plate, keeps the larger ratchet fast, and the unbending of the springs is then utilized to drive the clock.

In planting the click, draw a line from the centre of the wheel through the nose of the click ; another line drawn through the centre of the click-hole should form with it an angle of about 100°. If less than 90° the click will be liable to fly out.

For large clocks, some makers use a spiral spring coiled round a curved pin fixed to one of the arms of the great wheel, as shown in Fig. 6 ; an eye screwed to the larger ratchet compresses the spring. This answers well if the spring is of sufficient length.

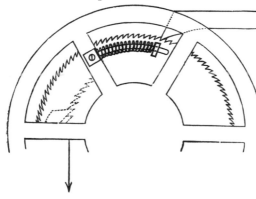

Fig. 6.

Maintaining work often fails for want of attention to the clicks, especially that of the larger ratchet wheel. The extreme point of the click should be the most advanced and sharp, to receive the pressure when it is in the tooth ; it may be finished with a moderately coarse file used across the width of the face of the click, which part should be hardened.

Endless Cord Maintainer.—See Huyghens.

Sun and Planet Maintainer.—In turret clocks there is often a weighted lever or segment brought to bear on the great wheel while winding, but this is open to the objection that it has to be renewed if the operation of winding takes long, and also because after winding, until the maintainer is removed, there is considerable extra pressure on the escapement. In some of Arnold's watches is a continuous maintainer, which also appears to have been invented by Harrison. Although not so suited for watches as Harrison's maintaining spring already described, it appears to be admirably adapted for turret clocks, which take some time to wind. The great wheel and barrel both ride loose on the arbor, to which is fixed a pinion represented by the smallest circle in Fig. 7. The largest circle represents a ring of internal teeth fixed to the side of the great wheel next to the barrel. There are two wheels, which gear with both the pinion fixed to the barrel arbor and with the ring of internal teeth on the great wheel, as shown. These two wheels run on studs in the end of the barrel. While the handle attached to the barrel arbor is turned as in winding, a continuous pressure is exerted on the internal teeth which really afford the resisting base in raising the weight. There is a ratchet wheel fixed to the

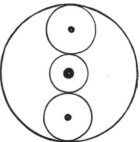

Fig. 7.

barrel arbor, with a click pivoted to the clock frame to prevent the weight running down, when the winding is completed. During the going of the clock the whole system turns with the barrel, so that there is no extra friction from the maintaining work. It is not absolutely necessary to have both of the wheels which run on studs at the end of the barrel, and sometimes one of them is omitted.

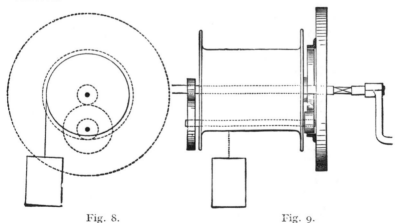

Fig. 8.   Fig. 9.

Mr. I. Herrmann patented the adaptation of this maintainer as keyless work for fusee watches and chronometers. The ring of internal teeth is on the great wheel, the stud pinions are on the fusee, which rides loose on the arbor, and the arbor carries a wheel which gears with a winding pinion at the pendant.

Messrs. Gillett & Co. have introduced another form of the

sun and planet maintainer for turret clocks with double gearing to give additional purchase in winding. Of this side and end views are shown in Figs. 8 and 9. As before, the great wheel and barrel both ride loose on the arbor. A pinion fixed to the barrel arbor gears with a wheel fixed to a spindle running in holes formed in the barrel near its circumference. On the other end of this spindle is a pinion which gears with the ring of internal teeth fixed to the great wheel. The extra winding wheels allow the ratchet and click to be dispensed with, but it would, perhaps, be safer to add them.

Maltese Cross.—[*Croix de Malte.—Das Malteserkreuz.*]— The name given to the star wheel of the ordinary going-barrel stop work.

Mandril.—[*Tour universal.—Die Klammerdrehbank.*]—A lathe having a face plate furnished with dogs for clipping watch plates and the like, and having a spring centre (pump centre) in the face plate with a fine conical point. English, and latterly, some Swiss mandrils are driven by a gut passing over a large pulley fixed either at the back or at the end of the lathe head, which pulley has a handle attached for turning it. A mandril is fitted with a slide rest, and sometimes with a back or "meeting" centre.

A good mandril with slide rest is one of the most perfect tools a watchmaker can possess. The English and Swiss tools have each advantageous features. The English mandril is worked by means of a hand wheel and gut, giving an easy, silent motion, considered by many to be superior to the Swiss wheel and pinion action. On the other hand, the Swiss slide rest is better than the English. The Swiss get the best effect by forming their dovetails at an angle of 45°, while ours are a great deal too upright. The Swiss slide rest main screw is finer cut, and has a projecting end with an adjustable stop for precise turning. The slides, too, are all adjustable, and the main slide is turned upside down, so that it can be capped to keep the dirt out. The Swiss meeting centre is also a great convenience in centring work from the front when the pump centre is not available. But perhaps the greatest convenience of all in the Swiss tool is an arrangement on the slide rest whereby the cutter can be raised or lowered.

Fig. 1 is a good example of a Swiss mandril with wheel and pinion driving action. There are square runners fitting the back centre head, for holding gravers or other tools required in jewel setting and similar operations.

Fig. 2 seems to combine the advantages of the English and Swiss styles. The head is crossed out, and the driving wheel most conveniently placed. I have not thought it worth while to again reproduce the meeting centre and hand rest.

In choosing a mandril, select one with a steel bar and crossed out face plate. See that the slide is free from end shake throughout its entire range, and then observe if the oil on the dovetail reproduces the pitch of the screw ; if it does it is a sure sign

SWISS MANDRIL.—Fig. 1.

that the screw is not true, but " drunk " as it is termed. See that the pump centre is free from shake and works easily in and out. Remove the dogs and pump centre. Examine the mandril to see that there is no end shake ; put on the slide rest, and, having cut a peg to a chisel-shaped point, put it in the tool post

Screw the rest out so that the point of the peg just touches the surface of the face plate near the edge. Now if the mandril is

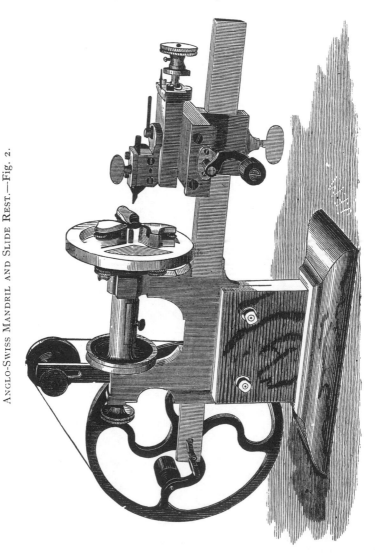

Anglo-Swiss Mandril and Slide Rest.—Fig. 2.

rotated it can be seen with an eye-glass whether the face plate is perfectly true in flat.

To test the dogs take them to pieces and, with a vernier gauge or a notch cut in a piece of brass, try all three, and if they are all exactly the same height alike they may be passed as correct. To test the truth of the hole for the pump centre get a long square box-wood stick small enough to enter the hole, round the corner off one end, and put on the hand rest ; bring it up close to the hole and rest the peg on it so that it touches the hole *only* at the *point*. If the hole is not true the end of the stick will rise and fall according to the amount of error. Wipe out the hole and replace the pump centre, the point of which can be tested by a peg in the way just described.

The meeting centre, as bought with the mandril, is seldom in a line with the pump centre. The top part of the head through which the meeting centre passes is made adjustable by screws, so that it can be corrected if necessary. To adjust it, put a piece of plate brass in the dogs, catch the centre with a graver and drill a very small hole in it, try it with a peg and see that it is perfectly true, put on the upright head and fix it in its proper position (there is usually a mark across the bar to indicate the proper position for it ; if there is no mark, place it about $3\frac{1}{2}$ inches from the front bearing of the mandril), bring the point of the meeting centre up to the hole you have drilled, and see if it goes correctly into it ; if so, loosen the head and draw it back so that with the meeting centre pushed completely through the head, the point will just reach the piece of brass in the dogs ; if it again goes correctly into the hole you may assume that it is right. As a final test, shift the piece of brass in the dogs a little and with the head in its first position, viz., near the face plate, push the centre gently through until it just touches the brass, at the same time rotating the mandril. If it makes a fine dot, remove the head to the second position and again try it ; if it again makes a dead centre, it may be passed with confidence. If the centre is not in a line with the pump centre, it can be adjusted by the screws at the side of the head to make it parallel with the bar. If too high the top part can be lowered by a stroke or two of a smooth flat file, and if too low it can be packed up to the required height by layers of tinfoil ; a moment's reflection will show the workman in which direction the alteration is required.

To use the tool make a good pointed cutter, leave it quite hard and gloss it, buff the extreme sharpness off the cutting edges, and proceed to turn as large a sink as possible on a flat piece of brass ; having turned the sink, make the cutter take the slightest possible cut, rotating the mandril wheel as fast as convenient, while you screw the cutter very slowly across the face of the sink. This should give good work without polishing and as a further trial rub it with a piece of flattened slate pencil wetted on

s

the tongue, when a rub or two ought to take all cutter marks clean out. If the screw pitch is repeated on the sink it indicates a bad rest. By varying the shape of the cutter, what at first sight seem inaccessible places may be got at. It is always well to consider if it is possible to shape a cutter to compass a difficulty, and with a stock of cutters it is surprising what a number of things can be done with a mandril. As a rule, pointed cutters suit watch-makers best. Watchmakers' tools are not so solid and the fixings so rigid and free from vibration as to make it practicable to turn with a flat face cutter, but a partially flat cutter like the sketch turns very clean ; the flat of the cutter must always be level with the centre of the mandril, and when it wants sharpening whet the sides, merely polishing the top. The front of the cutter is often by beginners formed to make too acute an angle with the top. The front should only be ground back sufficiently to just clear the face of the work.

Steel of a much smaller size may be used for the cutters, if they are held in a cutter-holder, such as is shown under the head of Lathes, p. 224. The cutter-holder swivels so that the cutter may be instantly adjusted to any required angle. Cutter bars are rapidly displacing the use of solid tools for lathes.

Marble Clock Cases.—[*Cabinets d'horloges en marbre.—Das Marmorgehäuse.*]—To polish marble clock cases, make a thin paste of best beeswax and spirits of turpentine, clean the case well from dust, etc., then slightly cover it with the paste, and with a handful of clean cotton wool rub it well, using abundant friction ; finish off with a clean old linen rag, which will produce a brilliant block polish. For light-coloured marble cases mix quicklime with strong soda-water, and thickly coat the marble. Clean off after twenty-four hours, and polish well with fine putty powder.

Mending Marble Clock Cases.—Plaster of Paris may be used, but it is better, especially if the mended part is visible, to soak the plaster of Paris in saturated solution of alum, and then bake it. It is used with water, may be mixed with any desired colouring material, and will take a high polish. Lime and white of egg make the best cement for closely fitting surfaces, but it requires using very quickly, as it soon sets. Marble case makers use a cement composed of Russian tallow, brickdust, and resin melted together, and it sets as hard as stone at ordinary tempera-tures.

Marine Chronometer.—[*Chronomètre de marine.—Das See Chronometer.*]—A chronometer for use on board ship. It is sus-pended on gymbals and enclosed in a wooden box. Greenwich mean time is of the utmost value to the navigator for readily determining the longitude of his ship. Observations of the sun are taken by means of a sextant (see Sextant), and as the time of apparent noon at Greenwich is known, it is clear that if the apparent noon at any other place is noted, the difference in time between the two will give the longitude of the place of observation, each degree of longitude being equal to four minutes of time. Every ship belonging to the Royal Navy carries three chronometers for comparison. Deck or assistant watches are used to note the observations, and these are checked by the chronometers, which are not removed from the chronometer room.

The following particulars of Marine Chronometers are given to me by Mr. T. Hewitt : —

Description of Chronometers.	Frame.		Barrel Cubical contents	Pull of Mainspring Inch-grains.	Balance.					Balance Spring, Bending Moment, Inch-grains.
	Diam. inches	Hgt. ins.			Diam. inches	Weight grains.	Mo. of inertia	Rad. of Gyration.	Arc of Vib.	
8 day	3·5	1·7	2·71	32666	1·15	143·3	·1064	·537	450°	16·80
2 day modern	3·275	·9	1·12	23556	1·273	170	·152	·586	470°	24·32
2 day 30 years ago.	2·875	8·75	·69	12486	1·115	89·7	·0615	·517	480°	9·92

Length of detent from shoulder to face of locking, 8 day ·53, 2 day ·54, old 2 day ·52.

Strength of detent deflected through an angle of 12° or ·2 radian, 8 day 14 grs., 2 day 15 grs., old 2 day 10 grs.

Diameter of escape wheel, 8 day ·50, 2 day ·54, old 2 day ·48.

Diameter of impulse, 8 day ·26, 2 day ·28, old 2 day ·25.

Radius of unlocking, 8 day ·0565, 2 day ·06, old 2 day ·045.

Angle of impulse, 8 day 43° 16·, 2 day 44° 18·, old 2 day 16·.

Numbers of wheels and pinions will be found under the head of " Train."

After chronometers have been at sea a cobwebby film collects on the bright steel work, quite marring its appearance. Kullberg found that this may be readily removed by an application of spirits of ammonia ; others recommend a solution of caustic potash in water for the purpose.

Mass.—[*Masse.—Die Masse.*]—The amount of matter contained in a body. It is often convenient to speak of the mass of a body in terms of its weight, and though as long as the body is the same distance from the centre of the earth mass and weight are convertible terms, it is necessary to distinguish between mass and weight, because while the mass of any body is the same in whatever part of the world it may be, its weight would vary in different latitudes. Mass at any particular place equals the weight divided by the accelerative force of gravity at that place, thus :—

$$m = \frac{w}{g}$$

Massey, Edward.—A Staffordshire watchmaker. Inventor of a form of lever escapement called the crank roller, in which the impulse pin is projected beyond the periphery of the roller something like the finger in the going-barrel stop work. Contact of the extremities of the lever with the edge of the roller formed the safety action. Massey also invented (Patent No. 3854, November, 1814) keyless winding for watches which worked by a pumping action of the pendant, and many other contrivances connected with watches and clocks. Mr. Massey died May 10, 1852, aged 82, and was buried at St. John's, Duncan Terrace, Islington.

Matted.—[*Mat.—Die Körnung.*]—The granular surface formed on watch plates and wheels prior to gilding is spoken of indifferently as matted or frosted. In English work this surface is obtained by immersing the parts in diluted sulphuric acid. The Swiss silver the work first, and then mat it by scratch brushing. (See Frosting.)

Mean Time.—[*Temps moyen.—Die mittlere Zeit.*]—The time shown by clocks generally. The ordinary day of 24 hours is the average of all the solar days in a year. (See Time.)

Meridian Dial.—[*Cadran méridien.—Mittagslinie Sonnenuhr.*] —A dial for showing when the sun is on the meridian. The following are nearly Ferguson's concise instructions for tracing a meridian line. " Make four or five concentric circles a quarter of an inch from one another, on a flat stone, and let the outmost circle be but little less than the stone will contain. Fix a pin perpendicularly in the centre and of such length that its whole shadow may fall within the inmost circle for at least four hours in the middle of the day. The pin ought to be about an eighth of an inch thick, with 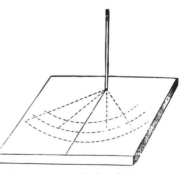 a round blunt point. The stone being set exactly level, in a place where the sun shines, suppose from eight in the morning till four in the afternoon, about which hours the end of the shadow should fall without all the circles ; watch the times in the forenoon, when the extremity of the shortening shadow just touches the several circles and there make marks. Then, in the afternoon of the same day, watch the lengthening shadow, and where its end touches the several circles in going over them make marks also. With a pair of compasses find exactly the middle points between the two marks on any circle, and draw a straight line from the the centre to that point which line will be covered at noon by the shadow of a small upright wire, which should be put in place of the pin. The reason for drawing several circles is, that in case one part of the day should prove clear, and the other part somewhat cloudy, if you miss the time when the point of the shadow should touch one circle, you may perhaps catch it in touching another.

Mercer's Balance.—Chronometers fitted with this balance have been very successful at the Greenwich trials. The balance itself is of the ordinary kind, the special feature being the auxiliary, which consists of a laminated arm of brass and steel fixed at one end to the central bar of the balance. The auxiliary may be arranged to act in either extreme of temperature. For low temperatures the steel would be the outer of the two metals composing the arm, and for high temperatures it would be the inner one. The free end of the arm carries two screws, the weight and position of which may be varied as required. A banking screw tapped through the rim of balance serves to regulate the action of the auxiliary.

Micrometer. — *Micromètre.—Das Mikrometer.*] —A measuring gauge of extreme exactness.

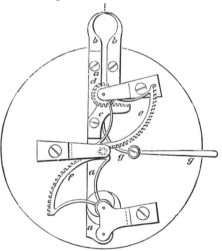

Fig. 1.

Grossman's.—The back of the dial of Grossman's micrometer gauge is shown in Fig. 1. The right hand limb *b* is fixed ; the other is the termination of a long lever *a* which carries a rack *c* pitched into a pinion fixed on the arbor of the rack *e*. This rack takes into the centre pinion, to which the index is fixed.

The rack f also takes into the centre pinion, which has a spiral spring round its arbor by which the jaws are kept closed and the index at zero. The lever g by pressing against the lever a opens the jaws. The dial may be divided so that each division shows either ·01 of a millimetre or ·0004 of an inch.

Horstmann's.—Fig. 2, which is the invention of Mr. G.

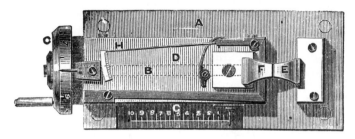

Fig. 2. Horstmann's.

Horstmann, measures to ·0001 of an inch. The plate A is supported on four feet. The jaw E is fixed. F is attached to the

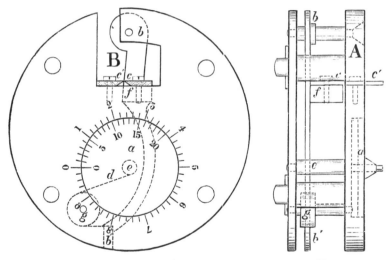

Fig. 3. Bonniksen's.　　　　Fig. 4.

slide D, which is actuated by the screw B, having 100 threads to an inch. The wheel C by which the screw is turned is divided into 100 equal parts, and each division consequently represents ·0001. To avoid the possibilty of delicate work being crushed between the jaws, F yields to pressure, and in doing so acts upon a bent lever which moves a pointer H along a graduated arc. When the undue pressure is removed, the pointer is brought back to zero and the jaw to its normal position by a bent spring as shown in the drawing. A standard inch is laid down at G, and an index attached to the slide registers the number of turns made by the handle.

Bonniksen's Micrometer.—By this simple gauge it is possible to measure to the $\frac{1}{10000}$ of an inch. Two plates A and B are rigidly connected by four stout pillars. A lever b b' turning on a fulcrum at b, carries on it a knife edge c ; another knife edge is attached to the top plate. A cord D is attached to the movable end of the lever b', the other end of the cord is attached and wound once round the arbor e. The arbor carries on it a circular disc a. In moving the lever from left to right, it will turn the disc in the same direction as the hands of a watch, and at the same time the two knife edges will separate. In the plate A, a sink is turned to receive the disc, and a balance spring which has to turn the dial and knife edges to their original position. The balance spring is not shown in the drawing, but it is placed underneath the disc, in the middle of the plate. On the plate round the edge of the disc are fifty equal divisions. These with the vernier on the disc indicate the one-thousandth of the circumference. The length of the lever from its turning centre to where the cord is attached is two inches. The distance from the turning centre of the lever to the knife edge $\frac{1}{2}$-inch ; therefore when the lever is moved $\frac{1}{10}$ inch, where cord is attached the knife edge has only moved $\frac{1}{10}$ inch. The circumference of the arbor e is made $\frac{1}{10}$ inch nearly (the thickness of cord makes it necessary to reduce the thickness of arbor to a little less than $\frac{1}{10}$ inch). When the lever is moved $\frac{1}{10}$ the disc has had to make one complete rotation, and the knife edges are separated $\frac{1}{10}$ inch ; and as the circumference is divided into 1000 parts, therefore $\frac{1}{10}$ is divided into 1000 parts, or 1 inch into 10,000. The plate A is left open in front of the two edges at B, so as to make it possible to measure a balance staff pivot.

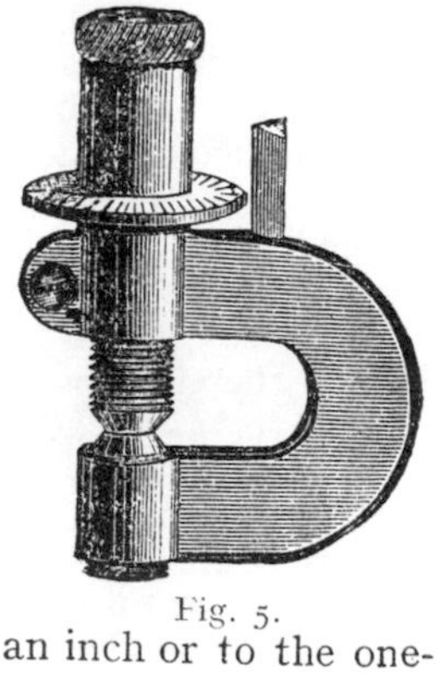

Fig. 5.

Mainspring Micrometer Gauge.— This, shown in Fig 5, is very handy for gauging the thickness of metals, etc. These gauges may be had to record measurements to the one-thousandth of an inch or to the one-hundreth of a millimetre.

Middle.—[*Carrure.—Der Gehäuserand.*]—The portion of a watch case to which the pendant, covers, and bezel are attached. The middle is made up of the band which occupies the centre and the edges on which the covers and bezel close.

Middle Temperature Error. [*Erreur de température moyenne.—Der Kompensationsfehler.*]—The element of irregularity known under this name is one of the most perplexing that trouble the maker of superior timekeepers. The object of applying a compensation balance to a watch or chronometer is that its vibrations shall be performed in the same time, notwithstanding that from changes of temperature the energy of the balance spring is varied. In the ordinary form of balance this is sought to be effected by causing the balance to contract with a rise of temperature when the balance spring is weaker, and to expand with a fall of temperature when the balance spring is stronger.

Airy demonstrated that the loss in heat from the weakening of a steel balance spring is uniformly in proportion to the increase of temperature. But the compensation balance fails to meet the temperature error exactly ; the rims expand a little too much with decrease of temperature, and with increase of temperature the contraction of the rims is insufficient ; consequently, a watch or chronometer can be correctly adjusted for temperature at two points only.

While the energy of the balance spring varies equally for equal increments and decrements of temperature, the number of vibrations made by the balance in a given time varies not inversely as the distance of its weights from the centre, but inversely as the SQUARE of the distance of the centre of gyration from the centre of motion. If a chronometer keeping mean time at 60° Fahr. were tried first at 30° Fahr. and then at 90° Fahr., and the compensation weight on the second trial approached the centre the same distance as they receded from it in the first trial, the chronometer might be expected to lose at 30° and gain at 90°. But the difficulty is actually, as already stated, too much expansion in cold, or too little contraction in heat. The action of the balance is complex. The weights do not move radially, and, with an increase of temperature, while the weights are moved inwards the central arm lengthens and the parts of the rim adjacent thereto move outwards.

However, as the adjustment can only be perfect at two points, chronometer makers generally arrange so that the compensation is right at the two extremes of temperature which the chronometer is likely to encounter, leaving the greater error at the middle temperature. The chronometer will then be found to gain at all temperatures within the extremes, and to lose if exposed to temperatures outside the extremes.

A marine chronometer is usually adjusted at 45° and 90°,

unless special adjustment is ordered to suit particularly hot or cold climates ; pocket watches at about 50° and 85°. In a range of 40° Fahr. there would be a middle temperature error of about 2 secs. in 24 hours with a steel balance spring. The amount of the middle temperature error cannot be absolutely predicated, for in low temperatures, when the balance is larger and the oil thicker, the arc of vibration is less than in high temperatures when the balance is smaller and the oil thin ; consequently its time of vibration is affected by the isochronism, or otherwise, of the balance spring and the action of centrifugal force on the balance. Advantage is sometimes taken of these circumstances to lessen the middle temperature error by leaving the piece fast in the short arcs.

To avoid middle temperature error in marine chronometers, various forms of compensation balances have been devised, and numberless additions or auxiliaries have been attached to the ordinary form of balance for the same purpose. Of these, some examples are given and will doubtless be of interest.

The late John Hartnup, director of the Liverpool Observatory, advocated the acceptance of the ordinary compensation balance, and proposed that navigators should be instructed to allow for the middle temperature error, which, after trial of 1000 chronometers, with steel balance springs, he found amounted to 1·5 sec. in 24 hours for a change of 15° above or below the temperatures at which the chronometers had been compensated. This can only be accepted as an approximation, because, as we have seen, the amount of the middle temperature error is to some extent dependent on other disturbing influences.

With palladium balance springs the middle temperature error is very sensibly less than with steel, which may be accounted for in the following way :—

There is a greater temperature error to be compensated with palladium than with steel springs, and the compensation weights with palladium have consequently to be moved more towards the free ends of the rim. With a rise of temperature the superior expansibility of the brass or outer metal of the rim not only carries the weights towards the centre, BUT, BY ELONGATING, CURLS THE RIM ; AND THE PATH OF THE WEIGHTS IS THEREFORE VARIED ACCORDING TO THEIR POSITION ON THE RIM. It happens that the movement of the weights at the point where the action of the palladium spring causes them to be placed, more nearly conforms to what is theoretically required than if they were farther back, as they would be with a steel spring. But the middle temperature error is not yet eliminated or provided for, and its amount, though reduced with a palladium spring, is even less constant than with steel.

Checking the outward movement of the rim in low temperatures, or causing a subsidiary weight to be carried inwards in high temperatures, may be taken to represent the two principles on which most auxiliaries are constructed.

Mill.—[*Meule.*—*Die Schleifscheibe.*]—A disc of steel or other material rotated in a lathe charged with some cutting or polishing medium. The article to be cut or polished is pressed against and moves over the surface of the mill. (See Diamond Mill.)

Millimetre.—[*Millimètre.*—*Das Millimeter.*]—A unit of measure used generally by French watchmakers.

Comparison of English and French Measures.

Millimetres	1	2	3	4	5	6	7	8	9	10
Inches.	·0303	·0787	·1181	·1574	·1968	·2362	·2756	·3149	·3543	·3937

NOTE.—10 Millimetres = 1 Centimetre ; 10 Centimetres = 1 Decimetre. A Millimetre equals the 1·25th of an inch nearly.

Scale of Centimetres and Millimetres.

Milling Cutter.—[*Fraise.*—*Die Fräse.*]—A circular cutter used for cutting the teeth of steel wheels, and also for finishing surfaces of various shapes where accuracy is required. The milling cutters, formed to correspond with the outline of the work to be finished, are serrated, so as to present a large number of cutting edges. Milling cutters have the advantage of retaining their sharpness for a considerable time, and they may be reset without lowering their temper by means of a small emery wheel rapidly rotated, and passed over the face of each cutting edge in succession. The emery wheel should be conical, so as to leave the cutting edge in advance. Mr. Curzon finds that milling cutters made from flat steel are more liable to distortion in hardening and fracture in use than others made from cheese-shaped pieces of steel cut from a round rod.

Minute Wheel.—[*Roue de renvoi.*—*Das Wechselrad.*]—The wheel driven by the cannon pinion. (See Motion Work.)

Minute Wheel Pin, or Stud.—[*Tenon de minuterie.*—*Der Wechselradstift.*]—The stud fixed to the plate on which the minute wheel revolves. (See Motion Work.)

Minute Wheel Pinion or " Nut."—[*Pignon de renvoi.*—*Das Wechseltrieb.*]—The pinion in the motion work of watches that drives the hour wheel. (See Motion Work.)

Mitre Wheels.—[*Roues coniques d'égale gradeur.*—*Gleichgrosse Räder mit konischer Zahnform.*]—Two bevelled wheels of the same size used for transmitting motion from one spindle to another at right angles to the first.

Modulus of Elasticity.—A constant, represented by E, which is used for ascertaining what proportion of its length material is strained when subjected to stress. If the body is stretched, the strain is a lengthening, and if it is compressed, a shortening of its original dimensions. In Young's formula, which is usually accepted, the stress in pounds per square inch of section, divided by E, gives the strain ; E being the force in lbs. that would stretch a rod one square inch in section to twice its original length, supposing its elasticity to remain perfect all the time. Young gives 29,000,000 as E for steel, but Mr. Robert Gardner considers this too high for the average quality of steel used in balance springs, and places it at 23,000,000.

Molyneux, Robert.—A chronometer maker who carried on business in King Street, Holborn (now called Southampton Row). Inventor of a compensation balance with auxiliary. (Patent No. 8418, March, 1840.) (See Auxiliary.)

Moment of Elasticity of a spring is its power of resistance. It varies directly as the modulus of elasticity of the material, and as the breadth and the cube of the thickness of the spring when its section is rectangular. $Mo = E\frac{bt3}{12}$ is a usual formula, E representing the modulus of elasticity, b the breadth, and t the thickness of the spring.

The moment of elasticity must not be confounded with the bending moment. The bending moment is a measure of the resistance a spring offers to bending, and of the amount of bending which has been produced, which varies directly as the angle wound through, and inversely as the length of the spring. $M = \frac{Ebt3A}{L12}$ is the usual formula for ascertaining the bending moment, E being the modulus of elasticity, b the breadth, t the thickness, and L the length of the spring, and A the angle through which it is wound.

This formula also determines the value of the force which has produced the bending, for if the forces are in equilibrium, the moment of the resisting force must be exactly equal to the moment of the bending force.

Moment of Inertia.—The resistance to change of velocity offered by a rotating or revolving body. The moment of inertia, which is generally represented by I, varies directly as the mass and as the square of the radius of gyration of the body.

Momentum.—[*Quantité de mouvement.—Momentum.*]—The quantity of motion in a body ; its mass multiplied by its velocity. (Momentum is occasionally confounded with the measure of the work stored up in a moving body, its kinetic energy, which varies as the mass and the square of the velocity.)

Motion Work.—[*Minuterie.—Das Zeigerwerk.*]—The wheels used for causing the hour hand to travel twelve times slower than the minute hand. The centre arbor of a watch rotates once in an hour, and on it is fixed friction tight a pinion with a long boss or pipe called the cannon pinion. The cannon pinion drives the minute wheel. which, together with the minute wheel pinion attached to it, runs loosely on a stud fixed to the plate of the watch. The last-named pinion drives the hour wheel, which has a short pipe, and runs loosely on the pipe of the cannon pinion. The minute hand is fixed to the pipe of the cannon pinion, and the hour hand to the pipe or body of the hour wheel.

The product obtained by multiplying together the number of teeth in the minute and hour wheels must be 12 times the product obtained by multiplying together the teeth in the cannon and minute wheel pinions.

Below are given 12 sets of motion work. If any other numbers are desired, the proportion is very easy to calculate. As before stated, the product obtained by multiplying together the minute and hour wheels must be twelve times that of the cannon and minute wheel pinions. Applying this to the first set in the table, we have 12 × 10 = 120, and 40 × 36 = 1.440, which

TRAINS FOR MOTION WORK.											
Minute Wheel Pinions	10	10	10	12	12	12	12	14	14	14	16	16
Cannon Pinions	12	14	16	12	14	16	18	14	16	18	16	18
Hour Wheels ...	40	42	48	48	48	48	54	56	56	56	64	64
Minute Wheels ...	36	40	40	36	42	48	48	42	48	54	48	54

is 12 times 120. Of course, the number of wheels or pinions given in the table may be transposed. For instance, if cannon is given as 12, and wheel pinion as 10, the cannon may be 10, and the wheel pinion 12. But there is an advantage in making the cannon the larger pinion which is often overlooked. With a small cannon pinion the oil is almost invariably drawn away from the centre wheel lower pivot. If a larger cannon is used, a square sink or recess may be cut in it, which will effectually cure the evil referred to. In keyless watches it is especially desirable to have the cannon pinion large so as to get an easy action for setting hands. In some instances the cannon pinion and the minute wheel are made of the same size as each other for this purpose, and then of course the minute wheel pinion and the hour wheel bear to each other a proportion of 12 to 1. (See Cannon Pinion.)

For 24–hour motion work suitable numbers would be 8, 10, 40, 48, or 8, 12, 48, 48.

In House Clocks the cannon and minute wheel have each the same number of teeth for the convenience of letting off the striking work by means of the minute wheel, which thus turns in an hour : consequently the hour wheel and minute nut bear a proportion to each other of 12 to 1. Generally a pinion of 6 and a wheel of 72 are used. The cannon wheel is loose on the centre arbor, and behind it is a bow-shaped spring, which at its centre is squared on to the arbor so that its ends bear on the face of the cannon wheel. A tapered pin passes through the extremity of the centre arbor and between this pin and the cannon is a washer which keeps the cannon wheel pressed against the spring sufficiently hard to drive the motion work, while the adhesion of the spring is easily overcome when the clock is set to time by turning round the minute hand with the finger.

In French Mantel Clocks the body of the cannon is thinned at one part and punched in to fit friction tight to the centre arbor.

In Turret Clocks, where the striking is not discharged by the motion work, the cannon pinion is tight on the centre arbor and the arrangement is similar to the motion work of watches.

Movement. — [*Mouvement. — Das Werk der Uhr.*] — The mechanism of a watch or clock without the dial or case. The plates and train of a watch without the escapement are also spoken of as the movement. (See " Train," " Quarter Clock," " Regulator," and " Watch Movement.")

Movement-covering Glass and Tray.— The sketch shows an improved kind of tray for the reception of the parts of a watch movement. It is divided into compartments, so that the pieces of the escapement, barrel work, &c., may each be kept separate and readily found.

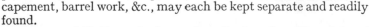

Movement-Holder.—A frame for clipping the pillar-plate of

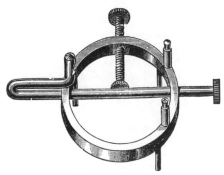

Movement Holder.

a watch when putting the watch together. There is no danger of soiling the plate by handling it, and as the holder with the plate in it may rest on the bench, both hands are at liberty to proceed with the work. The pattern shown in Fig. 1 is a good one. The movement is held between the two studs and the notch of the sliding piece which is firmly held in position by the binding screw. On page 35 is shown an adjustable holder capable of being turned in any desired way for position timing.

Mudge, Thomas.—Born 1715, died 1794. He was apprenticed to the celebrated George Graham, and from 1750 to 1766 he carried on business in Fleet Street, at first in partnership with Dutton, another apprentice of Graham. In 1771 Mudge removed to Plymouth. He invented the lever escapement, which he applied to a watch for Queen Charlotte, about 1765. This identical escapement is now in the possession of the Clockmakers Company. It is analogous in its action to the present form of double roller escapement, with two exceptions : the locking faces of the pallets are curves struck from the centre of motion, so that the pallets cannot be drawn into the wheel by pressure from the teeth ; and the impulse pin is divided, for the purpose of ensuring the safety action after the finger enters the crescent, and before the impulse pin is fairly in the notch, a result now attained very simply by having horns to the lever. Mudge devoted the best part of his life to the improvement of the marine chronometer. Mudge's chronometers are admirable as specimens of fine work and correct proportion of details ; but though clearly a man of inventive genius, he unfortunately clung to the principles on which Harrison's timekeeper was constructed, and allowed Earnshaw and Arnold to solve the problem by the introduction of the spring detent escapement.

Musical Box.—[*Boîte à musique.—Das Spielwerk.*]—As nearly every country watchmaker is at some time or another called on to repair these instruments, a few hints thereon will not be out of place. It may be premised that if a very large number of the pins on the cylinder are broken, the box had better be sent to an expert. But, assuming the job to be undertaken, it will be prudent first of all to remove the comb or key-plate. Then let down the mainspring and see that the driving mechanism runs well, and that the cylinder, though free, has no end or side shake. The fly depth is important, for unless this runs smoothly and easily the box will stop.

If there are one or more keys missing they may now be reinstated. At the point where the new key is to be toothed in, file a dove-tailed notch in the key-plate like the sketch. Then file up a key similar to the adjacent ones, but rather full at the point and with a heel to fit into the notch. Harden the key and temper it by boiling in oil. Drive it tightly into position, and to make it secure, slightly rivet it or run a little solder into the joint by heating the spot with a blow-pipe or heavy soldering bit. Heat the comb as little as possible, and confine the heat to the place under repair. Now the key may be tuned, leaving it half a tone too high ; for it is easier to lower than raise, and the damping spring will bring it down the half-tone or nearly so. Keys are

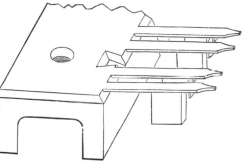

lowered in tone by weighting them with lead near the point, and raised by thinning a little on the underside behind the lead. To get at the underside to file it, have a rectangular brass stake as wide as the key, and with a little ledge as shown on one side, hardly so high as the key is thick. Rest the top of the key to be filed on this stake, holding the comb in the hand, so that there is enough weight resting on the stake to elevate that key above the rest, and then it can be filed in comfort, the ledge offering the requisite resistance to the file. If the key is near the middle of the comb, it may be necessary (to use a file with an *over* handle to it. When getting the point of the tooth to length, continually apply a glass surface plate or straight-edge

along the tips, for it is essential that all the tips should be exactly in line.

If only the tip of a key is missing, it will not be necessary to replace a whole key, but merely to file a slit in the stump and let in a new point which may be fixed by soldering. The tip may be let down a little by means of a blow-pipe to enable the file to cut, but care should most be taken not to soften the bending part of the key.

It is sometimes necessary to elevate or depress a key, or to make it point a little to the right or the left. Place the top of the comb on a steel stake or anvil, face downwards, and, to elevate a key, tap the under-surface gently with the hardened pane of a hammer so as to stretch it. In the same way, if a key is to be turned to the right, stretch the left edge. If a key is to be depressed, an expert will bend it with a smart blow of the hammer on the middle of the underside while it rests on the anvil, but this is risky and will often result in a broken key. It is better to stretch the upper surface of the key with light taps, even though the marks show.

Now just put the key-plate in position and see that the points of the keys are exactly in a line with the pins in the barrel, and if not, the cylinder bearings must be bent till this is right. Then see to the damping springs and supply new ones where necessary, fixing them quite tight with the old pins. It will be observed that the thicker keys for the lower tones have heavier damping springs than the intermediate keys, while the highest notes are without dampers. Occasionally some of the notes above the springs have dampers of quill. These are fixed with shellac dissolved in spirits of wine. The keyboard points not to the centre of the barrel, but above it, the proper elevation being about 15° from the horizontal line. The free end of the damping spring should be as close as possible to the point of the key without touching it, shaped as shown in the figure appended ; so that the pin in the barrel touches the spring first at about the point indicated.

If the damping spring is too thin, it will fail to stop the vibrations of the key soon enough, and if too thick will create a buzzing noise just as the key leaves the pin. A spring may be thick enough and yet fail to stop the vibrations, because it is not forward enough. The springs will be readily bent to position with a pair of tweezers.

To observe the action of the springs, place the key-plate in position and note first that it is the right height, as indicated by the dots on the cylinder. The shortest key should be on a level with the dots, and the longest one, which has more movement, about half a dot below it. To alter the height, the bearings of the cylinder may be raised or lowered as required.

To see if the key-plate is at the right distance from the pins, let the cylinder rotate slowly, and if the keys are not drawn up enough there will be but little sound, and the comb must be set a little closer to the cylinder by bending the feet of the base. If the sound is harsh and the dampers fail to stop the vibrations, the key-plate is already too close. If in playing a tune the notes at one end are produced too late, it shows that end of the key-plate is too close to the cylinder.

Let the box run through all its tunes, and if at any tune the pins do not pass exactly in the centre of the keys, the star wheel for that tune must be corrected. The drop from the highest to the lowest step of the snail in time causes an indentation in the latter, which may be filled by screwing into the face of the snail a piece of tempered steel to receive the blow of the pin. Any of the pins in the cylinder that are out of shape will be noted as the tunes run through, and carefully bent as required. New pins are formed with a pivot which fits tightly into the hole in the cylinder ; the pin is driven in up to a shoulder, the part projecting being rather larger in diameter than the pivot.

Sometimes a buzzing noise is observed while the box is playing. This is generally caused by something loose. To discover it, sound each key by striking it with a suitable pointer till the buzzing is heard ; then continue sounding that note while placing the hand on different likely parts of the box and mechanism till the buzzing is arrested, when an examination of the part will probably reveal a screw that requires tightening, or even the want of a drop of oil, which has been known to cause this disagreeable sound. In connection with these notes on musical boxes, I have to acknowledge the courtesy of Messrs. C. Paillard and Co., who have readily answered all my inquiries on the subject.

Name Bar.—[*Barrette à nom.—Die Namenplatte.*]—The bar carrying the upper end of a watch barrel arbor.

Nippers.--[*Pince à couper.—Die Beisszange.*]—A pair of jointed steel levers with tempered edges at one extremity for cutting wire.

Nippers are now to be had with removable jaws, a convenience appreciated if either of the cutting edges should become chipped, it is remarkable how very many nippers one sees with chipped edges ; this sometimes arises from bad or imperfectly tempered steel, although it is often caused by abuse, or improper use, of the nippers. It may be occasionally a temptation to cut larger wire than the tool is calculated to sever, but nothing can excuse the way that many people, and among them are some who pass for experienced workmen, handle their nippers. The tool and the wire should be held perfectly still when the cutting pressure is applied, but a majority of the accidents to the jaws are due to a wriggling twist given to the nippers at the moment of cutting, a movement which those who adopt it, erroneously suppose makes the cutting easier.

Non-Magnetizable Watch.—[*Montre non magnétisable.—Die antimagnetische Taschenuhr.*]—A watch in which the quick moving parts are made of some other metal than steel or iron. Aluminium bronze, which combines strength with lightness, is particularly suited for the lever and pallets. The balance spring may be of palladium alloy, or possibly of a nickel steel alloy. The steel balance staff, pallet staff, and escape pinion may be retained, their circumferential velocity being small. For the ordinary run of watches, a plain gold or brass balance is often used ; in view of its lower co-efficient of expansion the nickel

T

steel alloy may be preferentially employed. Many attempts have been made to devise a compensation balance in which the use of steel is dispensed with. Arnold and Dent used some of platinum and silver, which compensated fairly well, but were lacking in rigidity. Mons. Paillard has achieved considerable success with palladium, alloyed with silver, copper, and other metals. In some instances he appears to have used a palladium alloy for the inner part, and brass for the outer part of the rim, and in others to have formed both laminæ of different alloys of palladium. Good results have also been obtained by the Waltham Watch Company, but the nature of their alloy is not divulged. Excellent but costly balances in which the inner part was an alloy of platinum and iridium have been made. Steel alloyed with about 24 per cent. of manganese has also been used for the inner ring. Both the platinum, iridium and the manganese steel alloys are, I believe, very intractable under the cutters. Compensation balances of brass and nickel steel alloy have recently been used with success. If the springs are of nickel-steel alloy, requiring but little compensation, these balances are cut in the middle of the rims.

Oil.—[*Huile.—Das Oel.*]—The lubricants most in use for watches and chronometers are porpoise, olive, sperm, and neats-foot oil. Two or more of these are often mixed together, with the addition generally of a little mineral oil. The following processes, or some of them, are usually adopted for refining :—(1st) The removal of the solid portions when congealed, the operation being repeated until oil is obtained that will not readily freeze. (2nd) The stearine is removed by placing strips of lead in the oil, which is continued until it ceases to precipitate. (3rd) The resin and mucilage are taken out by washing with alcohol and afterwards with water, and the oil is finally filtered through animal charcoal. Latterly vaseline has found favour as a lubricant for the keyless work of watches, and will probably find more extended use. Oil should be applied sparingly. The effect of giving too much oil is that it is drawn over the plate, leaving the pivot dry.

Oiler.—[*Cheville à huile.—Der Oelgeber.*]—A wire instrument by which oil is conveyed to pivot holes. A pivot drill is a favourite medium, but a good form is a double-ended brass wire having a couple of light collars near its middle ; they prevent the ends touching the bench and so licking up the dust. One end is flattened and is the oiler proper, the other end is pointed, and is used to follow the oiler in holes with end stones, to ensure that the lubricant really reaches the end of the hole. In some of the American factories an ordinary steel pen is now used as an oiler.

Oil sink.—[*Huilier.—Die Oelsenkung.*]—A cavity turned in watch and clock plates around the pivot holes. Experience has shown that when the oil sink in chronometers and clocks where the plates are not gilt is thoroughly well polished, not only is the oil drawn to the pivot more freely, but it is less decomposed by contact with the metal than when the sinks are rougher. Oil sinks should be deep and small in diameter rather than shallow and wide.

Oilstone.—[*Pierre à huile.—Der Oelstein.*]—A stone upon which cutting tools are rubbed to give them a fine edge. Oil or some other lubricant is always used with the oilstone.

An oilstone thoroughly saturated with oil is often cast aside, but if it is soaked in benzine for two or three days it will be as good as ever. The ordinary animal and vegetable oils are not so suitable for use with the oilstone as petroleum, especially for setting small tools. A mixture of glycerine and alcohol is even better than petroleum **for** watchmakers' tools, or glycerine alone may be used. Glycerine has the advantage of neither evaporating nor clogging as oil is apt to do.

Oilstone Dust.—[*Pierre à huilepilée.—Das Oelsteinpulver.*]— Powdered oilstone used with oil for smoothing watch pivots, frosting keyless wheels, etc.

Oven.—[*Etuve.—Der Ofen für Temperatur Reglage.*]—A heated chamber in which watches and chronometers are placed when they are being adjusted for varying temperatures. An oven for marine chronometers is generally a square box of sheet copper or stout zinc with one or two shelves. There is a jacket or casing round it filled with water so as to keep the temperature uniform. The jacket may with advantage be covered with flannel or felt, except the bottom, which is slightly domed and heated by gas. One of the sides is formed into a door having double panes of glass let in so as to leave an air space between them. The oven is furnished with a self-registering maximum and minimum thermometer.

Hearson's Oven, shown in Fig. 1, is a good example taste- fully arranged as a fitting for a watchmaker's shop. A is the water jacket ; B and D plugs for filling and emptying ; B gas tap ; F shelves of coarsely woven wire on which the watches E are placed. The " thermostat " C is a very sensitive device on the principle of the aneroid, for keeping the temperature con- stant. It is a brass capsule, two inches square and quite flat, suspended from a tube running through the water jacket ; this is filled with ether ; when the ether boils the sides of the capsule bulge out, and by means of a rod (o), passing through the tube the gas is turned off at the tap. In this way a constant temperature of about 85° F. is maintained. There is a small bye pass for enough gas to keep the flame from being entirely extinguished. N is a thermometer.

W. G. Schoof describes a very cheap way of making an oven.
To the middle of a thin iron plate (A Fig. 2)
he fastens an iron tube (B), long enough to
go right through a wooden box, of the height
the oven is to be, the iron plate being under
the bottom of the box. The front of the
box is formed into a door with double panes
of glass as just described. A very small
flame from a gas or oil lamp suffices to
keep the highest temperature required for
compensation adjustment. The one objection to the use of iron in the construction of
an oven, is its liability, after hammering or
rolling, to retain magnetism.

A very handy form of oven large enough
for watches is shown in Fig. 3. There is
an outer ring of thin sheet copper or zinc.
18 inches high and 12 inches diameter, with
a horizontal partition 6 inches from the
bottom, and with three legs to raise it from
the floor. Inside is a cylindrical copper or
zinc box as shown by the dotted lines, 9
inches in diameter and 9 inches deep with a

Fig. 1.

rim to prevent it going too far down, and a hinged bezel with a glass.
A disc of wood is placed on the bottom to rest the watches upon.
The oven is heated by a paraffin lamp, and there are a number
of holes in the outer ring just below the partition as shown. Round the sides above the
holes are several layers of flannel with a
loose cover of the same for the top.

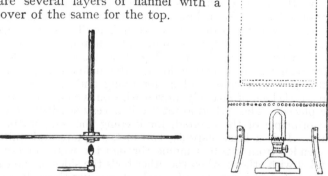

Fig. 2.

Fig. 3.

Gas Governor.—Fig. 1 is Kullberg's effective and ingenious
attachment to an oven heated by gas. The gas enters by the
right-hand pipe into the glass tube, and from there through a hole
in the left-hand pipe to the burner that heats the oven. It will be

observed that the left-hand pipe is prolonged to nearly the bottom of the glass tube to furnish an attachment for a laminated arm. This arm, made of a thick piece of mainspring and rather thicker brass outside, carries at its free end a weight, and just beyond a conical tit that acts as a stopper for the hole in the exit pipe. A nick is made in this hole to allow of the passage of sufficient gas to keep the burner just alight even when the stopper is pressed home. As the temperature rises the bending of the laminated arm causes the tit to approach the hole, and the reverse action takes place with a fall of temperature. By tilting the instrument to the right or left the weight retards or assists the action of the laminated arm so that any adjustment needed may be made. A screw with a large head nips the two pipes when the proper inclination has been obtained. There are also a pointer and circular scale for denoting the angle of inclination. The two pipes are of brass, but they may have flexible tube connections to the gas supply and burner respectively to allow of the tilting of the governor, the lower part of which is of course immersed in the oven.

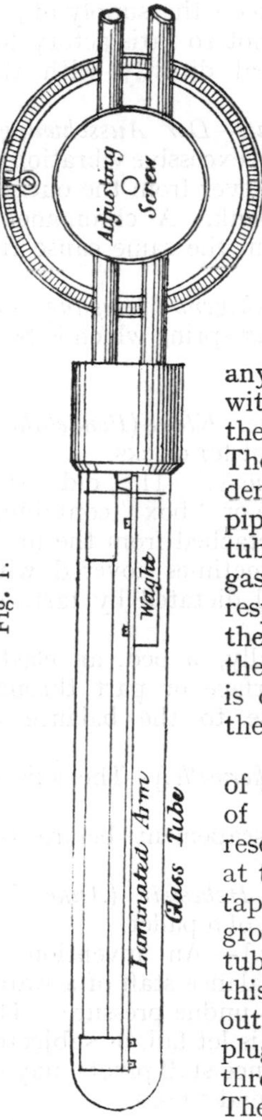

Fig. 2 is another form of a regulator, consisting of a glass tube with a reservoir of mercury (c) at the bottom. A hollow tapered glass plug is ground into the top of the tube. The gas enters this plug at A, and passes out of the bottom of the plug into the tube, and through B to the burner. The volume of gas allowed to pass depends upon the distance between the bottom of the plug and the top of the mercurial column. The screws allow the temperature to be adjusted by enlarging or restricting the space for the mercury.

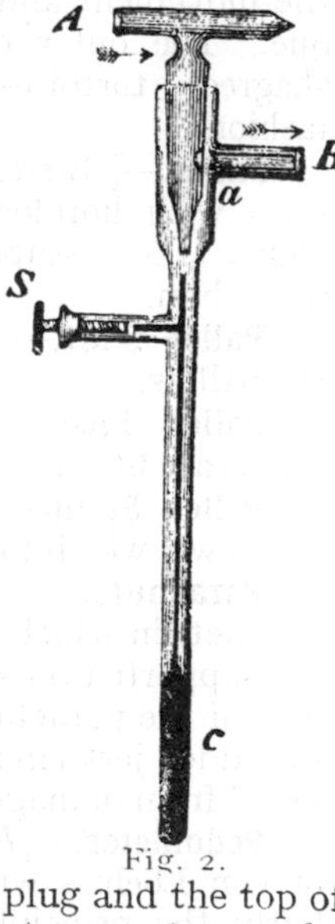

Fig. 1.

Fig. 2.

A minute hole A in the side of the plug allows enough gas to flow to just keep the flame alight irrespective of regulation.

There is another kind of gas governor in which a sensitive valve of Persian sheepskin or similar material is interposed in the supply pipe and weighted so as to keep the supply of gas at a desired pressure. This principle is not so satisfactory for watchmakers' ovens as those which deal directly with the temperature.

Overbanking.—[*Passage de renversement.*—*Der Ausschwung.*] —In a resilient lever escapement, when from excessive vibration of the balance the ruby pin pushes past the lever from the outside of it, the escapement is said to overbank. A chronometer escapement is said to overbank when from the same cause the escape wheel is unlocked a second time.

Overcoil.—[*Coudé.*—*Der anfwärts gebogene Umgang der Bréguet Spirale.*]—The last coil of a Bréguet spring which is bent over the body of the spring.

Oversprung.—(See Sprung Over.)

Pad.—[*Palette de recul.*—*Die Hebefläche. Klaue (Pendeluhr).*] —The pallet of the anchor recoil escapement for clocks.

Pair Case. —[*Boîte double.*—*Zweigehäusig.*]—The old style of casing watches with an inner metal case or " box " containing the movement and an outer case quite detached from the inner one. The outer case was of metal, sometimes covered with shagreen, tortoise-shell, or other material dictated by taste or fashion.

Pallet.—[*Ancre.*—*Die Palette.*]—Literally, a bed, an elastic term with horologists ; generally the surface or part through which the escape wheel gives impulse to the balance or pendulum.

Pallet Staff.—[*Tige d'ancre.*—*Die Ankerwelle.*]—The axis of the pallets.

Pallet Steel.—Pallets for a lever escapement before the jewels are fitted.

Pallet Stone.—[*Pierre de levée.*—*Der Hebestein (Anker).*]— The jewel which forms the rubbing surface of a pallet.

Parachute. — [*Parachute.* — *Parachûte.*] — An invention of Bréguet, in which the end stones of the balance staff of a watch are supported on springs so as to yield to undue pressure. The idea of the parachute is that if the watch is let fall, or subjected to sudden jerks in any other way, the balance staff pivots may be saved from damage by the yielding of the end stones.

Pedometer.—[*Pédomètre.*—*Der Schrittzähler.*]—The drawing of von Loehr's winding work (see Self-Winding) conveys very fairly the principle of the pedometer used for registering the number of paces walked. In the case of the pedometer the pawl is rigid and the last wheel of the train is planted in the centre of the movement. The centre wheel arbor carries a hand which traverses a dial on the other side of the plate. Instead of the lower banking pin there is a screw to adjust the amount of travel of the lever, and therefore the number of teeth of the ratchet wheel pushed forward at each vibration, to suit the stride of the wearer. By this means the divisions on the dial are made to represent approximately the number of miles walked.

Pedometer Watch.—[*Montre à secousses.—Die Perpetuale Taschenuhr.*]—(See Self-Winding.)

Peg Wood.—[*Chevilles de bois carré.—Das Putzholz.*]—Small sticks of wood, preferably dog-wood, used principally for cleaning out pivot holes of watches and clocks. The wood is pointed to enter the hole, and twirled round to remove the dirt. A fresh point is cut and the operation repeated till the point of the peg is not soiled when it is removed from the hole. The small size of watch work renders stocks of peg wood useful as pointers and for a variety of purposes. In fact, a watchmaker would be quite at a loss without a stock of peg wood. In speaking of peg wood many Swiss watchmakers use the term " Bois de fusain." Fusain wood is used for pegs, and also for making the sticks of charcoal for giving a circular grey surface to brass.

Pendant.—[*Pendant.—Der Gehäuseknopf.*]—A small neck and knob of metal, connecting the bow of a watch case with the band.

Pendulum.—[*Le Pendule.—Das Pendel.*]—The swinging body in a clock whose vibrations regulate the progress of the train.

A pendulum drawn aside from its point of rest and then released, is impelled by gravity to fall at once to its lowest possible point, and but for the momentum acquired in falling it would remain there at rest, but its momentum carries it up as far on the other side as it fell at first through the action of gravity. The length of the pendulum determines the time occupied in its vibrations ; a long pendulum moving slowly, and a short one quickly, because with a long pendulum the curve described at the centre of oscillation is flat, and with a short pendulum it is steep. The course of the pendulum bob attached to a long suspending rod, and that of one attached to a short suspending rod, may be compared to rolling a ball first down a slight decline, and then down a steep hill.

Gravity, a constant force, requires to act through four times the distance to impart twice the velocity to a body, and therefore a pendulum, to vibrate twice as fast as another, must describe a path four times as steep, and to attain this condition it must be but one-fourth the length of the first.

In estimating the time that a pendulum takes to vibrate, it must not be forgotten that after gravity has impelled it to its lowest position, it takes just as long to rise on the other side as it did to descend, so that each vibration takes twice as long as it might be expected to do if the influence of gravity on it, as it descended, were alone considered. Then as the pendulum swinging freely describes nearly a circular arc, the time in which a pendulum vibrates bears the same relation to the time in which a body would fall through a distance equal to half the length of the pendulum, as the circumference of a circle bears to its diameter.

Applying this to a pendulum 13 feet long : half its length is 6·5, and a body would fall through 6·5 feet in ·625 of a second nearly at the sea-level in London. And ·625 × 3·14159 gives as nearly as possible 2 seconds for the time in which such a pendulum would vibrate. As a matter of fact, the length of a 2-seconds pendulum at the sea-level in London is found to be 13·04 feet, or a trifle over 13 feet.

A clock has only to overcome the resistance of the air and the slight friction of the suspending spring in order to keep the pendulum going when it is once started ; and although a heavy pendulum requires a greater force to start it than a lighter one, the retardation afterwards, due to the extra resistance of the air on a larger surface, is insignificant. As inaccuracies in the clock work, currents of air, and other disturbing influences are likely to interfere with the regularity of a light pendulum, it is desirable always to use as heavy a pendulum as possible in reason.

The theoretical length in London of a seconds pendulum for mean solar time—that is the distance between the point of suspension and the centre of oscillation,* is approximately 39·14 inches : the length of pendulum for vibrating sidereal seconds in the same latitude is 38·93 inches.

As the force that gravity exerts on a body depends on the distance of the body from the centre of the earth, the length of a pendulum varies in different latitudes. A seconds pendulum is 39 inches long at the equator, and 39·206 inches at the poles.

* The centre of oscillation is that point in a vibrating body in which if all the matter composing the body were collected into it, the time of the vibrations would not be affected. In a straight bar suspended at one extremity, the centre of oscillation is at two-thirds of its length, and in a long cone suspended at the apex, at four-fifths of its length from the apex. From the irregular form of the pendulum, the position of its centre of oscillation is not easy to calculate, but it is always situated below the centre of gravity or centre of mass of the pendulum. In constructing a pendulum, it will be sufficiently near to assume the centre of oscillation to be coincident with the middle of the bob. The Board of Education has, I believe, recently decided to substitute " centre of motion " and " centre of percussion " for " centre of suspension," and " centre of oscillation." There seems to be no sufficient reason for the change, and I therefore venture to retain the original and admirably descriptive designations.

At   Rio Janeiro it is 39·01 inches.
At   Madras   ,,   39·02   ,,
At   New York   ,,   39·10   ,,
At   Paris   ,,   39·13   ,,
At   Edinburgh   ,,   39·15   ,,
At   Greenland   ,,   39·20   ,,

The usual formula for ascertaining the time of one vibration is $t = \pi \sqrt{\dfrac{l}{g}}$ in which π represents 3·14159, l the length of the pendulum in feet, and g the acceleration of a body exposed to the action of gravity (32·19 at the sea level in London).

The following is a simple rule often used for finding the length of a pendulum for a given number of vibrations, accepting the seconds pendulum of 39·14 inches as a datum :—

Let V = the given number of vibrations per minute.

Let L = the length required in inches.

Then $L = (375\cdot4 \div V)^2$

EXAMPLE.—The length is required of a pendulum to give 120 vibrations a minute.

375·4 ÷ 120 = 3·128, and 3·128 squared = 9·78, the length required.

If the length of a pendulum is given, the number of vibrations it would make a minute, may be ascertained as follows :—

Let L = the length given in inches.

Let V = the number of vibrations per minute.

Then $V = 375\cdot4 \div \sqrt{L}$.

NOTE.—Tables of Square Roots and Lengths of Pendulums are given at the end of the book.

Importance of Fixing.—Whatever kind of pendulum is used, it will not keep time unless it swings from a rigid attachment. Just as engineer clockmakers invariably make their escape wheels and other quick-moving parts too heavy, so many clockmakers seem afraid to put enough metal in their pendulum cocks and brackets, which have rarely enough base either. The beneficial effect of the heavy pendulum bobs, which it has been the custom recently to use for regulator and turret clocks, is often quite lost for want of sufficient fixing for the pendulum. For a regulator the pendulum should be supported on a cast-iron bracket with a base at least 10 inches square, bolted right through the back of the case, which should not be less than an inch and a quarter thick. For a turret clock a bracket of proportional size should be used, bolted to one of the main walls of the building, or, if attached to the clock frame, the rigid connection of the latter with the walls by means of girders or cantilevers should not be lost sight of. A timber frame fixing for a turret clock pendulum will never be satisfactory.

Length of Pendulums.—One-second pendulums are long enough for all but turret clocks, and longer than two seconds pendulums should not be used. The very long pendulums used by the old clockmakers for the turret clocks in order to get, as they expressed it, " a dominion over the clock," were very unwieldy and unsteady from the action of the wind and other causes. The requisite " dominion" is now obtained by making the bob heavier.

Circular Error.—The long and short vibrations of a free pendulum will only be isochronous if the path described at the centre of oscillation is a cycloid, which is a curve described by rolling a circle along a straight line. In the article on Wheels and Pinions is represented the method of describing an epicy-cloid for wheel teeth, which is done by rolling one circle on another. If the generating circle, instead of being rolled on another circle, were rolled along a straight edge, it would describe a cycloid. But a pendulum swung freely from a point travels through a circular path, and the long arcs are performed slower than the short ones. The divergence from the theoretical cycloid was of importance when the arc described was large, as it was of necessity with the verge escapement, and many devices were tried to lead the pendulum through a cycloid. With an arc of about 3° only, such as regulator pendulums describe now, the divergence is very small.

Escapement Error.—The kind of escapement used also affects the time of vibration. For instance, it is found that, while with the recoil escapement increased motive power and greater arc causes the clock to gain, the contrary effect is produced with the dead-beat escapement. The pendulum error may, therefore, be aggravated or neutralized by the escapement error. Again, with the Graham Escapement the pendulum requires less compensation than with a Gravity, because with increase of temperature the arc of vibration falls off with the Graham, the inference being that with heat the oil at the pivots becomes thinner, the train runs easier, and so the escape wheel teeth press harder on the locking faces of the pallets.

Barometric Error—With a decrease in the pressure of the air, and consequent fall of the barometer, the pendulum increases its arc of vibration ; with an increase in the pressure of the air, and consequent rise of the barometer, the pendulum diminishes its arc of vibration. In the Westminster clock the pendulum vibrates 2·75° on each side of zero, and Sir Edmund Beckett pointed out that with this large arc the circular error just compensates for the barometric error. Where the escapement is suitable, this is doubtless the best way of neutralizing the barometric error ; but it is not applicable to the dead beat, for extra run on the dead faces of the pallets or larger angle of impulse than usual is found to be detrimental as the oil thickens. Gardner's pendulum, which carries a column of mercury to counteract the effect of changes in the density of the air, is described on p. 305. The device adopted at Greenwich Observatory is shown on p. 53.

Rolling and Wobbling.—The path of a pendulum in plan should be a straight line. Any deviation from this will affect the regularity of its time-keeping. A want of squareness in the chops, or a twist in the suspension spring, will often cause rolling or wobbling. Many clockmakers fix the lower end of the spring with but one screw, so that the pendulum may hang plumb without danger of binding. If the pallet staff is not perfectly at right angles to the path of the pendulum, rolling may be caused by the oblique action of the crutch. This shows the necessity of care in adjusting the movement on the seat board in cased clocks, and is an argument in favour of attaching the pendulum of a turret clock to the frame of the movement instead of to a separate wall bracket.

Temperature Error.—With increase of temperature, a pendulum, in common with most other substances, lengthens, and the clock loses ; with decrease of temperature the contrary effect is produced. The object of the compensation pendulum is to meet the error arising from change of temperature by keeping the distance between the point of suspension and the centre of oscillation constant. As related on page 24 the researches of Dr. Guillaume have shown that an alloy of steel and nickel may be produced which is nearly insensible to temperature changes, and therefore would be admirably suited for pendulum rods. But with precision timekeepers, whatever material may be adopted for the rod, some provision for adjustable compensation is desirable.

With an Invar rod there may be a very short brass tube resting on the rating nut, the lower part of the bob being recessed to pass over the tube.

Zinc Tube Compensation.—Fig. 1 shows the construction of a zinc tube compensation pendulum, similar in principle to those employed in the Westminster clock and in the Standard sidereal clock at Greenwich. In the Greenwich clock the central rod and the outer tube are of steel and the bottom half of the hole in the bob is enlarged to bring the seat to the middle of the bob as in Fig. 1 ; the idea being that as the bob by reason of its greater bulk answers more slowly to changes of temperature than the other members of the pendulum, it should be neutral as far as the compensation is concerned. In the Westminster clock the rod and the outside tube are of iron, and the bottom surface of the bob rests upon the collar of the outer tube. Preference is now generally given to the centrally supported bob, and I will therefore give a detailed example of a seconds pendulum on this plan, but with an iron outer tube as a steel one is not so easy to obtain. The rod of steel should be ·3–inch diameter and 46 inches long from the top of the free part of the suspension spring to the bottom of the rod ; with a screw of 40 threads to the inch for a length of four inches from the bottom to receive the rating nut. On the top, to receive the suspension spring, is screwed a cap with a pin through it for safety. A drawn zinc tube, just large enough to slip over the rod, ·18 inch thick, and 28 inches long. This is a trifle long according to the table of expansions given at the end of this book ; still, it would be prudent to start with the tube rather long, for with a heavy bob resting on a zinc tube the tube in course of years shortens perceptibly. The Westminster clock was kept to mean time by adding small weights to the pendulum above the bob ; such a number of these weights accumulated that it has been found desirable to cast one large one instead. It was, I believe, concluded that the zinc tube had shortened and so lowered the bob. This tube rests in a square sink formed in a thick washer or collar just above the rating nut. Provision should be made to prevent this collar turning with the rating nut. The simplest way is to file the pendulum rod flat for about five inches up, and to put a pin through the collar just free of the flat part of the rod as in the appended sketch. Outside the zinc tube slipping freely over it is a thin iron tube, having at the top a cap recessed to fit the end of the zinc tube, with a hole in the centre of a size to slip freely over the central rod, and at the bottom an outer collar to form a seat for the bob. From the recessed surface of the cap to the upper surface of the bottom collar should be 24 inches. Slotted holes with semi-circular ends are sometimes made at intervals in the tube to allow the air to get freely to the zinc tube. There is a thin kind of iron tube, with a brazed joint sometimes used, but a long brazed joint seems to me to be objectionable, and I would prefer a solid drawn tube ; ¾-inch gas barrel would be right in the bore, and it could be turned on the outside, and have screw threads at the ends to receive the cap and collar.

The bob is of lead, 3·5 inches in diameter and 9 inches long, and weighs about 29 lbs. The upper part of the hole just passes over the iron tube. Half way up from the bottom the hole is enlarged so as to pass over the collar on the outside of the iron tube. By this means a seat is provided in the middle of the bob for it to rest on the collar.

If the bob is to be supported at the bottom, it may, perhaps, with advantage be smaller in diameter and longer, and the lengths of the tubes must be reconsidered. The co-efficient of expansion of lead and zinc is nearly the same, and with a lead bob supported at the bottom, the zinc tube should be shortened by half the length of the bob. Taking a bob three inches in diameter and ten inches long, then the zinc tube would be 23 inches instead of 28. Cast iron expands but little less than wrought, so that with a cast iron bob of any reasonable length, the length of the zinc tube would be 28 inches, whether the bob were supported at the bottom or in the middle.

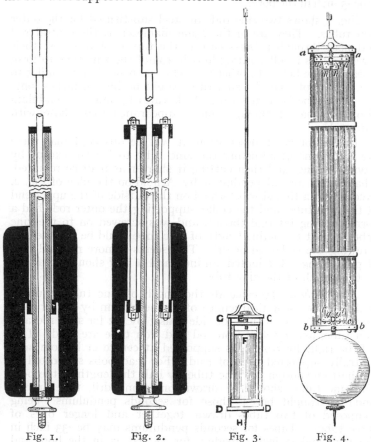

Fig. 1.   Fig. 2.   Fig. 3.   Fig. 4.

The suspension spring is usually about ·5 inch wide, from ·006 to ·008 in. thick, and the free part from an inch to an inch and a half long. The crutch should reach nearly to the bottom of the clock plates. With a short crutch a shorter pendulum spring must be used, which is a disadvantage, for the shorter the spring the nearer the path of the pendulum approaches a circular arc. On the other hand, with a very long spring and a short crutch the spring is apt to buckle and the pendulum arc fall short of what it should be.

For a lighter pendulum, with a bob 8 inches long and 2·5 inches diameter, weighing 14 lbs., a central rod ·25 inch in diameter may be used with a zinc tube ·1 thick, and an outer tube ·5 inch in the bore.

Fig. 2 shows two side rods of steel substituted for the outer iron tubes. They are of the same diameter as the central rod and pass through a brass cap at the top as shown. At their lower ends is a collar fitting loosely round the zinc tube to form a seat for the bob. In other respects, the pendulum is similar to Fig. 1, except that the zinc tube would be but 26 inches long. The zinc tube is coated with black varnish, and the pendulum altogether has, I think, a nicer appearance than those with outer tubes.

In the foregoing description, it will be observed that there are no means of adjusting the zinc tube to length, except by having it long, and then cutting it shorter as may be required. I have seen an old pendulum by Arnold on the plan of Fig. 2, except that a thread was formed on the outside of the upper end of the zinc tube, and the collar supporting the outer rods had a corresponding internal thread, and was screwed on to the zinc tube, so that the acting length of zinc tube could be adjusted by screwing the collar on or off. This seems a more rational way of proceeding. If adopted, an inch and a half should be added to the length of the zinc tube.

Zinc Tubes.—To calculate the length of zinc tube required, multiply the theoretical length of the pendulum by ·67 for steel centre rod and steel tube (or side rods) ; by ·72 for steel rod and iron tube, and by ·8 if iron rod and iron tube are used. This will be right for iron bobs supported at centre or bottom, and centrally supported bobs of lead. For lead bobs supported at the bottom shorten the zinc tube by half the length of the bob. The zinc tubes should be drawn on a mandril, to ensure a smooth, straight bore ; those for seconds pendulums being composed of two tubes drawn together and longer ones of three tubes. Tubes for seconds pendulums may be ·33 inch in the bore and ·7 inch outside ; for 1¼ sec. ·7 in the bore and 1·15 inch outside ; for 1½ sec. ·85 inch in the bore and 1·4 outside ; for 2 secs. 1·1 in the bore and 1·9 outside. These dimensions will permit of the use of standard sizes of gas barrel for the outer tube.

Long Zinc Tube Pendulums.—For turret clocks, rods and outer tubes of iron are generally used with the zinc tubes. The appended details may be useful :—

Theoretical length.			Rod: Diam.	Rod: Length.		Zinc tube.		Bob iron.			
ft.	in.		in.	ft.	in.	ft.	in.	in.		ft.	in.
5	1	= 1¼ sec., 48 beats a min.	⅝	6	4	4	0½	8	×	1	1
7	4	= 1½ sec., 40 beats a min.	¾	8	10½	5	10	9	×	1	3
13	0½	= 2 sec., 30 beats a min.	1	15	0	10	5	11	×	1	6

The length of the rod is from the point of suspension to the bottom.

Under the head of Electric Clocks is shown Murday's ingenious method of compensation which acts on the suspension spring, and appears to answer well.

Mercurial Compensation.—In the mercurial pendulum with a glass jar, as originally constructed by Graham, the mercury does not answer so quickly to a change of the temperature as the steel rod ; there is difficulty, also, in obtaining glass jars perfectly true. Preference is therefore now generally given to thin metal jars for precision clocks, although the elegant appearance of the glass jar in a stirrup causes it to be retained for regulators in many instances when a showy appearance is desired. In Fig. 3, page 301, is represented Graham's arrangement, with a little addition by Adam Reid for regulating the

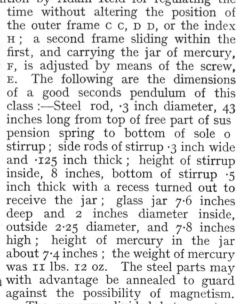

Fig 5.

time without altering the position of the outer frame C C, D D, or the index H ; a second frame sliding within the first, and carrying the jar of mercury, F, is adjusted by means of the screw, E. The following are the dimensions of a good seconds pendulum of this class :—Steel rod, ·3 inch diameter, 43 inches long from top of free part of suspension spring to bottom of sole o stirrup ; side rods of stirrup ·3 inch wide and ·125 inch thick ; height of stirrup inside, 8 inches, bottom of stirrup ·5 inch thick with a recess turned out to receive the jar ; glass jar 7·6 inches deep and 2 inches diameter inside, outside 2·25 diameter, and 7·8 inches high ; height of mercury in the jar about 7·4 inches ; the weight of mercury was 11 lbs. 12 oz. The steel parts may with advantage be annealed to guard against the possibility of magnetism.

The mercury divided between two jars answers quicker to changes of temperature. Fig. 5 is an arrangement of this kind, in which the lower part of the pendulum rod is square to receive a brass sleeve, the bottom of which is a platform with two cups to suit the bottoms of the jars. A brass clip supports the tops of the jars, which may have glass or metal covers. The brass work has a very nice appearance if it is silvered or nickelled.

Precision clocks with mercurial pendulums may have jars larger in diameter than 2 inches, made of cast iron enamelled on the inside, or of steel. I have seen one recently made by Messrs. Charles Frodsham & Co. for the Lick Observatory with a 40lb. pendulum, in which the jar of steel is three inches in diameter inside and $9\frac{1}{2}$ inches deep, the height of the mercury being rather over $8\frac{1}{2}$ inches. The upper end of the jar has a screw on the outside, of 20 threads to the inch, by means of which a cover with a corresponding internal screw is attached. There is a boss in the centre of the cover drilled and tapped to receive a screw of ·6 of an inch in diameter and 30 threads to the inch. The screw is pinned to the pendulum rod 7 or 8 inches from the bottom, so that the rod may dip well into the mercury. The upper end of the screw terminates in a square, through which the pins pass, and there is a wing on each side of the square to enable one to hold the rod steady while screwing the jar up or down for adjustment. There is a lock nut for tightening after the adjustment is perfect. A smaller boss on the top of the cover is drilled and tapped for a plug, and has a conical mouth to facilitate the pouring in of additional mercury if such should be required. For removing small quantities of mercury a dipper is used, that is, an open glass tube, over the upper end of which the finger is placed after it is dipped into the mercury, so that a partial vacuum being formed in the upper part of the tube as it is withdrawn, a portion of the mercury remains inside the tube, and is discharged as soon as the finger is removed.

Great care should be taken, when filling the mercury jar, to avoid air bubbles. The best plan is to push the centre of a good silk handkerchief into the jar and pour in the mercury through a long box-wood or other funnel with but a mere pinhole for the outlet. When the whole of the mercury is poured in, carefully draw up the handkerchief by its four corners. The jar of mercury, with a piece of bladder tied over the top, may then be subjected to a temperature of about 120° for a week or two.

It is important to get the mercury as pure as possible for a pendulum. A good way of removing impurities is to add sulphuric acid to the mercury and shake the mixture well. The metal is then washed, and afterwards dried on blotting-paper. Another method of purifying mercury is to put it in a bottle with a little finely-powdered loaf sugar. The bottle is stoppered and shaken for a few minutes, then opened and fresh air blown in with a pair of bellows. After this operation has been repeated three or four times, the mercury may be filtered by pouring it into a cone of smooth writing-paper, the apex of which has been pierced with a fine pin. The sugar and impurities will be retained by the cone. Some filter mercury by squeezing it through a piece of chamois leather. In dealing with mercury, care should be taken to avoid the injurious vapour which rises from it even at the ordinary temperature of the air, and of course more freely at higher temperatures.

Gardner's Pendulum.—The appended engraving is an elevation of the bottom part and a sectional plan of a one second

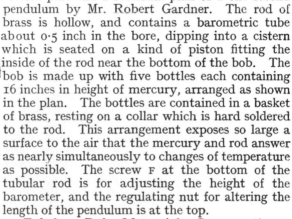

pendulum by Mr. Robert Gardner. The rod of brass is hollow, and contains a barometric tube about 0·5 inch in the bore, dipping into a cistern which is seated on a kind of piston fitting the inside of the rod near the bottom of the bob. The bob is made up with five bottles each containing 16 inches in height of mercury, arranged as shown in the plan. The bottles are contained in a basket of brass, resting on a collar which is hard soldered to the rod. This arrangement exposes so large a surface to the air that the mercury and rod answer as nearly simultaneously to changes of temperature as possible. The screw F at the bottom of the tubular rod is for adjusting the height of the barometer, and the regulating nut for altering the length of the pendulum is at the top.

Tubular Rod Mercurial Compensation.— Edward Troughton about 1790 formed a one-second pendulum of a glass tube about the size of a barometer tube holding with the bulb 45 ounces of mercury which reached up to about the middle of the tube. Surrounding the bulb it had a lenticular bob weighing 9 pounds.

S. Riefler patented in 1899 (No. 14259) a pendulum with which he has achieved remarkable results. It beats seconds and consists of a steel tube with a bore of 16 mm. and a thickness of 1 mm. The tube has no bulb and contains mercury for about two-thirds of its length. There is a lenticular bob, and below threaded on the rod are collars for adjustment.

Harrison's Gridiron Pendulum. — This, shown in Fig. 4, page 301, is still the form of compensation adopted in many foreign regulators.

U

It is composed of nine parallel rods, five of steel and four of brass, the total length of each kind being nearly as 100 to 60, that being the ratio of expansion of the two metals. Depending from the cross frame A are two rods of steel *a a*. The frame B, to which they are fixed at their lower extremities *b b*, carries also two brass rods *c c*, which at their upper ends, *d d*, are carried in the frame C together with two other steel rods *e e*. Those at their lower extremities *f f*, are fastened in the frame D, which also carries the brass rods *g g*. The frame F carries the upper ends of this last pair of brass rods at *h h*, and also the central steel rod to which the bob is attached.

Wood Rod and Lead Bob.—A cheap and good compensated pendulum may be made with a wood rod and lead bob. For a seconds pendulum the rod should be ·5 inch in diameter, of thoroughly well-seasoned straight-grained deal 45 inches long, measuring from the top of the free part of the suspension spring to the bottom of the bob.

A slit for the suspension spring is cut in a brass cap fitting over the top of the rod, to which it is secured by two pins. A bit of thin brass tube is fitted to the rod where it is embraced by the crutch. The rating screw, ·25 inch in diameter, is soldered to a piece of brass tubing fitting over the rod and secured by a couple of pins. Wooden rods require to be coated with something to render them impervious to the atmosphere. They are generally varnished or polished, but painting them answers the purpose well. Mr. Latimer Clarke recommends saturating them with melted paraffin. The bob 2·25 inches diameter and 12 inches high, with a hole just large enough to go freely over the wood rod, rests on a washer above the rating nut.

Shorter pendulums for chime and other clocks are made of teak, mahogany, and ebony, simply because in such small sizes deal does not allow of sound attachments to the ends. These pendulums have generally lenticular shaped bobs. Such rods cost scarcely any more than brass or iron, and are infinitely preferable.

It is essential that the grain of a wood pendulum rod should be perfectly straight, for if the grain is not straight, the rod is likely to bend, causing the clock to go very irregularly.

Pendulum Spring.—[*Ressort de suspension.*—*Die Aufhäng-ungsfeder.*]—(1.) A short ribbon of steel used to suspend the pendulum of a clock. The lower end of the spring is fixed in a slit in the pendulum rod, and the upper end between two clamps or chops as clockmakers call them. An arbor runs through the chops, and rests on the pendulum cock. It is most essential that the ends of the spring should be clasped perfectly tight. Some makers prefer to make the slit in the pendulum rod cap wide enough to admit of filing or milling it true ; packing strips are then inserted with the spring to make up the width. Generally but a single screw is used to fasten the spring in the cap, so that the pendulum may hang plumb without distressing the spring, which it might do if fastened slightly awry, with more than one screw. In some French clocks two widths of spring are used, with a space between them, with the idea of keeping the pendulum from wobbling, and though this plan is occasionally followed in seconds pendulums, it does not find much favour. In large clocks the spring occasionally breaks close to the chops, and the liability to failure is increased if the edges of the chops are not smoothly rounded off, or do not form a right angle with the length of the spring. Some makers use for the suspension a piece of steel thick at the ends, and made with a gradual curve to the desired thinness in the middle. This plan lessens the liability to break, and permits of a sounder attachment. No chops are needed at the upper end. A hole is drilled through the steel to admit the suspending spindle, the nuts of which clasp the steel on each side. The lower end is fitted to a slit in the pendulum head, and secured by a single bolt. In small clocks the pendulum spring is often too stout, and an increased vibration of the pendulum may be obtained by thinning it. It occasionally happens in mantel clocks that when the clock is brought to time the pendulum is just too long for the case. This may be remedied by thinning the pendulum spring at the top of the free part : the better plan is though, I think, to cut a piece off the bottom of the bob.

Phillips' Spring.—[*Spirale à courbe Phillips.—Die Spirale mit theoretischen Endkurven.*]—A balance spring with terminal curves formed on the lines laid down by M. Phillips, an eminent French mathematician, who gave rules for tracing different forms of isochronous curves with which the centre of gravity of the spring would always be on the axis of the balance, and springs so made would in action not exert lateral pressure on the pivots. The engraving is one of a variety given by M. Phillips in his work *Sur le Spiral Réglant*. Though many of the curves used by English watchmakers very closely agree with Phillips' prescription, the term Phillips' spring is rarely used here. (See Balance Spring, page 18.)

Pillar.—[*Pillier.—Der Pfeiler.*]—The pillars of a watch are the three or four short pieces of brass which serve to keep the two plates of the movement in their proper relative positions. In the watches of Mudge and other old English masters, great attention was paid to the form of the pillars, which were either pierced or richly engraved.

Pillar Plate.—[*Grande platine.—Die Pfeilerplatine.*]—The circular plate of a watch movement to which the pillars are fixed.

Pillar File.—[*Lime à pilier.*]—A flat file, narrower than a potance file. Though larger ones are made, a pillar file is generally understood to mean one three inches and half long, measured from the point to the end of the cut, and three-eighths of an inch wide.

Pinion.—[*Pignon.—Das Trieb.*]—A small toothed wheel. The smaller of toothed wheels which work together. A pinion is generally understood to be a wheel that has not more than twenty teeth. (See Wheels and Pinions.)

Pinion Gauge.—[*Mesure aux pignons.—Das Triebmass.*]—A

gauge used by watchmakers for taking the height of pinion shoulders and other measurements. It is something the shape of a pair of compasses ; the legs are kept apart or brought together by a screw.

Pinion Height Tool.—[*Outil pour hauteur de pignons.—Das Triebhöhenmass.*]—The exact distance between the jaws of this tool corresponds to the exterior measure of the feet. If, therefore, the gauge is adjusted so that the feet pass freely between the plate, or the plate and the bridge, the exact length of pinion is

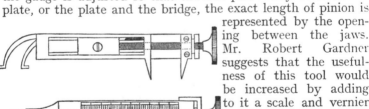

represented by the opening between the jaws. Mr. Robert Gardner suggests that the usefulness of this tool would be increased by adding to it a scale and vernier

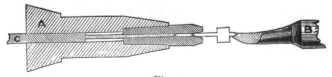

as shown in the lower view.

Pinion Leaf.—[*Aile de pignon.—Der Triebzahn.*]—The tooth of a pinion. An ingenious method of holding pinions while cutting which is adopted in the American factories, is shown in Fig. 1.

![Fig. 1 illustration]

Fig. 1.

The pinion is held in the tapered mouth of the spindle A, which carries a division plate with notches corresponding to the number of leaves to be cut. The other end of the pinion rests in the beak of the runner B. C is a pusher to remove the pinion from the tapered-mouthed holder. Three circular milling cutters are used mounted in a triangular frame, so that each in turn can be quickly brought to bear on the work. The first is a saw or slitter, the second roughly shapes the leaf, and the third finishes it. The last two cutters, formed of steel disc about five-eighths of an inch in diameter, have each ten slits, equidistant about the circumference, to give cutting edges, and from one bottom corner of each slit a saw-cut is made, going obliquely towards the centre, as shown in Fig. 2. The circumference of the disc is then squeezed in till the surfaces where the saw-cut started are in contact, and a clearance equal to the width of the saw-cut is thus obtained. This idea was patented by Mr. Church, of the Waltham Company.

Fig. 2.

Pinion Wire.—[*Acier à pignons.—Der Triebstahl.*]—Steel wire drawn with corrugations resembling pinion leaves, from which pinions are made.

Pin Pallet Escapement.—[*Ancre à chevilles.—Die Stiftenank-erhemmung.*]—An escapement used mostly in French clocks, in which it is often placed in front of the dial. The pallets are formed of semicircular stones, generally carnelian.

This excellent escapement (invented by M. Brocot), rarely seen except in small French clocks, appears to be worthy of more extended use. The fronts of the teeth of the escape wheel are sometimes made radial, as shown in the engraving; sometimes cut back so as to bear on the point only, like the "Graham"; and sometimes set forward so as to give recoil to the wheel during the motion of the pendulum beyond the escaping arc. The pallets, generally of carnelian, are of semicircular form. The diameter of each is a trifle less than the distance between two teeth of the escape wheel. The angle of impulse in this escape-ment bears direct reference to the number of teeth embraced by the pallets. Ten is the usual number, as shown in the drawing. The distance between the escape wheel and pallet staff centres should not be less than the radius of the wheel × 1·7. This gives about 4° of impulse measured from the pallet staff centre.

English clockmakers rather object to this escapement on account of the difficulty of keeping oil in the pallets, which is aggravated if there is much space between the root of the pallet stone and the face of the wheel. The effect of the want of oil is much more mark d if the pallets are made of steel instead of jewel. Any tendency of this escapement to set is generally met by flattening the curved impulse faces of the pallets as indicated by the dotted line across the right-hand pallet.

BROCOT'S PIN PALLET ESCAPEMENT.

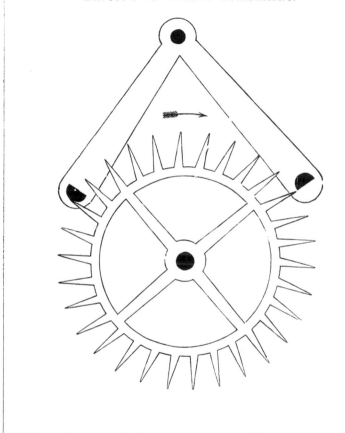

Pin Tongs.—[*Mandrin.—Die Stiftenzange.*]—Clams opening with a spring in which small pins are held while being shortened or otherwise altered.

Pin Vice.—[*Etau à queue.—Der Stielkloben.*]—A small vice held in the hand, chiefly used as a holder for pins and small pieces of work while they are being filed.

A much improved pin vice, hollow throughout its length, and

made on the principle of a split chuck, is shown in the sketch.

The wire to be held is instantly gripped perfectly central by giving a turn to a serrated nut with the thumb and finger. Two sizes are made, commanding between them a wide range of opening. In the smaller size the conical grip is slit into three, and in the larger one into four. As a drill holder, and for many other purposes where a rapidly adjusted handle is required, these vices will be found useful.

Pin Wheel—[*Roue à goupilles.*—*Das Hebestiftrad.*]—A wheel in the striking train of a clock, in which pins are fixed to lift the hammer.

(2). The escape wheel of a pin wheel escapement.

Pin Wheel Escapement.—[*Echappement avec roue à goupilles.* —*Der Stiftengang.*]—Invented by Lepaute about 1753. A clock escapement analogous in its action to the " Graham." The impulse is given by nearly half-round pins standing out from the face of the escape wheel. The one advantage over the Graham is that the pressure on the pallets is always downwards, so that excessive shake in the pallet staff hole, which may be looked for in course of time, especially in large clocks, would not affect the amount of impulse.

This escapement (see next page) is used principally in turret clocks. The chief objection to it practically is the difficulty of keeping the pins lubricated, the oil being drawn away to the face of the wheel. To prevent this a nick is sometimes cut round the pins, close to the wheel, but this weakens the pins very much. The best plan is to keep the pallets as close as they can be to the face of the wheel without touching.

Lepaute made the pins semicircular, and placed alternately on each side of the wheel so as to get the pallets of the same length. This requires double the number of pins, and there is no real disadvantage in having one pallet a little longer than the other, provided the short one is put outside, as shown in the drawing. Sir Edmund Beckett introduced the practice of cutting a piece off the bottoms of the pins, which is a distinct improvement, for if the pallet has to travel past the centre of the pin with a given arc of vibration before the pin can rest, the pallets must be very long unless very small pins are used.

The escaping arc is generally 2°, and the diameter of the pins is then 4° measured from the pallet staff hole.

Then with a given diameter of the pin, to find the mean length of pallets, divide the given diameter by ·069.

Or if the mean length of pallets is given, the diameter of pins may be found by multiplying the given length by ·069.

The opening between the extreme points of the pallets == 2°, that is, half the diameter of the pins.

With an escapement arc of 3° the mean length of the pallet arms is ten times the diameter of the pins.

The angle of impulse is divided between the pins and the pallets, and care must be taken that the pallets are not cut back

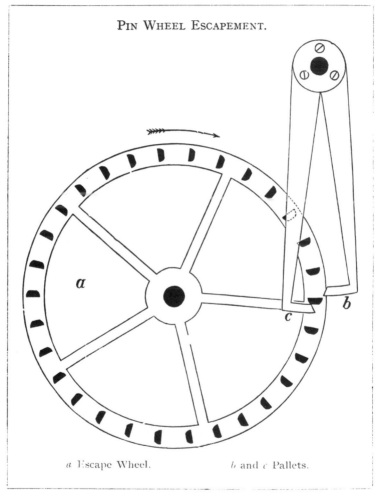

PIN WHEEL ESCAPEMENT.

a Escape Wheel.   *b* and *c* Pallets.

too much. When a pin escapes from one pallet, the bottom of the succeeding pin must fall safely on the rest of the other pallet.

It is best before finishing the impulse planes to place the pallets in position, and mark them off with reference to the pins. The thickness of the two pallets and one pin contained between them equals, less drop, which is very small, the space between two pins from centre to centre. The pallets are of steel hardened at the acting parts, and screwed to a collar on the pallet staff. The rests are slightly rounded, so as to present less surface to the pins, and the curves struck from a little below the pallet staff hole so as to be hardly " dead." The pins should be of gun-metal or very hard brass, or aluminium bronze, round when screwed into the wheel, and cut to shape in an engine afterwards.

Pipe.—[*Canon.—Das Rohr.*]—Any long boss through which a hole is bored. (2.) Of a watch key, the steel part which fits on to the winding square.

Pitch. — [*Distance des dentures. — Die Teilung.*] — The " pitch " of wheels and pinions is the portion of the circumference of the pitch circle between the centre of one tooth and the centre of the next.

Pitch Line. Pitch Circle. — [*Ligne d'engrenage. — Die Eingriffslinie.*]—A circle described from the centre of a wheel or pinion as a basis for constructing the teeth. The pitch circles of a wheel and pinion working together should touch but not intersect each other.

Pith.—[*Moëlle de sureau.—Das Fliedermark.*]—The pith used by watchmakers to clean their work is the pith of the elder.

Pivot.—[*Pivot.—Der Zapfen.*]—Strictly the end of a rotating arbor on which it is supported, but watch and clockmakers call any part of a rotating arbor that runs in a support a pivot.

Pivot File.—[*Lime à pivots.—Die Zapfenfeile.*]—A file used for forming pivots.

Pivot Gauge.—[*Filière à pivots.—Das Zapfenmass.*]—A steel plate with tapered slit for measuring pivots. It is chiefly used with the Jacot tool. The numbers on the side of the gauge correspond with the numbers stamped on the pivoting beds, so that if a pivot equal to, say, number 30 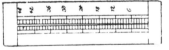 on the gauge is desired it can be formed in the bed of that number without any fear of its being made too small.

Pivoted Detent.—[*Bascule.—Die Wippe (Chronometerhemmung).*]—A form of detent mostly seen in French chronometer escapements, which moves on pivots instead of through a weak spring as in the English escapement.

Plate.—[*Platine.—Die Platine.*]—The plates of a watch are the discs of brass which form the foundation of the movement. The chief plate, called the pillar plate, lies underneath the dial ; the side of it next the dial is recessed to contain the motion work ; on the other side the pillars are fixed. Unless the watch has a " bar movement " there is another plate, kept a little distance from the pillar plate by the pillars, called the top plate. In full-plate watches this plate, like the pillar plate, is circular. In three-quarter-plate watches there is a piece cut out of the top plate, sufficiently large to allow the balance to move in the same horizontal plane as the top plate. In half-plate watches the fourth wheel arbor is cut short, and its upper end carried by a cock, so as to permit of the use of a larger balance than would otherwise be the case. The plates of a clock are two pieces of brass which receive the pivots of the train.

Pliers.—[*Pinces.—Die Flachzange.*]—Small tongs. Watch-makers use pliers of various shapes. The jaws of those used to handle the steel work of watches are generally lined with brass, which is not so likely to bruise or magnetise the work as steel

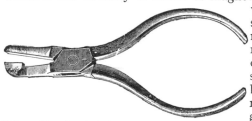

would be. But as absolute contact is not necessary to impart magnetism, it is evident that the pliers should be wholly of brass to secure immunity from danger in this respect. Pliers lined with ivory are used by some for handling highly polished work. Pliers with jaws annealed and polished are most suitable for gripping tempered steel pieces. The pliers shown in the engraving have one jaw jointed, so as to give a parallel grip. They are especially handy for keeping two pieces together while soldering them, or for holding a plate while shellacing a piece to it.

Combined Plier, Nipper, and Joint Tool.—This is especially useful for removing joint pins in watch movements, brooches, and other articles of jewellery. The jaws of a pair of flat-nosed pliers are, towards the joint on one side, formed with cutting edges. On the other side, which is visible in the sketch, a raised hollow stud stands on one jaw, and on the other jaw is a projecting pin in a line with the hole in the stud. When the joint-pin of a brooch or other article is to be removed, the joint is brought between the hollow stud and the pin, the latter bearing

against the joint-pin of the brooch, and on closing the plier the joint-pin is forced out of the joint into the hollow stud. The extremity of one handle of the pliers is straightened and pointed, to push home the new pin of a joint; the extremity of the other handle is also straightened and flattened to act as a screw-driver.

Pneumatic Clocks.—[*Horloges pneumatiques.*—*Pneumatische Uhren.*]—Clocks actuated by air. In the Popp-Resche system for some years used in Paris for public clocks, air, dried by pressure of about 40 lbs, per square inch. From these reservoirs a constant pressure of 10 lbs. per square inch is maintained in a closed vessel from which the air is used for driving the clocks. Behind the dials of the public clocks is a ratchet wheel fixed to the arbor of the minute hand, as in appended sketch. A click, working into this ratchet wheel, is pivoted to a lever whose extremity is attached to a bellows of thin metal. Pipes are laid to all the public clocks, and at the completion of every minute a master clock at the central station opens a communication between the supply vessel and the bellows, the pressure of air expands the bellows, and the ratchet wheel is advanced one tooth.

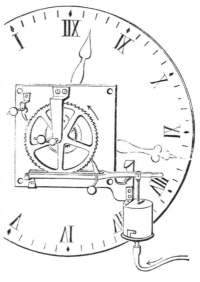

Compressed air may possibly be reliable in its action, and the mechanism needed is simple, but the time occupied by the transmission of signals appears to preclude its use for clocks at any considerable distance from the central station.

Pocket Chronometer.—[*Chronomètre de poche.*—*Das Taschen Chronometer.*]—A watch with chronometer escapement.

Poising Tool.—[*Outil à équilibrer.*—*Die Unruhwage.*]—A tool for ascertaining if the metal in a watch or chrono neter balance is evenly disposed round the axis. In the opinion of many testing in the callipers does not afford a sufficiently exact in-dication of the truth of a compensation balance for fine work, and recourse is had to a special tool, of which the one in the drawing (Fig. 1), by R. Bridgman, is a very fine example. The movable standard is kept in contact with the bed of the tool by springs, so as to be independent of the truth or otherwise of the traversing screw. The pivots of the balance rest on knife edges formed of sapphire, which may be adjusted to the desired distance by means of a screw. Screws on each side of the standard serve to level the knife edges. Fig. 2, the " Gross-

Fig. 1.

mann," is a more usual and less costly pattern. (See Callipers.)

Polishing.— [*Polissage.— Polieren.*] —The methods employed for producing the beautiful polished and square surfaces to be found in watchwork may be divided into two general principles. First, where the work is rigid and receives a reproduction of a previously squared surface ; and, secondly, where the work is " swung " or arranged so as to yield to unequal pressure in polishing.

Fig. 2.

Polishers of steel are either of soft steel, iron, bell-metal, tin, zinc, lead, or box-wood. They must in all cases be formed of softer material than the object to be polished ; for instance, bell-metal, which brings up a good surface on hard steel, is unsuited for soft.

Polishers for brass are generally of tin or box-wood, with willow for finishing.

The polishing medium is either emery, which is used for grey surfaces, carborundum, oilstone dust, red-stuff, or diamantine, used with oil, or Vienna lime, which is mixed with spirits of wine. ALL POLISHING MATERIALS MUST BE PROTECTED FROM DUST, OR THEY WILL BE SPOILED.

Brass surfaces are generally "stoned" preparatory to polishing, that is, rubbed to a surface with bluestone or water of Ayr stone and water or oil.

Mixing Polishing Material.—Red-stuff should be thoroughly beaten up on glass or a polished steel stake to a stiff paste with very little of the best oil that can be obtained. It is very poor economy to use inferior oil. Far too much oil is often used, and the mixture left thinner than it should be. Olive oil is not suitable, and, if used, the polishing stuff becomes sticky in a day or two. Refined sperm oil, such as is used for watches, answers well. Diamantine should be mixed on glass with a glass beater in the same way, as dry as possible, so that when it is used the polisher is only just damped with it. If diamantine is brought into contact with metal in mixing it turns black. Vienna lime, which is much favoured in the American factories, is very quick in its action. It may be used with water, but is much better mixed with spirits of wine, and used with either a tin lap or a boxwood polisher. It should be kept in stoppered bottles away from the light, and mixed just as it is required to be used.

Polishing Watch Wheels.—Escape wheels are generally fixed to a small brass block. The block is heated in a bluing pan, and a piece of resin passed lightly over it so as to leave a very thin varnish only, which is quite enough to make the wheel adhere ; there should be circles marked on the face of the block as a guide for fixing the wheel as nearly central as possible, or else a small pin in the centre of the block to go through the hole in the wheel with the same object. The wheel fixed to the block is first rubbed till quite flat on a piece of bluestone having a true face, which is kept moistened with water ; it is rubbed with a circular motion by means of a pointer (generally a drill stock) pressed down on the middle of the back of the block, which is hollow, as shown in the sketch (Fig. 1). The wheel is thoroughly cleaned, and then polished on a block of grain tin with sharp red-stuff and oil well beaten up previously. The block of tin rests on a leather pad. When one side of the wheel is finished, it is placed again in the bluing-pan. The old resin is cleaned off, and the finished side of the wheel fixed to the block. After both

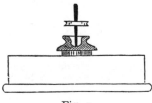

Fig. 1.

sides are polished, the wheel is placed in spirits of wine to remove any resin adhering to it.

Solid Train Wheels are placed on a piece of cork screwed in the vice, and rubbed with a piece of bluestone (previously rubbed on a stone to an even face) moistened with water. When the wheel is rubbed flat, it is put in a pair of turns with a ferrule screwed on the arbor, and revolved with a hair bow, and rubbed at the same time with a thin piece of bluestone moistened with oil, and having a perfectly flat face. This process must be continued till the wheel presents a perfectly smooth face, free from all marks and scratches. This point is important, as deep marks cannot be removed by subsequent polishing, except by the sacrifice of much time, and also of the squareness of the work. The wheel, having been well cleaned with bread (it should be seen that the bottoms of the teeth *are* quite clean), is next rubbed with a piece of flat boxwood, and the unction or paste obtained by rubbing two bluestones together, with clean oil, and afterwards, if brightness be desired, with a bit of cedar or willow ; but a true, flat surface, the teeth also flat and not rounding at the sides or end, is more to be sought after than a mere polish. Pierced wheels are first rubbed flat on a cork either with a bluestone or water of Ayr stone. After cleaning they are polished with a soft tin polisher,* and moderately sharp red-stuff, using a slightly circular stroke. Instead of a plain cork some finishers use a hemisphere of cork resting in forks cut in another cork, as in Fig. 2. When quite flat and smooth, the

Fig. 2.

* *Grain Tin Polishers.*—Owing to the extreme softness of this metal, making a polisher at once light and rigid is a task of some difficulty. If it is to be of tin alone, the smallest size that will be of use in polishing wheels will be of about seven-eighths of an inch broad by five-sixteenths thick, and even this size will require great care in filing and use to avoid bending. A plan recommended by Mr. Gray is to file up a bell-metal polisher about one-eighth of an inch thick, and of the required width, and to tin the face with a copper bit, muriatic acid, and solder, and making a mould for half the length of the polisher in plaster of Paris, cast on a layer of grain tin, previously heating the bell-metal to rather more than the melting point of tin. A polisher made in this way is far lighter and more rigid than any solid tin one. Circular blocks should be cast with a flange on the bottom, and considerably thicker than required. By means of the flange you can grip them in the mandril and surface them, taking a considerable portion off the top. If there are impurities and grit in the metal, they have a tendency to rise to the top, and are thus removed. Tin, previous to being cast into blocks, etc., should be carefully melted several times in a *clean* iron ladle, and each time poured from a height while in a melted state into water, thus breaking it up into very small particles, and enabling you to wash it thoroughly to remove all grit, etc. The file used for finishing the face of tin polishers should be an old smooth-cut one, *well worn.* A new file is useless for this purpose, as it clogs and cuts, leaving deep scratches in the metal. The file marks should be stoned out, and finally the face carefully burnished with a flat burnisher.

wheels are washed in soap and water, and burnished on a clean hard cork with a burnisher well rubbed on a board with rotten stone or red-stuff. To ensure success in polishing wheels the greatest cleanliness must be observed, and the polisher frequently filed to keep a flat, clean surface. If the operator cannot get on after a little practice, failure can generally be traced to a want of the scrupulous cleanliness absolutely necessary throughout the operation. In fixing escape wheels to the brass block there must be but a smear of resin; too much will be fatal. A mixture of equal parts of beeswax and resin is preferred by some, as it has the advantage that it may be removed from the block with hot water.

Large Steel Pieces, such as indexes and repeater racks which are not solid, and springs, should be shellaced to a brass block, and polished underhand; a flat surface is first obtained by rubbing with fine emery on a brass plate; afterwards with coarse red-stuff on a bell-metal block. The work is then finished off with diamantine on a zinc or grain tin block. The diamantine should be well beaten up on a glass with as little oil as possible. Such parts as rollers and collets are polished in the same way. Levers are pressed into a piece of willow held in the vice, and polished with a long, flat bell-metal or zinc polisher, moving the polisher instead of the work. There is nothing like diamantine for giving a good black polish. It is, however, very quick in its action, and requires some little experience to avoid overdoing it and making the work foxy. The work, polishers, etc., must be kept scrupulously clean.

Pinion Leaves may be polished by means of a tool as shown in Fig. 3, which consists of centres for holding the pinion and guides through which the runner carrying the polisher works. The polisher is of some soft material, usually lead. Where pinions are made in large quantities the polisher is

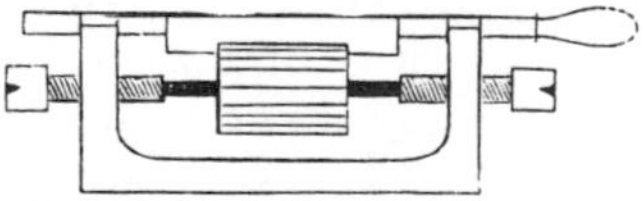

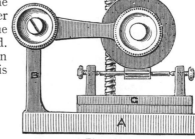

Fig. 3.

Fig. 4.

actuated by a " Wig Wag" if it has a to-and-fro motion. Pinions may also be polished by means of a rotating willow cylinder carried between two arms pivoted to a frame (A B, Fig. 4). A slide rest (G) holding the pinion between centres is set under the roller, not exactly at right angles to it, but sufficiently askew to cause the pinion leaves to cut a screw in the cylinder, which then rotates the pinion. During the operation the slide rest is worked to and fro. There is an adjusting screw to keep the roller from going too low.

Facing Pinions.—The facing tools should be made of steel, to be used with coarse stuff for squaring up the face, and bell-metal with diamantine for finishing. The pinion arbor must have only moderate play inside the tool, but its *end* must be kept perfectly free, or the points of the leaves will not come *up*. To ensure this freedom, the tool should be drilled from the back with a large drill, so that the smaller part of the hole shall not embrace more than half the length of the pinion arbor. Convenient tools may be made by providing a tube of a proper size tapped on the inside, into which stoppings of steel or bell-metal (to form the facing tool) are screwed. This plan has the advantage of giving ample play to the end of the arbor, and the polisher when worn out can be easily renewed. To face pinions one end is supported against the back centre of the turns, and the pinions rotated by means of a screw ferrule and hair bow. (See Fig. 5.) The tool is pressed against the pinion with the end of the forefinger, and the bow held at the *top*, moved with long steady strokes. The tool must be kept flat by frequent filing, care being taken to keep it upright, so as to present a true face to the work. If it gets out of truth, it should be put on an arbor and turned true. When the tool is

Fig. 5.

of proper size the pinion will " speak " (make a squeaking noise) as the red stuff is dying off ; if it does not speak it may be taken for granted that something is wrong, most probably an improperly sized or badly shaped tool. The steel should be used till the face is quite square at the ends and sides of the leaves and with the rings well formed at the bottoms. Before finishing with the bell-metal tool, the pinion must be thoroughly cleaned with bread and by a peg being passed through it, so as to remove all dirt. The hole in the tool must also be cleaned in the same way. In finishing, the tool must be filed with a very old smooth-cut file, or rubbed with water of Ayr stone, and only slightly damped with diamantine. The bow must be moved smoothly and steadily ; if a jerky motion is given to it, perfect flatness of the face will not be obtained. For a similar reason the finishing should be very quickly done, for if the tool is at all worn the face will be rounding.

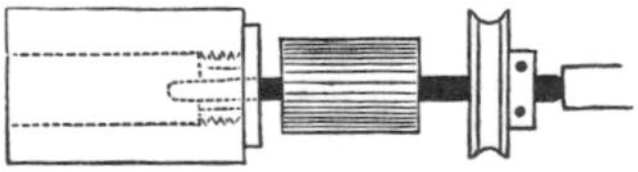

Fusee Hollows may be polished with a cup shaped polisher made of iron or sheet copper, and held in a pair of clams. The tool is made by first stamping the cup in a die : a hole of the proper size is then made in the centre of the cup, and the tool held by the sides is slightly turned in the mandril to true it. The inside of the cup must also be chamfered a little to give the tool effective play behind the fusee shoulder. Care must be taken to turn the hollow of a proper shape, not too deep, and to give the requisite concave form, the edge of the graver should be curved on the oilstone. The fusee pivot is supported against the back centre of the turns, the clams held in the left hand, and the fusee rotated with a bow. After turning the hollow, the tool must be tried with a little red stuff, and if it does not touch the hollow all over, that part or parts of the hollow touched by the tool must be again turned slightly. This must be repeated till the hollow fits the tool exactly, when but very little polishing will be required. When the tool fits the hollow properly, it will speak in the same way as a pinion during the process of facing. When the hollow is smooth from coarse red stuff, it may be finished with fine red-stuff or diamantine. If the hollow should have a milky appearance from the fine red-stuff, it may be improved by rubbing it with a clean peg and clean fine red-stuff mixed wet. Set-square hollows and the hollows of centre pinions may be polished in similar manner. Since the introduction of diamantine the hollows of small pinions are often polished *in* the turns with a small zinc polisher held in the fingers, which quickly accommodates itself to the shape of the hollow, or a bit of copper wire filed to the proper shape may be used in the same way with fine diamond powder.

For Solid-Faced Pinions, barrel arbor shoulders, the backs of rollers and the like, the work is fixed in a swing tool of the form shown in Fig. 6. The work to be polished is mounted on an arbor which is held between the screw centre at bottom and the runner at the top of the tool. The face to be

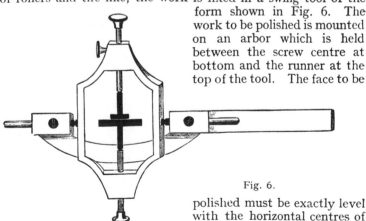

Fig. 6.

polished must be exactly level with the horizontal centres of the swing tool, or the work, instead of being flat, will be rounding.

V

Sometimes this swing tool is held in turns, but it is preferable to have a frame as shown in the sketch, which allows free motion of the polisher. The end of the frame is held in the vice. The arbor on which the work is mounted must run freely in the centres of the swing tool, so that the motion of the polisher causes the work to rotate ; at the same time the arbor must have no shake. The proper adjustment of the arbor between the centres must be seen to occasionally during polishing.

Lever End and Roller Crescent Tool.—The top part of this

tool (Fig. 7) is used for polishing crescents in rollers for lever escapements and similar pieces. The roller is secured by means of a clamp screwed to the body of the tool. There are steady pins to keep this clamp always in true position, and a short spiral spring round the body of the lower screw presses the clamp open as the screws are relaxed, so that the work is at once released. A steel rest for the polisher, fixed to the body with screws passing through slotted holes to allow of adjustment for height, runs down the whole width of the other end of the tool. The lower clamp is for holding

Fig. 7.

evers and other pieces that are polished on a block underhanded ; the steel piece with the slotted holes for adjustment, just referred to, being rounded on the end to act as a former, the shape of the work of course corresponding to the curve of the former.

Lever Edges, chronometer detents, and many other parts

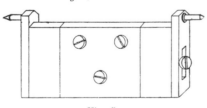

are polished in a swing tool, like Fig. 8, care being taken as in the previous case to get the surface to be polished level with the swinging centres. The two upper screws in the cover plate are tapped into the body

Fig. 8.

of the tool. The lower screw tapped into the cover plate is used to keep the plate parallel with work of different thicknesses. The teeth of chronometer escape wheels are also polished in a very similar tool, each tooth in succession being brought to the swing level.

Roller Edges for lever and chronometer escapements are often polished by means of a rotating disc or mill of bell-metal. The roller on an arbor is fixed to the slide rest of the lathe in a pair of turns or specially adapted holder, as shown in Fig. 9. When brought into contact with the mill, it is turned with a bow or the thumb and finger, and the slide rest traversed the while so as to move the roller in a plane parallel with the face of the mill. After the edge is polished, if the corners of the roller are to be chamfered, the holder is turned first one way and then the other to an angle of 45°.

Another way of polishing roller edges is to place the roller to be polished on an arbor between two flint hard rollers of the same diameter as the one to be polished is to be when finished, and polish in the turns.

Yet another way of finding a rest for the polisher is by means of a broad-headed screw put through the runner of the turns, the roller is placed near the end of the arbor, and varying diameters are accommodated by raising or lowering the screw.

Stud Holder.—Fig. 10 is a little holder for studs and similar pieces. It turns on a universal joint at the bottom, so as to bring all the angles of the stud in turn within range of the polisher,

Fig. 9.

which works to and fro on a steel rest attached to the body of the tool. There are screws for tightening the joint in both directions if it is not stiff enough. For watch studs of steel, the jaws of the holder are of brass ; for chrono-

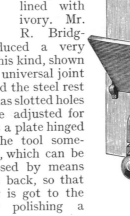

Fig. 10.

meter studs of brass they are lined with ivory. Mr. R. Bridg-

man has introduced a very superior tool of this kind, shown in Fig. 11. The universal joint is very strong, and the steel rest for the polisher has slotted holes to allow it to be adjusted for height. There is a plate hinged to the body of the tool something like a desk, which can be raised or depressed by means of a screw at the back, so that when the holder is got to the right angle for polishing a

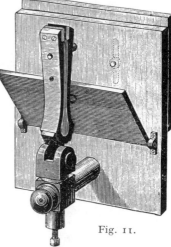

Fig. 11.

surface, this flap is brought up to be in contact with it, and the holder may be shifted to examine the work with the certainty of being able to bring it back to the same place.

Polishing Screws.—Finishers generally use the old English screw-head tool for producing the beautiful " tallow top " screws used in English work. This tool is a mandril running in one bearing, with an overhanging ferrule and a rest for the polisher, like Fig. 12. The screw whose head is to be polished is screwed into a chuck, of which there must be a sufficient variety to suit

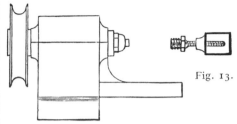

Fig. 13.

Fig. 12.

all of the ordinary run of screws. Tool marks are usually first removed with a slip of oilstone, and the polishing finished with red-stuff or diamantine. A chuck like Fig. 13 is used for polishing the taps and the points of the screws. Here the screw is slipped into a hole in a narrow-faced " lantern " from the *back*, and kept in position by a screw which runs through the boss of the lantern and jambs the screw head against the back of the face. With a long bow, screws are polished very expeditiously in the English tool, but it is not so handy for jobbing, where so many different kinds of screws are met with.

Fig. 14.

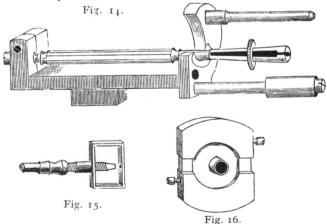

Fig. 15.

Fig. 16.

For jobbing, the cheap form of Swiss screw head tool, shown in Fig. 14, is still used. There are a number of different sized holders lined with brass so as not to bruise the taps of the screws.

It is important to use one of the right size, for if too small in the hole the screw is sure to be marked. The holders are sprung open, and a sliding thimble serves to nip them together sufficiently to grip the screw to be polished. The holders

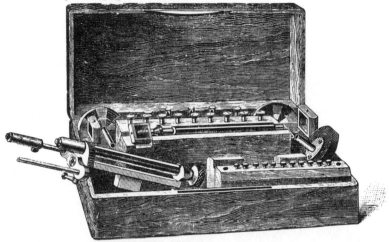

Fig. 17.

are rotated by rubbing the palm of the left hand to and fro over the octagonal body. The polisher rests on a roller to permit of smooth running. The upper arbor seen projecting from the body of the tool is to receive a lap, which is pressed against the work and slighly rotated to and fro by thumb and finger of the right hand, while the work in the holder is rapidly rotated with the left. Two nearly semicircular laps are generally screwed to one holder, as shown in Fig. 15, one of soft steel and the other of gun-metal. Fig. 16 shows a lantern for polishing the ends of fusee squares. Smaller ones of the same kind are used for holding screws so that the taps and points may be polished as described in speaking of the English tool.

A more modern tool by Boley is shown in Fig. 17. The holder has a round corrugated body, and it is furnished with split chucks of brass and steel useful for holding other pieces of work besides screws ; when the lap is not in use its arbor may be swung out of the way.

Meek's Pattern Screw Head Tool (Fig. 18) is especially adapted for English work. It has one arbor fitted with brass lanterns large enough for fusees, another with small steel lanterns for jewel screws, and a third with brass split chucks of various sizes which may be instantly opened or closed perfectly central by means of a large milled nut. The laps are of a good form.

326

The Screw Point Tool (Fig. 19) is a holder which is rotated in a hole formed in a block gripped in the vice.

Fig. 20 is a little triangular plate used for polishing screwheads flat underhand. In two of the holes are screws with hardened points, and in the third, head downwards, is the screw to be polished. It is obvious that small pieces other than screws may be shellaced to a tool of this kind and polished underhand.

Conical Pivots. —The cone should be an easy curve dying away into the pivot proper, which runs in the hole ; this part must be perfectly straight and parallel. The pivot having been turned to a little over the required size, its end is laid on a bed formed in the runner of

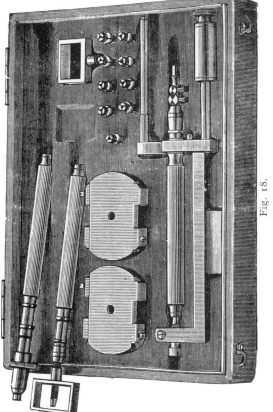

Fig. 18.

Fig. 19.

the turns. (See Turns.) Every time work is examined the bed of the runner must be cleaned and the runner adjusted to a slightly different length, so that it does not bear on the same part of the pivot. If this is neglected the pivot is sure

Fig. 20.

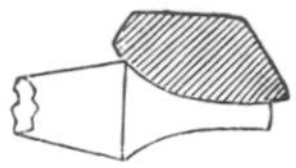

Fig. 21.

to be marked. A soft steel polisher made to suit the pivot, as shown in Fig. 21, is then used with either oilstone dust or red-stuff. It should be used with a backward and forward as well as a rolling motion till the pivot is reduced so that the hole, when taken up on the tip of a damp finger and applied to the acting part of the pivot, fits without shake, but with just sufficient freedom to fall off. The pivot is then finished with a very smooth burnisher and oil. Instead of the soft steel polisher, some prefer to use a hard steel burnisher roughened on a piece of lead with emery, which makes an equally good pivot. For rounding the end of the pivot, a thin edged runner to allow the end of the pivot to come through is used. The pivot is rounded by passing the burnisher from the body of the pivot over the end. If the burnisher is used *from* the point towards the body of the pivot a burr may be formed. There is a little difference of opinion as to the proper shape for the ends of balance staff pivots. Many manufacturers say the watches time better if the ends are left nearly flat, as shown in Fig. 21. This is not accepted by others, who prefer the pivot ends left rounder.

The size of the ferrule is not a matter of indifference. While a large size is an advantage in facing a pinion, as small a one as possible should be used in polishing a pivot. The chief difficulty of beginners is to get the polishers and burnishers of exactly the same shape from end to end, so as to follow each other properly.

This uniformity may be secured by drawing the polishers and burnishers from end to end across a piece of hardened steel, the edge of which is of the shape it is desired the pivots should be shaped, till they coincide with the form of the steel. (See Fig. 22.)

The best plan is to file up two pieces of steel together, slightly bevelling the edge of one to give it a cut, and leaving the other

quite square. With only one piece there is a tendency to cut the polishers too deeply, leaving them ridgy.

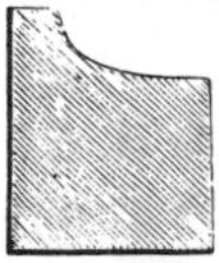

Strai ht Pivots with square shoulders are polished with a steel polisher slightly curved along the edge

Fig. 22. that acts against the shoulder of the pivot, as shown in Fig. 23. This edge is also dovetailed a little so as to form rather less than a right angle with the bottom of the polisher in section,

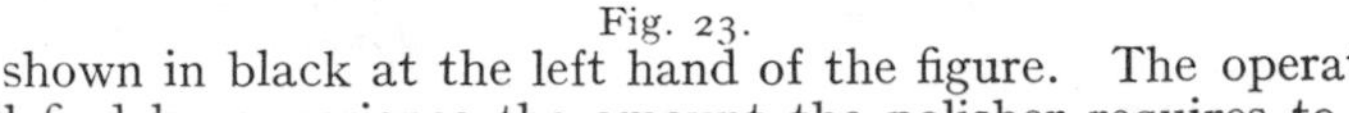

Fig. 23.

as shown in black at the left hand of the figure. The operator will find by experience the amount the polisher requires to be curved. It is rarely that one man can use another's polisher so well as his own. If the edge of the polisher is too much dovetailed, it will produce a wavy shoulder to the pivot. The pivot must be turned nearly to its right size, with the shoulder quite square. During polishing the *end* only of the pivot must rest on a suitable bed on the end of the runner. A piece of paper may be placed underneath the pivot to reflect the light. The light so reflected must be divided equally on each side of the shoulder during the process of polishing, and uniform pressure exerted along the pivot. The polisher must be used with a backward and forward motion, and with a slightly lateral motion also, to prevent ridges being cut in the pivot.

Those who have not had much experience in polishing may with advantage use a lap for straight pivots and shoulders. The lap and pinion are rotated in opposite directions by means of two bows held in the right hand, the lap being centred in the back limb of a depthing tool and the pinion in the front one. An arm (*a*, Fig. 24) is fixed to

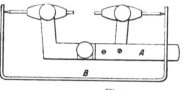

Fig. 24.

the depthing tool to hold it in the vice, and a piece of brass wire (*b*) clasps the runners of the front limb, so that the operator can move the pinion to and fro with his left hand. A soft steel lap at first, and afterwards a zinc one, are generally used. They should be turned true on the edge and the face slightly undercut. One of the laps with ferrule is shown at *c*.

Lately there has been introduced the " Compact " mill carriage and holder, Fig. 25, which fits into the tee rest pillar of the turns. There is a spring pressing against the end of one of the runners of this carriage, so that the pressure of a finger against

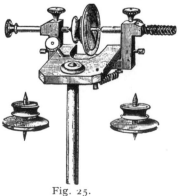

Fig. 25.

Fig. 26.

the end of the other runner is sufficient to give a to-and-fro motion to the mill. The carriage is mounted on a hinged platform, and by turning a set screw the free end of it is raised or lowered, causing the mill to recede from or approach the centre of the turns. Besides a wide range of polishing, this tool is adapted for snailing.

A neat lap-holder, shown in Fig. 26, accompanies the " Hopkins " and other American lathes. This may be angled in any direction, and is also suitable for snailing. Other forms are shown under the head of " Lathe."

Machine Polishing.—Fig. 27 shows the principle on which the polishing is conducted in the American factories. The pinion

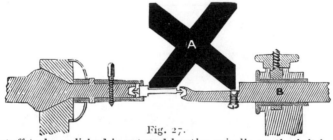

Fig. 27.

or staff to be polished is rotated by the spindle on the left hand, the other end of it being centred in the runner B. The polisher, to which a rapid to-and-fro motion is given by a " Wig-Wag,"

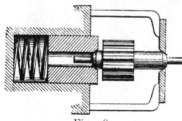

Fig. 28.

has four acting faces, each of which in rotation can be quickly brought to bear on the work in hand. By means of a jewelled screw in the runner B for the edge of the polisher to run on, the taper of the staff or pinion can be adjusted to what is required.

For polishing the pivots, the pinion is held in a cage by the slight chamfers at the shoulders of the pivots. The construction of this holder will be understood from the sketch, Fig. 28. The back sink is formed in a sliding block, which is kept against the pinion sufficiently tight during the process of polishing by a spring. This spring yields and allows the block to be pushed back when a fresh pinion is inserted. The pivot is quite unsupported, and is found to be quite strong enough to sustain the pressure of the wig-wag polisher. A thin plate containing the front sink for the reception of the chamfer through which the pivot to be polished projects, is shellaced to the cage so that it may be readily attached and set true with the hole when it has had to be changed for a different size of pinion.

A spring holder of this kind would form a very useful adjunct to a lathe for many purposes. The front plate might preferentially be

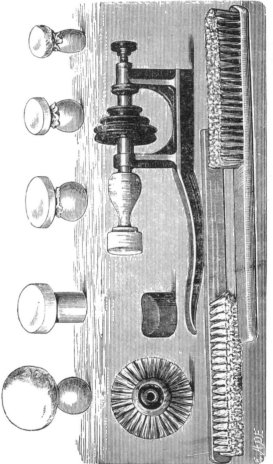

Fig. 29.

Fig. 31.

Fig. 30.

fastened by a couple of sunk screws instead of with shellac.

Watch Case Polishing.—The outsides of watch cases, when not engraved or engine turned, are polished by rubbing them on the leg of an old worsted stocking stretched on a board and charged with rotten-stone at first and afterwards with red-stuff. The final polish is given with the palm of the hand and rouge. For the insides, wooden dollies of suitable shape are rotated in the lathe. The dollies are covered first with felt, and for finishing with the finest doeskin ; the beads, if any, are polished against a rotating brush. Boley provides a very useful lathe and requisites for polishing watch cases, jewellery, etc., as shown in Fig. 29.

For polishing rings, buffs covered with felt or cotton (Figs. 30 and 31) are useful.

To Polish Mother of Pearl, Ivory and Horn.—Rub with wet linen cloth dipped in finely-powdered pumice stone, finish off with putty powder and a wet rubber. Ivory which has become discoloured may be restored by exposure to the sun's rays *under glass*. If not placed under glass the surface will crack.

Clock Plates.—Brass clock plates after being filed flat are surfaced either by stoning or " shaving " ; the latter operation being performed with a sharp tool held in a long handle under the arm. They are polished with emery, rotten stone or oilstone dust and oil, used with a wooden wad or a soft brush or a polisher of willow wood. Chronometer plates are polished in the same way, and afterwards with red-stuff and oil.

Polishing Stake.—[*Bloc à polir.*—*Die Polierdose.*]—A square polished surface of steel on which red-stuff and other polishing material is mixed. It is usually enclosed in a box with a lid, to exclude dust and dirt.

Poole, John (born 1818, died 1867), a clever chronometer maker, and inventor of an auxiliary compensation which has met with much favour. (See Auxiliary.)

Potence.—[*Potance.*—*Der Gangkloben.*]—A hangdown bracket used for supporting the lower end of the balance staff in full-plate watches. Any hangdown bracket used in watch-work.

Potence File.—[*Lime à carrelotte.*]—A flat file, wider than a pillar file. The size of potence file most generally used is four inches long, measured from the point to the end of the cut, and nine-sixteenths of an inch wide.

Potential Energy.—[*Die aufgespeicherte Kraft.*]—Power to do work. A mainspring when wound possesses potential energy. The pressure multiplied by the distance travelled in winding would be the measure of its potential energy. The potential energy of a raised clock weight is equal to its weight multiplied by the distance through which it can fall. The potential energy in foot-pounds of any raised weight $= wh$, where w is the weight in pound and h the vertical height in feet.

Precious Stones.—[*Pierres précieuses.*—*Die Edelsteine.*]—
Beauty of colour, hardness, and rarity are strictly the criteria of precious stones, which include only the diamond, ruby and sapphire, though the term "precious" is often applied to the opal, which lacks hardness, and to the pearl, which can hardly be regarded as a mineral. The term "Gem" is by archæo ogists restricted to engraved stones such as intaglios and cameos, though more generally applied to precious and semi-precious stones when cut or polished. "Jewel" is applied to a stone only after it has been mounted.

The *Diamond* is pure carbon, crystallized in the isometric system, and is usually found as an octahedron or some modification of that form. Diamonds, as a rule, are cut into one of two forms, Brilliant or Rose. Large stones are, if the shape permit, cut as brilliants, that is, with a flat table on the upper surface, surrounded by small angular planes or facets. The under-surface is also cut into facets, and terminates nearly in a point called the collet or cullet.

Thin stones do not allow of this style of cutting, and are cut as roses, having the top surface covered with small facets and the underside left flat.

The *Sapphire* is alumina and stands at the head of the corundum class. It varies much in value and is most highly prized when of a rich velvet blue or dark red colour.

A stone of the latter kind is generally known as a *Ruby*, although of precisely similar composition to sapphires of other colours. Watch jewellers who use both light-coloured sapphires and rubies, assert that the former are as a rule of a higher degree and more uniform hardness than the latter.

The *Topaz* is a fluo-silicate of alumina.

Spinel is composed of alumina and magnesia. The red variety of this stone is sometimes called the Spinel Ruby.

The *Chrysoberyl* is an aluminate of glucina. The true cats-eye is a variety of this stone.

The *Beryl*, *Emerald*, and *Aqua-Marine*, the last-named of which is a pale bluish green in colour, may be classed together, being composed of silicates of alumina and the rare earth glucina.

The *Zircon*, *Hyacinth*, and the *Jargoon* are silicates of zirconia.

Garnets are divided into three classes, the iron, alumina and the chrome garnet. They vary much in value, the commoner kinds being used for jewelling the cheaper grade of watches.

Rock Crystal, a crystallized transparent form of silica, also known as "Bristol," "Welsh," "Cornish," or "Irish" diamond, is also used by watch jewellers.

The *Amethyst* and the *Cairngorm* are also varieties of crystallized transparent silica.

Chalcedony, Agate, Onyx, Sardonyx, Carnelian, Chrysoprase, Heliotrope, and *Jasper* are forms of silica either amorphous, translucent, or opaque.

Tourmaline is a boro-silicate with alumina, iron, magnesia, and fluorine.

Moon-Stone, Sun-Stone, Amazon-Stone, and *Aventurine* are forms of felspar.

Colour, Degree of Hardness, Specific Gravity, and the kind of Refraction (Single or Double) of the Principal Gems :—

	Colour-less	Red	Yellow	Green	Blue	Violet	Brown	Black	Hardness	Specific Gravity	Refraction
Diamond	*	*	*	*	*		*	*	10	3·5	S
Sapphire	*	*	*	*	*	*			9	3·9—4	D
Chrysoberyl			*	*					8·5	3·8	D
Spinel		*		*		*			8	3·7	S
Topaz	*		*	*	*				8	3·5	D
Zircon	*	*	*	*			*		7·5	4·6	D
Beryl	*			*					7·5	2·7	D
Emerald				*	*				7·5	2·6	D
Rock Crystal	*		*				*		7	2·7	D
Amethyst						*			7	2·7	D
Chrysolite				*					6—7	3·4	D
Garnet		*		*		*	*	*	6—7	3·8	S
Tourmaline	*			*	*	*	*	*	6—5	3·1	D
Turquoise				*	*				6	2·6—3	
Lapis Lazuli					*				5·5	2·4	
Opal	*	*	*	*					6	2·3	

NOTE.—To ascertain the specific gravity of gems in the absence of a proper hydrostatic balance, place a small glass partly filled with distilled water under one end of the beam of a pair of sensitive scales. Attach a horsehair with a loop to the scale beam over the water, and add weights to produce equilibrium. Tie the stone to the horsehair, and note its exact weight in the air. Then lower the horsehair so that the stone is immersed in the water, and again note its weight. Divide the weight in air by the difference between the weight in the air and the weight in water, and the quotient will be the specific gravity.

Pearls.—Pearls will not tarnish nor lose their brilliancy if kept in dry magnesia, instead of cotton-wool as they sometimes are in jewel cases.

A *Carat,* used as a unit of weight for diamonds, is divided into four grains ; these grains are sometimes called " pearl grains," to distinguish them from the grain troy. The carat itself, unfortunately, varies in different countries. In England it equals 3·17 grains troy ; in Holland, 3·0 ; and in France, 3·18. In the United States the English weight is mostly used.

The word carat, as used by refiners, though understood to mean a weight equal to 4 grains troy, simply expresses a $\frac{1}{24}$. Thus, " 18-carat gold " would imply $\frac{18}{24}$ of pure gold.

The *Grain Troy* is the unit basis also of Apothecaries' and

Avoirdupois weights.   The pound Troy and the pound Apoth-
caries' weight each contains 5,760 such grains ; the pound
Avoirdupois contains 7,000.

The *Ounce Troy*=480 grains or 20 pennyweights is now by
legal enactment the standard weight to be used in the buying and
selling of gold, silver, and platinum, and is to be decimally
divided ; so that the grain and pennyweight, though from long
custom still familiarly used, are not recognized in law.

Self-closing Diamond Tweezers.—The self-closing tweezers
shown in the engraving are useful for holding a stone
which it is desired to examine or display.   Their action is similar
to that of the well-known universal watch key.   The jaws of the

tweezers open to receive the stone when the projection at the end
of the handle is pressed, and clasp it sufficiently tight when the
pressure is removed.

Prest, Thomas.—Patentee of keyless action for watches.
(No. 4501, October, 1820.)   Prest was foreman to Arnold.   His
mechanism was for winding only ; it contained no provision for
setting hands, and is said to have been used as an adjunct to
watches for the blind.

Pump Centre.—[*Pompe à centrer.—Die Zentrierspitze.*]—The
centre of a mandril which is projected by a spring and kept back
by a bayonet joint.

Pump Cylinder.—[*Mesure à piston.—Das Höhenmass (für grosse
Arbeit).*]—A sliding telescopic gauge used by chronometer makers
for taking heights.   A helical spring expands the gauge, and a
set screw tightens the sliding piece to secure a measurement.

Push Piece.—[*Poussitte.—Der Druckknopf.*]—The knob pushed
in from the pendant to open the case of a key-winding watch.
In pendant-winding hunting watches, the button and winding
stem form the push piece.

(2.)   A little projection on the band of the case in pendant-
winding watches, which is pushed in when it is desired to set the
hands of the watch.

Quare, Daniel (born 1649, died 1724).—The invention of
repeating watches is claimed for Quare as well as for Barlow.
There is no doubt that Quare made repeating watches of his own
design about 1680.   Prior to that, he had applied motion work
for driving concentric hands of watches and clocks.   He was
buried in the Quakers' ground at Bunhill Fields.

Quarter Clock.—[*Horloge à quarts.—Die Vierteluhr.*]—A clock
that strikes or chimes at the quarter hours.   (See Striking Work.)

Quarter Rack.—[*Pièce aux quarts.*—*Der Viertel-rechen.*]—The rack that regulates the striking of the quarters in the clock or repeater. The sketch shows a quarter rack for a repeater. (See Striking Work and Repeater.)

Quarter Screws.—[*Vis réglantes.*—*Die vier Regulierschrauben (Unruh).*]—The four screws in a compensation balance that are used for getting the watch to time.

Quarter Snail.—[*Limaçon des quarts.*—*Die Viertelstaffel.*]—The snail used in the quarter part of clocks and repeating watches.

Rack.—[*Rateau.*—*Der Rechen.*]—A straight bar or segment of a circle with teeth. (See Cremaillere, Hour Rack, Quarter Rack, Rack Lever, etc.)

(2.) A wooden stand with wedge-shaped notches for the reception of watches while they are being timed in positions.

Rack Lever.—[*Ancre à rateau.*—*Der Ankergang mit Rechen und Trieb.*]—A watch escapement patented by Peter Litherland, in 1791. There was no impulse pin as in the present form of lever escapement, but the lever terminated in a circular rack, which worked into a pinion on the balance staff, so that the balance was never detached. Watches with this escapement were made in large quantities at Liverpool during the early part of the present century. The rack lever is said to have been invented by the Abbé Hautefeuille, sixty years before the date of Litherland's patent.

Rack Hook.—[*Grappe.*—*Die Einfallschnalle.*]—In a striking clock the lever with a click that takes into the teeth of the rack. (See Striking Work.)

Ratchet. Ratchet Wheel.—[*Rochel.*—*Das Sperrad.*]—A wheel with pointed teeth fixed to an arbor to prevent its turning back. The fronts of the teeth of a ratchet wheel are radial, and the backs are straight lines running from the tip of one to the root of the next tooth. Ratchet wheels are right or left-handed according to the direction of the teeth. A pawl or click is a necessary adjunct to a ratchet wheel. This pawl or click is a finger, one end of which fits into the teeth, and the other is pivoted on a tangent to the wheel ; so that the force of the wheel, in striving to turn back, is transferred to the pivot, which in its turn must be fixed to some resisting base capable of withstanding the pressure. In watches, chronometers, and clocks with mainsprings, a ratchet wheel is fixed to the barrel arbor to prevent the mainspring flying back after it is wound ; the click being pivoted to the plate or bar in which the barrel arbor works. In going-barrel keyless watches this wheel has epicycloidal teeth to gear with the winding work, but it is still spoken of as a ratchet. The teeth of lever escape wheels when pointed are sometimes spoken of as ratchet teeth.

Rating Nut.—[*Vis réglante.—Die Regulierschraube (Pendel).*]— A round nut with a milled edge screwed to the pendulum rod of a clock. It supports the pendulum bob, and by turning it to the right or to the left, the bob is raised or lowered, and the time-keeping of the clock altered. In the finest clocks a scale is engraved round the rating nut to serve as a guide to the amount it is turned.

Rat Tail File.—[*Lime queue de rat.—Die Rundfeile.*]—A tapered round file.

Recoil Escapement.—[*Echappement à recul.—Die Anker-hemmung mit Rückfalt.*]—An escapement in which each tooth of the escape wheel, after it comes to rest, is pressed backward by the pallets. (See Anchor Escapement.)

Red-Stuff.—[*Rouge.—Das Stahlrot.*]—Sesquioxide of iron used with oil for polishing brass and steel work. Crystals of sulphate of iron are subjected to great heat, and then graded into polishing stuff of various degress of fineness. The more calcined part is of a bluish purple colour, coarser and harder than the less calcined, the finest of which is of a scarlet hue. Clockmakers use the bluish purple under the name of " Crocus " ; and the scarlet, known as " Rouge," is esteemed for the polishing of silver plate, watch cases, and the like. Watchmakers use four grades; the coarsest, known as " Clinker," is used for giving a surface to steel after it is tempered. " Coarse " is used next for steel and for polishing brass. " Medium " is used to finish steel that has to be blued, and " fine " for polishing bright steel. " Fine " red-stuff must not be used for steel that has to be blued, or the colour will not be even. (See Polishing.)

Regulator.—[*Régulateur.—Die Normaluhr.*]—(1.) A standard clock with compensated pendulum. (2.) The lever in a watch by which the curb pins are shifted.

A few words on the construction of a watchmaker's regulator would not, I am told, be unacceptable. A good arrangement is shown in Figs. 1 and 2. The plates for the movement are, when finished, 8·75 in. by 7·5 in. by ·15 in. thick. They are made of brass, well hammer-hardened, flattened, and filed to a true surface. The plates are pinned together by two small pins near the top and bottom respectively, and the edges finished. A line is then struck on one surface from top to bottom in the middle of the plate, and another line crossing it at right angles 4·4 inches from the bottom will mark the spot for the centre arbor. From this point the centres of the other wheels are set off by the depthing tool*, and

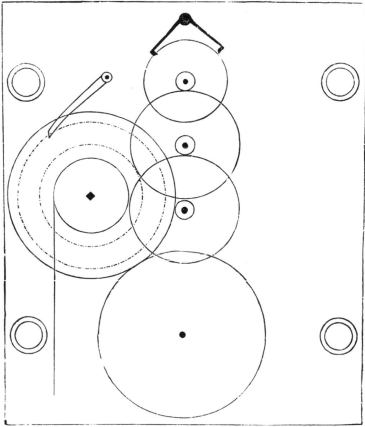

Fig. 1.—REGULATOR MOVEMENT.

Great Wheel—144 teeth—Pitch diameter, 3·4 inches.
Hour   do.   144 do.   do.   do.   3·4 do.
Centre do.   96 do.   do.   do.   2·26 do.
Third   do.   90 do.   do.   do.   2·11 do.
Escape do.   30 do.   do.   do.   1·75 do.
The Pinions have all 12 leaves.

* In the absence of a depthing tool, the correct depths of the wheels and pinions may be ensured by the tool described under the head of Clock Repairing, page 101, or, failing that, by marking off with the compasses on a strip of brass the proper distance of centres approximately, taking care that the centre distance is short rather than long. Then drill holes in the brass to receive the arbors, and stretch the brass a little at a time by hammering it, trying the depth after each stretching, till the depth is perfect ; the holes in the strip of brass are then the template for drilling the holes in the clock plates.

w

the pivot holes and pillar holes drilled and broached before the plates are taken apart. For the lighter arbors oil sinks may be formed out of the thickness of the plate, but the holes for the

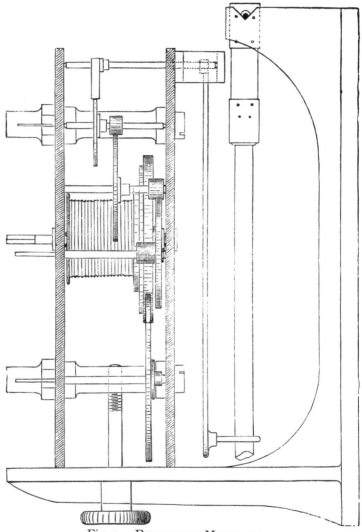

Fig. 2.—REGULATOR MOVEMENT.

barrel arbor should be bushed to get a longer bearing. The arbor of the wheel below the centre wheel, which is driven by the great wheel, carries the hour hand, and the escape wheel arbor the seconds hand, so that no motion work is needed. The train wheels are all of the same pitch, those requiring to be of extra strength being thicker than others that have less strain on them.

Using wheels of the sizes and numbers given throws the centre of the seconds hand and the centre of the hour hand each $2\frac{1}{2}$ inches from the centre arbor, which allows of seconds and hour circles as large as can be conveniently arranged on an eleven-inch dial, and bold enough for a twelve-inch one if such a size is desired. The barrel is an inch and a-half in diameter, and goes round once in 12 hours, making therefore 16 rotations in 8 days. Assuming that the weight has a pulley, a case allowing 4 feet of fall would be ample. The weight pulley should be about 3 inches in diameter. There would be less friction in working if the barrel were on the other side of the centre arbor and the cord fell between the centre pinion and the centre of the barrel, but then the weight would be closer to the pendulum and would affect its vibration more, so that the barrel is much better planted as shown. The late Astronomer-Royal felt so strongly the desirability of keeping the pendulum as far away from the influence of the weight as possible, that he not only had the weight of the Standard Sidereal Clock at the Observatory carried to the extreme corner of the case, but had a wooden shoot made to enclose it. The grooves in the barrel are better cut so that the cord winds from the front to the back of the case, for the weight as it descends is then carried more to the front of the case, and still further from the pendulum bob. The pillars should not be planted at the extreme corners of the plates, but equidistant from the centre arbor and from each other, and also as close to the arbors as is practicable. The pillars are half an inch in diameter with enlarged collars to give a good bearing surface on the plates, and each end beyond the collar about ·3 inch diameter and ·1 inch long is carefully turned down to fit the holes in the plates. The front pillar screws have hexagon heads with a round prolongation to serve as a seat for the dial, and the dial screws are then symmetrically placed where they do not interfere with the figures. The bottom pillars are square in the middle where the screw holes for fixing the movement to the seat board are tapped. The barrel is hollow, having the smaller ratchet cast on it, and is bored out to receive the front end, which is cast separately. It should be squared on to the arbor, and then turned on it. The barrel arbor has a solid collar at the back end to keep the great wheel and the larger ratchet wheel in their places. Details of the maintaining work may be gathered from the drawing under the head of " Maintaining Power." It will be seen that the bracket carrying the pendulum serves also as a seat board for the movement. This bracket is of cast iron, and the heavier it is the better.

The seconds hands of a regulator should not be turned back, for by so doing the backs of the escape wheel teeth scrape against the pallets, and so re-oil them, which is very likely to alter the rate of the clock.

When about to wind a regulator, or indeed any clock with maintaining power, it is always prudent to first press the winder the reverse way, to ensure that the maintaining click is properly in the ratchet, otherwise it may happen when the click is just on the corner of a tooth, it is jerked out by the action of winding, and the clock momentarily stops.

Reid, Thomas (born 1750, died 1834), a celebrated Edinburgh clockmaker. Author of *A Treatise on Clock and Watchmaking*, containing the experience of a practical and clever man.

Remontoire.—[*Force constante.—Die Hemmung mit konstanter Kraft.*]—A spring or other device which is wound by a clock and discharged at regular intervals. The function of a remontoire is generally either to impart impulse to the pendulum or to cause the hands of the clock to jump through certain spaces. Though this word comes from the French, it is not now used in that language, except in the sense of a stem winder.

Repeater.—[*Montre à répétition.—Die Repetieruhr.*]—A repeating watch. A clock or watch in which mechanism can be set going to denote the time approximately by hammers striking on bells or gongs. In a quarter repeater the last hour is struck, and afterwards the number of quarters that have elapsed since. A minute repeater in addition strikes the number of minutes since the last quarter. Half-quarter repeaters, instead of giving the minutes, strike one additional blow if the half-quarter has passed. Five-minute repeaters give after the hour one blow for every five minutes past it.

Repeating Watches.—Repeaters were first made about 1686. The honour of the invention is claimed for Daniel Quare and for Edward Barlow, a clergyman, who was undoubtedly the inventor of the rack striking work for clocks, which was originally applied by him simply as repeating work.

The engraving is a very fair representation of half-quarter work. The parts are sometimes differently arranged but the principle of all is the same.

The small mainspring which supplies the power for repeating is wound up by the wearer pushing round a slide that projects from the band of the case. This slide is the extremity of a lever which presses against a pivoted rack engaging with a segment on the barrel arbor. There is underneath a segment of greater radius containing twelve ratchet teeth. The number of hours to be struck is regulated by the position of the hour snail in precisely the same way as is the striking work of a clock. At twelve o'clock the lowest step of the snail is presented to the stop, so that the rack can be traversed its full extent. In returning, each one of the twelve ratchet teeth in turn lifts the hammer which strikes the hours. To sound each quarter, two blows are given in quick succession on different gongs, the quarter rack being furnished with two sets of three teeth each; one set lifts the hammer which gives the first note of the quarter and the other set lifts a separate hammer for the

HALF-QUARTER REPEATING WORK.

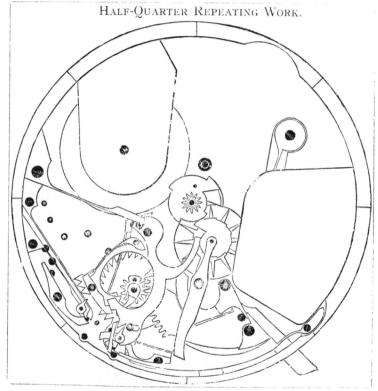

second note. As the slide is moved round, the all-or-nothing piece, as it is called, releases the quarter rack, and as a spring is constantly pressing towards the quarter snail it moves in that direction till it is stopped by the quarter snail. After the hours are struck a curved finger, or gathering pallet, on the barrel arbor presses the quarter rack to its original position, and, in passing, each of the ratchet teeth, by pushing aside a pallet fixed to the same arbor as one of the hammers, causes a blow to be struck. Whether one, two or three quarters are struck depends, of course, on the position of the quarter snail.

The half-quarter rack, with but one ratchet tooth, is placed on top, and works with the quarter rack. Between each quarter and seven minutes past the tooth of the half-quarter rack is exactly coincident with the extreme right-hand tooth of the quarter rack and is kept in that position by a spring catch, so that it does not act.

The quarter snail, attached to the cannon pinion, is of two thicknesses, with steps just dividing each quarter space, so that after the half-quarter the half-quarter rack gets round a little nearer to the centre of the snail than the quarter rack. The spring catch is sprung out, releasing the quarter rack, which a slotted hole permits to move to the position shown in the drawing, and then, after the quarters have struck, it lifts the hammer and strikes one more blow.

Seeing the difficulty that the half-quarter action presents to many repairers, I will give another example of it, in which the result is attained, with slightly different mechanism, for which I am indebted to Mr. Etchells.

Here, the spring catch locks the rack in the position for giving the half-quarter and is pushed out by the pressure of the rack tail on the snail when the quarter alone is to be sounded. The engraving shows in separate pieces the quarter rack, the spring catch, and the half-quarter rack ; to the right of these the parts are shown together ; on the extreme right is an upper and an under view of the quarter snail, the last-named showing clearly the divided steps on each quarter space.

The hour snail is mounted on a star wheel, as shown on page 341, and the star wheel is moved by a pin in the quarter snail, or rather in the loose surprise piece underneath, which flies out to the position shown in the drawing directly the star wheel is moved. The surprise then prevents the quarter rack reaching any step of the quarter snail, and consequently no quarters are struck. When the pin in the surprise piece comes round to the star wheel again, the pressure of the pin on a tooth of the star wheel causes the surprise piece to retire, so that at the proper time the third quarter and half-quarter can be struck, but as the star wheel jumps forward the succeeding tooth flirts out the surprise.

The hammer arbors go through the plate, and the hammers are on the other side. The gongs of steel wire, fixed at one end to the plate, curl round it and lie between the plate and the band of the case.

There is also on the other side of the plate a train of runners for regulating the speed of striking. The centres of the wheels are indicated by dots on the left hand of the barrel. The last pinion is sometimes furnished with a collet to act as a fly, and it is usually mounted in an eccentric plug, by means of which its depth can be increased or made shallower. This is found to be sufficient regulation, though latterly an escape wheel and pallets have been applied at the end of the train of runners to regulate the speed in some repeaters. This is, perhaps, more scientific than making a bad depth, but the pallet staff holes are found to wear very much if not jewelled.

Minute repeaters are a most troublesome and delicate complication. Secured to the quarter snail is a four-armed surprise piece (Fig. 7) actuated by a click, so that, during the interval between the completion of each hour and one minute past, the rack is kept away from the four-armed minute snail (Fig. 8). This snail is secured to the set arbor, below the cannon pinion. Each of its arms has a set of 14 steps to give the number of minutes past each quarter hour.

Fig. 7.   Fig. 8.

Five-minute repeaters seem likely to grow in favour by reason of their simplicity, for they evidently involve but little more work in construction than quarter repeaters, and yet give the time more closely than half-quarters. In place of the quarter snail there is, on the centre arbor, one with 11 divisions to act with a rack having the same number of teeth for giving from 1 to 11 blows.

Rusty Gongs give a bad sound. They may be improved by polishing first, with a half-round steel polisher, and then finished with box-wood polisher on cork.

Examination.—If the repeating train will not run down when the slide is pushed to its farthest, see if the slide moves freely in the case ; if the lever moved by the slide is free and not liable to jamb ; that a particle of dust is not stopping one of the small wheels ; that none of their pivot holes have become too wide and caused a false depth ; that the gathering pallet is free to move on or with its arbor.

The fly pinion is a prolific source of trouble. If its flat shoulder is in contact with a hollow sink it will cut a groove in the sink and in time stick. Sinks for pinion shoulders should always be finished with a flat cutter ; if the pinion is a little out of round or badly sized it may stop the train when its depth is altered to regulate the speed ; if it has a weight the weight may foul ; see that the pinion end shakes and freedom are correct ; that the weight is free even when out of upright, to alter the depth ; listen to its running, if faster in one position than another it indicates either weight fouling, bad fitting, a dry hole, or want of upright. A weighted pinion may be set upright, and then, if it runs down too fast, the weight may be made to offer more resistance by forming it like a balance spring collet with flats on the round part. If it run too slowly shorten the weight rather than lessen its diameter, for shortening gains freedom, and you retain the power of forming flats if afterwards it is required to run slower.

In taking the movement out of the case if the gongs cannot be first removed, be careful and not distort them ; do not mix the screws ; be sure and not bruise the edge of the case where the slide works ; if a new screw has been tapped into the plate see that no burr remains ; see that the main wheel of repeating train turns freely on its arbor ; that other wheels are secure on their pinions ; that the quarter snail is firm on the cannon pinion and perfectly flat ; that all studs are perfectly upright ; all steady pins secure ; no burrs anywhere ; no bristle left in a tooth, under a stud or in any other crevice. Failure may be caused by the mainspring being bad and unadjustable, or binding in the barrel, or through oil that has become too thin. The mainspring should be fully set up to obtain the most uniform striking that is possible.

Half-quarter repeaters often fail through the pivot holes of the hammer arbors wearing. A tooth or teeth in these racks may be short even when the arbors fit the holes. The rack teeth may then be filled up again, carefully trying each tooth of both racks.

If any one of the springs fails to take home well the piece it drives, do not set it immediately, but first ascertain that all the parts in connection with it are well free, clean, oiled where needful, and without burr. If certain the spring is too weak, set it on, *but only just enough* as oversetting may derange the whole motion. For instance, if the half-quarter rack does not fall well home and its spring be set on to make it reach the snail properly, it is quite likely that the extra drag will be too much for the mainspring, and the rack will not be carried back as it should.

Too much oil is often applied, and very often in the wrong places. The studs and acting spring ends are essential places for the smallest portion of oil ; but care should be taken that no oil flows between the racks, or between the surprise piece and the snail, and none should be given to the case slide.

The " Astra " Repeating Mehanism.—Under this title there has been recently introduced a new form of mechanism for quarter-repeating watches, which appears to be of a simple and effective character.

The usual driving barrel, with its coiled mainspring and accessories, is dispensed with ; there are but one gong, one hammer, and one striking pallet.

The running train will be found in Fig. 1, under the bridge q. It consists of three wheels, z, z^1, z^2, and the piece z^3. The pinion t, which engages directly with the teeth of the hour rack, is attached to the wheel z by means of a ratchet and click, so that it may turn in one direction by itself.

The hour rack s (Figs. 1 and 5) has 20 teeth and is movable on a pivot c at its axis s^4 through the arm s^2 which passes through the band of the case, and to start the mechanism is pushed in the direction of the arrow. The spring p, held by the screw p^2, presses against the projection s^1 of the hour rack, and so drives it back. The rack by its teeth moves the pallet e, e^1, which is in connection with the hammer a, as well as actuating the running train. On the hour rack is a lever b (Fig. 4), which has two round holes, b^1, b^2, and the screw hole b^3. By the hole b^1 the lever is fitted to its pivot, the head of the screw c, the screw c^1, which is fixed firmly in the hour rack, passes through the hole b^2, which allows of a little movement. The movement of the lever on the rack will be distinguished in Fig. 5, where the second position is indicated by dotted lines. In the hole b^3 a screw d^1 is fixed forming a pivot for the click d, which has a hook d^3 at one end and a spring d^2 at the other. As the rack is carried round by moving the arm s^2 in the direction of the arrow, the point b^4 of the lever is pressed against the hour snail h, and the head of the screw c^1 presses against the click, and so releases the quarter rack, which is pushed home to its snail, m, by the spring f, acting against the pin v^2. The quarter rack moves on a pivot v^3 in the same plane as the click d. This rack has three pairs of teeth, as shown in Fig. 2. Until the quarter rack is pushed aside the curved end v^4 presses against the striking pallet at e^1.

The hammer a (Figs. 1 and 3) is under the plate, moves round its axis a^1, and is furnished with two long pins, a^2 and a^3, projecting through the curved passage w of the plate. The counter spring k by acting on the pin a^2 regulates the loudness of the striking.

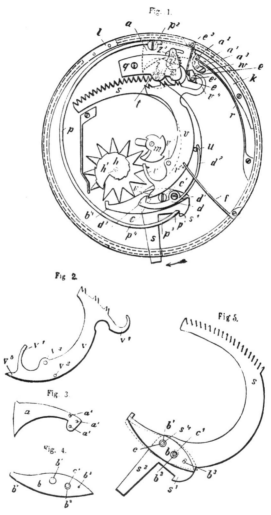

Fig. 1.

Fig 2.

Fig 5.

Fig. 3.

Fig. 4.

The pallet pivoted to the plate by means of the screw e^2, **puts** the hammer in motion by the finger e^1, which presses against the hammer through the pin a^3; the other end of the **pallet** engages with the teeth of the hour rack, by which it is actuated, one blow being given for every tooth, till the last tooth of the hour rack has passed the pallet. The striking of the hour is then completed, but the hour rack continues its motion and the pin *u* presses the quarter rack back, so that as each pair of its teeth passes the pallet, two blows in quick succession are given, the hammer not being lifted so far as for the hours, the sound of the quarters is unmistakably different. The spring e^3 ensures the return of the pallet to the position of repose as in Fig. 1.

Repeating Rack.—[*Rochet des heures.—Der Repetierrechen.*]— A rack in a repeating watch which is shifted one tooth for each blow that is struck.

Repeating Slide—[*Tirage de répétition.—Der Repetitions-schieber.*]—The slide on the band of a repeating watch case that is moved round to set the repeating work in motion.

Resilient Escapement. — [*Echappement résilient. — Ausweichende Prellung (Ankerhemmung).*]—A form of the lever escapement, in which, when the impulse pin presses on the outside of the lever, it yields and allows the pin to pass. (See Lever Escapement.)

Reversed Fusee. Left-handed Fusee—A watch movement arranged so that the pull of the chain takes place on the same side of the centre of the fusee as the force is communicated

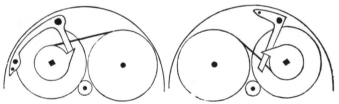

Fig. 1.   Fig. 2.

to the centre pinion, whereby the friction on the fusee pivots is lessened. Fig. 1 shows the parts planted in the usual way, and Fig. 2 the reversed fusee. I must confess that the titles for this arrangement are singularly unfortunate, as the fusee is neither reversed nor left-handed. The hooking in of the mainspring is reversed, and the barrel pulls from the other side of the same fusee. (See Stopwork.)

Ridged-back File.—A file, the cross-section of which is a very flat triangle, whereof the base forms the cutting surface.

Riveting Stake.—[*Outil aux trous.—Das Nietstöckchen.*]—A cube of steel having a polished face and pierced with a series of different-sized holes for the reception of arbors, so that the pinion or collet to be riveted finds a resting-place round the edge of a hole of suitable size (Fig. 1). A jointed stake made in

halves and hinged at one end, so that it may be opened to admit
an enlarged part and then closed round the arbor, is handy in

some cases. Boley has
provided a bolster for
riveting stakes together
with interchangeable
stakes of steel and brass

Fig. 1.

of the form shown in the figure. The ability to at once substitute
a brass for a steel stake, or *vice versa*, is a convenience.

The "*Staking*" or *Riveting Tool*, Fig. 2, has a shifting table
or stake round which holes of various sizes are arranged in a
circle, so that any particular hole may be brought exactly under
a suitable punch moving in a vertical guide. Punches for every
operation in the ordinary run of watch repairs are provided, and
there are a number of small stumps or stakes which fit into the
largest hole in the table. These are adapted for riveting

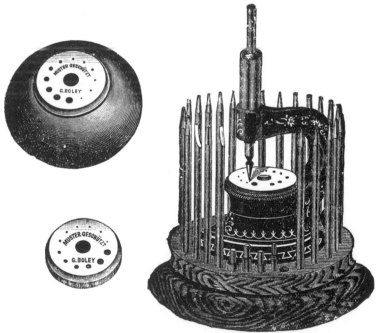

Fig. 2.

stoppings into sinks in watch plates and other purposes. One
great advantage of this tool is that a direct action of the punch
is ensured, so that irregular riveting and accidents arising from

an awkward blow in the ordinary way are avoided. A somewhat similar tool to that shown in Fig. 2 is used for closing

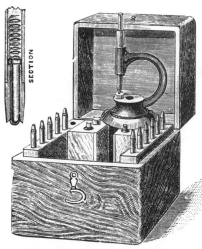

holes in watch plates. Fig. 3 is a good example of this kind by Boley. There are no holes in the table, and the punches are cupped so as to stretch the material of the watch plate towards the hole. The punches are formed with a pump centre, to ensure their acting concentrically round the hole in the plate, as shown in the enlarged section. Fig. 4 is a pedestal stand and an assortment of riveting stakes. With this useful collection is supplied an ingenious thimble for holding a punch or other tool by means of the forefinger. This leaves

Fig. 3.—Hole Closer.

the rest of the left hand free to support the work, and so enables

the operator to drive a minute hand and accomplish many other operations single-handed, and more easily than he could do otherwise.

Rocking Bar. Yoke.— [*Plateforme bercante.—Die Wippe.* (*Bügel aufzug*).] — A variety of keyless mechanism. The steel bar which carries the intermediate

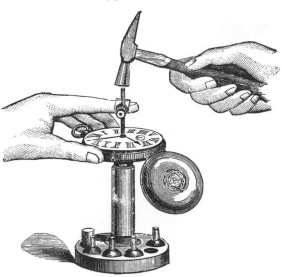

Fig. 4.—Riveting Thimble, Stand and Stakes.

wheels in going-barrel keyless mechanism. (See Keyless.)

Roller Remover.—More than one special tool has been introduced for removing a roller from a lever balance staff by drawing it forward with a screw. Such a method of procedure is not without danger to the staff in the case of a tight roller,

and a roller may always be more easily and safely removed by impact than pressure. Mr. Plose recommends a steel table having a taper slit, and made sufficiently thin round the slit to clear the collet as a bed for the roller, as shown in the sketch. With the roller resting on this table, the staff may be safely driven out by means of a hollow punch as depicted under the head of " Staff Punch."

Rose Cutter.—[*Tasseau fraise à trou.—Die Hohlfräse.*]—This adjunct to a lathe, which is fixed to the spindle in the same manner as a chuck, is exceedingly useful for quickly reducing pieces of wire for screws, &c., to a gauge. For screws, the wire would be of a proper size for the screw heads, and a cutter selected with a hole the size of a finished screw. The point of the wire is rounded to enter the hole of the cutter against which it is forced by the back centre of the lathe, the serrated face of the cutter rapidly cutting away the superfluous metal, the part

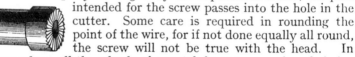

intended for the screw passes into the hole in the cutter. Some care is required in rounding the point of the wire, for if not done equally all round, the screw will not be true with the head. In cases where all the wheels of a watch have too much end shake, a rose cutter may be used to all the pillars in succession, to reduce their height.

Rouge.—See Red-Stuff.

Rouge Leather.—[*Peau chargée de rouge.—Das Waschleder mit Rot.*]—A chamois leather permeated with rouge.

Round Broach.—[*Alésoir.—Die Glättahle.*]—See Broach.

Rounding-up Tool.—[*Machin à arrondir.—Die Wälzmaschine.*] —An engine for slightly reducing the size of watch wheels. This ingenious tool, which is shown in the figure, is one of the most useful to watch jobbers. By its aid, a wheel may be almost instantly reduced in diameter ; corrected if out of round, or have the form of its teeth altered as may be required.

The cutters are serrated, and for a little over half a circle concentric in form ; they terminate in a screw-shaped adjustable guide, which conducts each space in succession to the action of the cutter. While one end of the guide meets the cutter, the other angles a little, so that instead of meeting the other extremity of the cutter when the circle is completed, it leaves a space equal to the pitch of the wheel to be cut. By this means, after the cutter has operated on a space, the wheel is led forward one tooth by the time the cutter arbor has completed its revolution.

Some little practice is required to select exactly the cutter required. Care must be taken not to use one too thick, or the teeth will of course be made too thin, and the wheel probably bent. When the guide is adjusted to the pitch, it will be well to see that it enters the space properly before rotating the tool quickly. The wheel should be fixed firmly, but not too tight between the centres, which should rest well on the shoulders of the pinion. The rest piece for the wheel should be as large as possible to keep the wheel from bending, to give it firmness, and to ensure a clean cut.

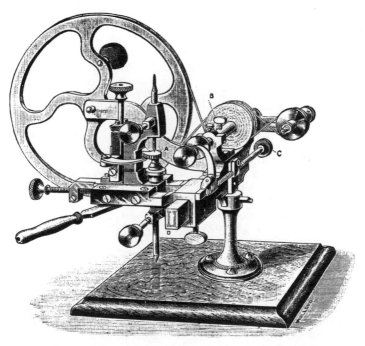

The slide which carries the cutter is generally adjusted by the eye. This is unsatisfactory, and occasionally results in error, even with experienced workmen, for, owing to the position of the top runner, it is difficult to get a good view of the cutter and the work, and, as the cutters are often more or less out of flat, their action can only be guessed at. Mr. Edward W. Webb has devised the registering arrangement shown in Fig. 2, which is a plan of the lower slide and part of the table in front of the cutter. To the table is fixed a small, circular scale. A corresponding vernier is carried at one end of a lever, which works in jewelled holes, and is accurately fitted. There is a spindle

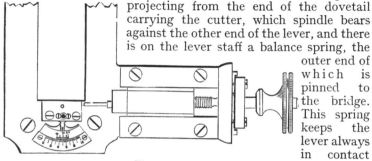

projecting from the end of the dovetail carrying the cutter, which spindle bears against the other end of the lever, and there is on the lever staff a balance spring, the outer end of which is pinned to the bridge. This spring keeps the lever always in contact with the

Fig. 2.

spindle projecting from the dovetail. All that is needed now is to note the position of the vernier when each one of the cutters is adjusted for certain work, to be sure of being always able to adjust it again in the same place, and of obtaining the same result. This will be found especially of value in fixing cutters for cutting one side of the teeth of wheels, and for altering the form of the teeth.

Recently a larger rounding tool suitable for clock and telegraph instrument wheels has been introduced.

Ruby Pin.—[*Cheville de plateau.—Der Hebestein.*]—The impulse pin in a lever escapement.

Ruby Roller.—[*Rouleau en rubis.—Die Steinrolle (Duplexhemmung).*]—The roller in a duplex escapement against which the teeth of the escape wheel are locked.

Run.—[*Chemin perdu.—Der verlorene Weg. (Ankergang).*]—In a lever escapement, the movement of the lever towards the banking pins, after the wheel has " dropped " on to the locking, caused by the inclination of the escape wheel teeth and the locking faces of the pallets. (See Draw.)

Runners.—[*Les broches.—Die Spitzen (Drehstuhl).*]—(1) The round pieces of steel or brass in a pair of turns by which the work is supported.

(2) A small train of wheels used in repeating watches to regulate the speed of striking.

(3) An idle wheel is also sometimes called a runner.

Rust.—[*Rouille.—Der Rost.*]

To Keep Steel Articles from Rusting.—Cover them with

powdered quicklime.   If they must be exposed, place near them a small open vessel containing chloride of calcium.   Immerse rusted steel articles for a few minutes in a strong solution of cyanide of potassium, and they will clean much easier.

Steel or iron that has been immersed in caustic soda will resist rust for a long time.

Spots of rust on chronometer springs and other steel pieces are generally rubbed with a piece of brass, but Mr. Kullberg tells me the best plan after cleaning the spots is to apply a little spirits of ammonia to them.

Safe Edged File.—[*Lime à bord non taillé.*—*Feile ohne Seite hieb.*]—A file with a smooth edge.

Safety Pin.—See Guard Pin.

Safety Pinion.—[*Das Sicherheitstrieb (Grossbodenradtrieb).*]— In a going barrel watch a centre pinion which allows a barrel to recoil when the mainspring breaks.   The Waltham Watch Company screw the pinion head on to the arbor, and backward pressure then causes the pinion head to rotate on the screw.   Mr. Alex. Edwards has a thin ratchet fixed to the centre pinion, the click for which is pivoted on the centre wheel.   The centre wheel being loose on the arbor, the pinion and barrel are free to move backwards.

Savage, George.—A watchmaker who, in the early part of this century, did much to perfect the lever escapement by good work and nice proportion, besides inventing the two-pin variety. He spent the early part of his life in Clerkenwell, but in his old days emigrated to Canada, and founded a flourishing retail business in Montreal, where he died.

Scorper.—[*Der Schaber.*]—A kind of graver used for squaring the corners of sinks, easing watch bezels, and other purposes.

Screw.—[*Vis.*—*Die Schraube.*]—The standard recommended by a Committee of the British Association and adopted by many watch manufacturers is a **V** thread of $47\frac{1}{2}°$, rounded top and bottom through 2-11ths of the height, and the pitch P is directly related to the diameter D by the formula $D=6P\frac{g}{5}$.   To formulate a standard series the successive powers of 0.9 mm. are used for the pitch.   The index of the power is taken as a designating number for the screws : thus the pitch of No. 6 screw is got by raising 0.9 mm. to the sixth power, the pitch being therefore 0.53 mm. and the diameter 2.8 mm.   From the figures so obtained in decimals of a millimetre a series is got in decimals of an inch.   No rule can be framed for proportions of screws as the best under all circumstances.   For large clocks the Whitworth thread of $55°$ is employed.   Obviously where the plate or nut to be threaded is very thin a fine pitch is desirable, but a fine pitch is in more danger of being stripped than a coarse one, and fine pitched screws are liable to overturn.   In

x

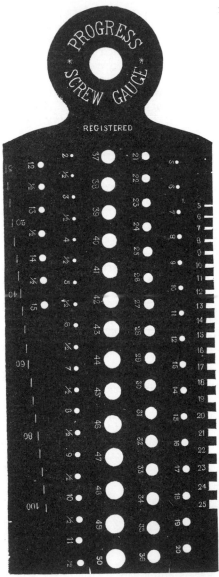

retapping a stripped hole a coarser thread gives more chance of success. The subjoined engraving represents a gauge issued with the " Progress " metrical system of screws adapted for small mechanism.

The notches at the right-hand edge are for measuring the thickness of the head. The series of holes consisting of three rows coming next, are for the diameter of the head ; the second series for the size of the body or screwed part ; and the short lines rising in steps from the left-hand edge are for gauging the length of the body. It will be understood that this last measurement is obtained by placing the other side of the head against the edge of he plate, and running the screw along until its point coincides with one of the lines. All the numbers on the gauge represent tenths of millimetres, one-tenth of a millimetre being accepted as the unit. For instance, the first of the notches, marked 5, is five-tenths of a millimetre in width ; the first hole in the second row, marked 21, is twenty-one tenths of a millimetre in diameter ; and the last short line, marked 100, is 100-tenths of a millimetre from the bottom edge of the gauge. These screws are to be had either grey or polished, and finished taps of the complete range of screws are also supplied.

To Extract Broken Watch Screws.—Make a cramp like the drawing below, large enough to reach across a watch plate, very strong just at the bow, so as to stand screwing up without springing. Provide two or three sets of steel screws with different-sized *hardened* points. To use it, tighten that screw of the cramp which is against the *point* of the broken screw, and when you have firm grip turn the whole tool round, and the broken screw will invariably be drawn out.

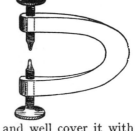

OTHER METHODS.—The cramp would be of no avail if the screw hole is drilled only partly through the plate. In such a case slightly warm the plate and well cover it with beeswax. Be careful not to let the wax touch the broken screw ; then make a solution of oil of vitriol—one part of oil of vitriol and 4 of water. Let it stand until quite cold, then put the plate in, and in a few hours the acid will dissolve the screw. The wax may be removed by warming it in olive oil, and washing it in hot soap and water.

To avoid the trouble of coating the plate with beeswax, many watchmakers prefer to boil it for a few hours in a strong solution of alum, which will be found to loosen the screw, and not affect the plate.

A Screw Rusted in may be loosened by placing the flat extremity of a red-hot stick of iron on it for two or three minutes. When the screw is heated, it will be found to turn easily.

Screw Arbor.—[*Arber-à-vis.—Der linke Drehstift.*]—A tool used for holding a wheel concentrically in order to turn it. Fixed to an arbor having centres is a shoulder or disc, against which the wheel is pressed to keep it flat, and on a reduced part of the arbor a sliding cone moves. The cone enters the hole in

the wheel to be operated on, and is kept in position by a nut. Screw arbors may, of course, be used without the cone, the object to be turned being simply gripped between the shoulder and the nut. A ferrule is fixed to the arbor so that it may be rotated with a bow or otherwise.

Mr. Plose states that by turning another cone on the back end of the conical bush and forming a corresponding recess in the nut as shown in the enlarged sketch, the object to be turned will be secured more truly concentric than would otherwise be the case.

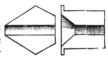

Screw Driver.—[*Tourne vis.*—*Der Schraubenzieher.*]—A screw driver for watchmakers' use should be as light as possible consistent with strength, properly proportioned to the work, with well-polished point of a width nearly equal to the diameter of the screw heads to be operated upon, and of a thickness to fit the slits with only sufficient taper to secure it from breaking. A tool with blunt taper will ruin the best of screws. Screw drivers made from pinion wire collect dust in the ridges, and are therefore objectionable. A better form is readily made from square steel twisted while hot.

A kind of double screw driver, useful for turning the timing *nuts* of a compensation balance, may be made from a spare pinion gauge by dressing the points to the required shape.

Screw Ferrule.—[*Cuivrot à vis.*—*Die Schraubenrolle.*]—See Ferrule.

Screw Head Tool.—[*Outil aux têtes de vis.*—*Die Schraubenpoliermaschine.*]—A tool in which screws are rotated while their heads are being polished. (See Polishing.)

Screw Plate.—[*Filière aux vis.*—*Das Schneideisen.*]—A plate of steel in which holes are drilled of sizes varying by regular gradations. In each hole a thread is formed by means of a tap, and the plate is then hardened and tempered. In use a piece of steel, softer than the plate and of a proper size, is passed through one of the holes by carefully turning the plate round and round. A thread corresponding to the one in the hole is thus produced on the steel, partly by cutting and partly by squeezing the softer metal. The larger-sized holes are slit to give a cutting edge to the thread in the plate.

To Make a Left-Handed Screw Plate.—Screw a piece of steel of the desired size in an ordinary right-handed screw plate. Then file it away to a feather-edge, something the shape of the shaded part in the sketch annexed, and harden it. A good left-handed screw plate may now be cut with the tap thus made if it is turned the reverse or left-handed way.

Screw Tap.—[*Taraud. Der Schneidbohrer.*]—A tool used to produce a screw thread in holes. Taps for watchmakers' use are made by running a piece of steel through a screw plate. Longitudinal slits are then filed through the screw, in order to give a cutting edge (unless the tap is very small), and the piece of steel, hardened and tempered, becomes a tap. Very small taps are filed square or three-sided after the thread is formed in order to give a cutting edge. The upper part of a tap is generally larger than the screw part, and in turning the tap care should be taken to make a long curve between the two, for if a square shoulder is turned there the tap is very likely to break in use.

Seat Board.—[*Support de mouvement.*—*Das Postament.*]—In a long case clock, the shelf that supports the movement.

Seconds Hand.—[*Aiguilles de secondes.*—*Der Sekundenzeiger.*] —The small hand that goes round once in a minute on the dial of a watch or clock.

Seconds Pivot.—[*Pivots de secondes.*—*Der Sekundenzapfen.*]— The prolongation of the fourth wheel arbor to which the seconds hand of a watch is fixed.

Secret Springs.—[*Ressorts secrets.*—*Die Gehäusefedern.*]—The fly and lock springs of a watch case. (See Case Springs.)

Sector.—[*Compas de proportion.*—*Der Proportionszirkel.*]—A proportional gauge consisting of two limbs joined together at one end ; used principally for sizing wheels and pinions. This invaluable and simple tool appears to be but rarely understood. The measuring of wheels and pinions by its means is but one of its many uses. The sector is really a proportional measuring gauge, suited for nearly all requirements of the watch and clock maker.

Construction of the Sector.—The length of the sector is quite unimportant : it may be made of any size considered to be most convenient for handling. It consists of two brass limbs carefully jointed at one end so that the centre of the joint pin is in a line with the inner faces of the limbs. The inner edges should be perfectly true and parallel. By means of a brass arc and thumb-nut at the other end of the sector, the limbs may be fixed in any desired position. The following are the dimensions of one of Jump's sectors :—

Length of limbs	...	...	...	...	16·5	inches.
,, ,, from centre of motion				...	16·125	,,
Width of each limb	...	...	...	...	·8	,,
Length of slot in arc	...	...	...	...	4·625	,,
Width of slot	...	...	...	...	0·15	,,
Thickness of limbs ...	...	...	...	...	0·1	,,
Thickness of cover plates for joints		...	...	0·18	,,
Thickness of arc	...	...	...	...	0·05	,,
Diameter of steel joint pin		...	...	...	0·1	,,

In order to get the joint pin in a line with the faces of the limbs, a knuckle projects beyond the face of the right-hand or moving limb, and a corresponding piece is cut out of the left-hand or fixed limb. The cover plates for joints are strongly riveted to the fixed limb.

To Divide the Sector.—The first mark (120) having been made somewhere near the top of the limbs, the distance between this mark and the centre of the joint pin is divided into six equal parts, marked 100, 80, 60, 40, and 20 ; the last or zero is, of course, the centre of the joint pin. The part from zero to 20 is devoted to the sizing of pinions, and the division of it will be referred to presently. Each of the other five divisions is divided

358

into 20 equal parts as shown, at every fifth part a longer stroke is made, and at every tenth part a number, so that these points may be easily recognized. The numbers are placed on the fixed limb and may be repeated on the movable limb, or represented by a dot as shown. The limbs should be opened so that the distance between them at the 100 mark is exactly one inch, and a line drawn across the arc coincident with the inner face of the movable limb. This line is marked 1 inch. The limbs are then closed till the space between them at the 100 mark is half-an-inch, and another line drawn across the arc in a similar way. The process is repeated at ·25 and ·125 of an inch. If larger measurements than an inch are to be taken, other datum lines can be drawn across the arc as may be desired.

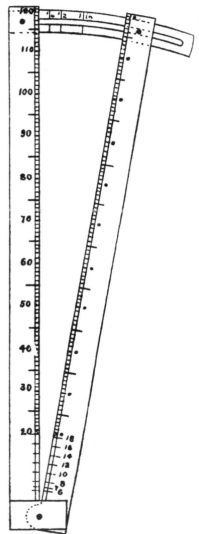

For the sizing of pinions, the edges of the limbs up to the Figure 20 are chamfered away to give a thinner edge. The 6 mark, instead of being $\frac{6}{20}$ of the whole distance between zero and 20, is placed $\frac{1}{30}$ higher, then from 6 to 8 is $\frac{2}{20}$ of the whole distance between zero and 20 ; 8 to 10, 10 to 12, 14 to 16, and 16 to 18, the same. The 7 mark is not midway between 6 and 8, but is placed $\frac{1}{20}$ of the distance between 6 and 8, nearer to 6. These are Jump's proportions, and they are pretty generally adopted. Some few consider the pinions should be larger, and therefore place the 6 mark higher than Jump. After all, the proper

The Sector of Proportional Gauge.

size of pinions depends upon the shape of the addenda. (See Article on Wheels and Pinions.)

The sector would certainly be much handier, especially for taking very small measurements, and for obtaining the pitch diameters of pinions when the centres and numbers only of a wheel and pinion are known if the limbs were divided equally all through. There does not seem to be any difficulty in doing so if the marks referring to the sizing of pinions were taken right across the limbs and the figures relating thereto placed on the outer edge.

Use of the Sector.—If the movable limb of the sector is fixed at the one inch line on the arc, the distance between the limbs at the 100 mark measures one inch ; at the 90 mark ·9 inch ; at the 80 mark, ·8 inch, and so on downwards till the 20 mark is reached, when the distance is ·2 inch. But it may be desired to measure something smaller than ·2 inch. Then the limbs are set at the half-inch or one of the other datum lines on the arc. If at the half-inch line the 90 mark would show ·45 ; the 80 mark, ·4 ; the 70 mark, ·35 ; the 60 mark ·3, and so on. If the limbs are set at the ·25 line the 90 mark would show ·225 ; the 80 mark, ·2 ; the 70 mark ·175 ; the 60 mark, ·15 ; the mark 50, ·125 and so on. In each case the distance between one mark and the next is $\frac{1}{10}$ of the distance between one of the figures and the next, so that whatever datum is selected it is divided decimally. For instance, it may be convenient to take the measurements in millimetres ; it is only necessary to set the limbs so that the distance between them at the 100 mark is, say, 10 millimetres, and draw a line across the arc so that the limbs may be fixed at the same spot on another occasion.

In the foregoing examples the sector takes the place of the ordinary slide gauge or other measuring tool. It is especially useful as a proportional measurer. For instance, in the article on Mainsprings, it is stated that taking the diameter of the inside of a barrel as 100, then with the mainspring in, the unoccupied part of the barrel should show a diameter of 74, and the barrel arbor 33. Set the limbs so that the 100 mark corresponds to the inner diameter of the barrel, and the 74 mark shows the size of the spring, and the 33 mark the size of the arbor.

For Sizing Wheels and Pinions.—Suppose a pinion of 8 is required for a wheel of 75 teeth ; the wheel is placed between the limbs at the 75 mark, and the proper size for the pinion is then the distance between the limbs at the 8 mark. Of course if the pinion is in hand and the size of the wheel is required, the operation is reversed ; the pinion is placed between the limbs at the 8 mark, and the distance between the limbs at the 75 mark gives the size of the wheel.

The numbers of a wheel and pinion and their distance apart from centre to centre being given, their respective pitch diameters may be obtained by means of the sector, provided it is equally divided all through as recommended in a preceding paragraph. Suppose a wheel of 60 and a pinion of 8 are to be planted ·75 apart ; open the sector so that at 68 (which is the sum of the wheel and pinion teeth) the width between the limbs is double the distance of centres—that is, 1·5 inch. Then the width between the limbs at 60 will represent the pitch diameter of the wheel, and at 8 the pitch diameter of the pinion. The full diameters may be obtained by means of the tables given under the head of "Wheels and Pinions." Or the full diameters may be obtained at one operation instead of the pitch diameters by adding 3 to the number of the wheel teeth and 1·25 to the pinion, if it have circular, or 2 if epicycloidal, addenda. Say it is a circularly rounded pinion, the sector would then be opened so that at 72·25 the width was 1·5 inch, and the width at 63 would represent the full diameter of the wheel, and the width at 9·25 the full diameter of the pinion.

Self-Centring Chuck.—[*Tasseau centrant de soi-même.—Das selbstzentrierende Futter.*]—A chuck that can be made to close concentrically on the work to be gripped. (See Lathe.)

Self-Winding.—[*Se remontant de soi-même.—Selbstaufziehend.*] —A watch or clock fitted with apparatus for winding it automatically.

Fig. 1 shows an arrangement of mechanism by M. Lebet for winding a watch by the action of closing the hunting cover. There is a short gold arm projecting beyond the joint. This arm is connected by means of a double link to a lever, one end of which is pivoted to the plate. To the free end of this lever is jointed a scythe-shaped rack, which works into a wheel with ratchet-shaped teeth on the barrel arbor. A weak spring fastened to the lever serves to keep the rack in contact with the wheel teeth. Instead of the ordinary fly spring there is a spring fixed to the plate and attached by means of a short chain to the lever. As this spring pulls the cover open, the teeth of the rack slip over the teeth of the wheel on the barrel arbor. Each time the wearer closes the cover, the watch is partly wound. By closing the case eight or nine times the winding is completed. The ordinary method of hooking in the mainspring would be clearly unsuitable with this winding work, because after the watch was fully wound the case could not be closed. M. Lebet places inside the barrel a piece of mainspring a little more than a complete coil with the ends overlapping, and to this piece the mainspring hook is riveted. The adhesion of the loose turn of mainspring against the side of the barrel is sufficient to drive the watch, but when the hunting cover is closed after the watch is wound, the extra strain causes the mainspring to slip round in the barrel.

The method of winding just described can be applied only to a hunting watch. Fig. 2 represents an invention of Herr von Loehr, in which the motion of the wearer's body is utilized in

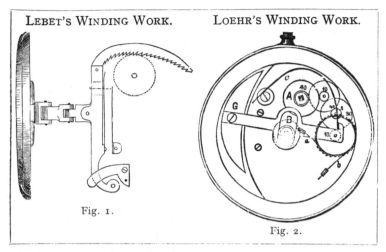

LEBET'S WINDING WORK.   LOEHR'S WINDING WORK.

Fig. 1.

Fig. 2.

winding. There is a weighted lever (G) pivoted at one end, and kept in its normal position against the upper of two banking pins by a long curved spring so weak that the ordinary motion of the wearer's body causes the lever to continually oscillate between the banking pins. Pivoted to the same centre as the weighted lever is a ratchet wheel with very fine teeth, and fixed to the lever is a pawl (*a*) which engages with the ratchet wheel. This pawl is made elastic, so as to yield to undue strain caused by the endeavour of the lever to vibrate after the watch is wound. A is the barrel arbor, and the connection between it and the ratchet wheel is made by a train of wheels as shown. *b* is a second pawl to prevent the return of the ratchet wheel. For setting hands there is a disc (B) which has a milled surface slightly cupped to suit the point of a finger.

Clocks with a kind of windmill wheel attached for winding by means of a current of air, the invention of Mr. Dardenne, were to be seen in London a few years ago. Pond's motor, described under " Electric Clocks," also comes within the category of self-winding.

Horstmann's Self-winding Clock.—In this ingenious arrangement, invented by the late Gustave Horstmann, the expansion and contraction of a liquid are used to wind the clock.

A strong metal vessel, A in the figure, is filled with an easily expanding fluid, such as benzoline, mineral naphtha, etc. Con-

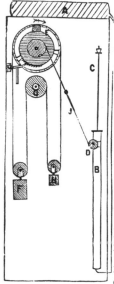

nected to this vessel by a strong tube with a very small bore are a cylinder and piston, B and C. Owing to the fact that most expanding fluids are incapable of driving a piston, being too volatile and thin, the cylinder and tube are charged with a thicker and more of a lubricating fluid, such as glycerine. The vessel containing the expanding fluid is on a higher elevation than the piston and cylinder. This is done to prevent them mixing, as benzoline is lighter than glycerine, and, therefore, rises to the top. It is easy now to see how that when the temperature rises, the expanding liquid will force the piston upward, and, by means of a slight counterforce, the piston will fall on the temperature lowering.

The piston terminates in a cross-bar, to each end of which is attached a steel ribbon like a wide watch mainspring. These two bands are brought down over pulleys at D, fixed on each side of the cylinder, and then carried direct to the winding mechanism, E, of the clock, which is all fixed on the back of the case, and independent of the movement. The two bands join into one a little before they reach the winding. A large pulley, E, is fitted on a stud at the back of the case, and is driven by means of a ratchet and click. The pulley E has a flat groove, and is studded with short pins. Over these runs a steel ribbon to drive the pulley G which actuates the hands and escapement.

Set-Hands Arbor.—[*Chevillot.*—*Die Zeigerwelle.*]—The arbor in a three-quarter plate winding watch by which the hands are set.

Set-Hands Dial.—[*Cadran indicateur.*—*Das Indicationsziffer-blatt.*]—A small dial attached to the movement of a turret clock for the convenience of observing the time when regulating the clock.

Set-Hands Square.—[*Carré de rapport.*—*Das Zeigerviereck.*]—The square in key-winding watches by means of which the hands are set to time.

Sextant.—[*Sextant.*—*Der Sextant.*]—This, the anglemeter of mariners, is used in conjunction with the chronometer for ascertaining the longitude of a ship at sea. Its construction may be explained with the aid of the subjoined Fig. 1. At 1 is a mirror pivoted into the frame of the instrument, but attached to an index arm which is free to travel round a brass graduated arc on the frame at R. On the frame at H is the horizon glass, the half of which next to the frame is a mirror, and the other part clear glass. If a ray of light from the sun or other object at s impinges on the mirror I, it will be reflected on to the horizon

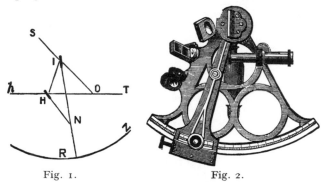

Fig. 1.   Fig. 2.

glass. At o T is a telescope. On looking through the telescope at T, the horizon *h* may be viewed through the clear part of the horizon glass by direct vision, and the index arm may be moved round till the reflected image from the mirror coincides with the horizon. The angle N between the mirror I and the horizon glass H will then be half the angle s o *h*. The arc R, though really but 60° of a circle, is divided into 120°, so that the reading correctly denotes the angle s o *h*. When the index arm points to zero, **z**, the faces of the mirror and horizon glass are parallel. For the engraving of the finished instrument (Fig. 2) I am indebted to Mr. John Short.

When using the sextant to ascertain the longitude by observation of the sun, the most usual plan is to take the sun's altitude when he is near the east or west point, and has an altitude exceeding 8° or 10°, and to compute the apparent time at the ship from that altitude ; the application of the equation of time, corrected for the Greenwich date, gives mean time at the ship : while mean time at Greenwich is obtained from the chronometer. (See Marine Chronometer.)

Shaping. Sinking.—[*Former. Noyer.—Zurichten, Fräsung, Aussenkung*.]—A kind of drilling tool, shown in Fig. 1, carrying a milling cutter in the vertical spindle, is useful for shaping, cutting sinks in plates and dies, and a variety of similar work often accomplished by hand cutting and filing at ten times the cost. There is a double action slide for moving the work, which is fixed to the circular table, and the table also turns on its centre, so that curves of any size within the limits of its radius may be cut. Fig. 2 is an adjunct to the American lathes for doing the same kind of work.

Contouring.—Fig. 3 is a shaping machine in which a circular cutter is used for milling outlines similar to a pattern. The pattern is fixed to the table so that a plain hardened rubber in the dumb spindle on the left hand bears against it. The article to be contoured is fixed under the live spindle on the right hand. The slides have no screws, and their motion is governed by the operator, who, holding the two handles, keeps

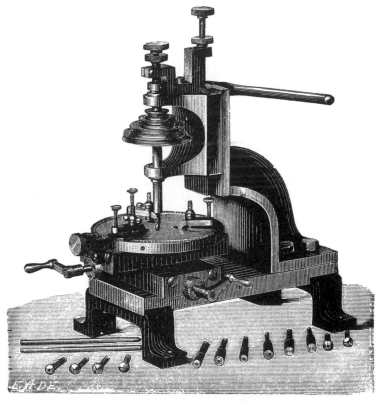

Fig. 1.

the rubber pressed against the edge of the pattern as he feeds it along. The pattern being fixed to the table, and the rubber brought to bear on its edge, the path taken by the edge of the cutter will indicate the position for fixing the work to be operated upon. Whatever radius of cutter is used, the rubber must

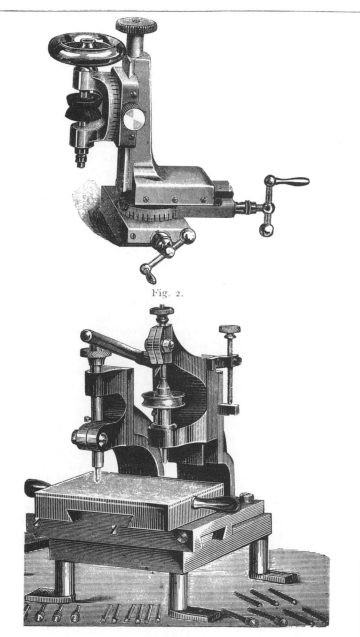

Fig. 2.

Fig. 3.

be of exactly the same radius in order to obtain an exact reproduction.

Ship's Timepiece.—[*Pendule de navire.—Die Schiffsuhr.*]—A " dial " timepiece controlled by a balance. A pendulum would clearly not vibrate correctly if subjected to the motion of a ship. A ship's timepiece has usually a lever escapement.

Shutting Off.—[*Dégrenage en fermant la boîte.—Die Ausschaltung der Aufzieheinrichtung.*]—A term used to describe the operation of throwing the winding wheels out of action by the act of closing either the hunting cover or the outer bottom of the case of a fusee watch. The most usual device is a push piece with a bevelled projection, above the seat of the cover, which acts on it as on a case spring.

Sidereal Time.—[*Heure sidérale.—Die Sternzeit.*]—The standard used by the astronomers. The sidereal day is the period occupied by the earth in making one rotation. (See Time.)

Silver.—[*Argent.—Das Silber.*]—Tarnish may be removed from silver goods by rubbing them with a leather or brush dipped in a paste of cream of tartar and water or in a solution of caustic soda or carbonate of soda and ammonia. A thin varnish of collodion will resist the action of the atmosphere and prevent tarnish. Paper in which silver goods are to be kept should be soaked in a solution of caustic soda and ammonia.

Restoring Silver Jewellery.—Silver ornaments which have merely become oxidised by exposure in a sulphurous atmosphere, and not by repeatedly cleaning, are simply restored by brushing with a clean brush and a little carbonate of soda. To restore the original dead or lustrous whiteness of silver goods, lost by having been too often and carelessly cleaned, they should be, if not soft soldered or very fragile, first annealed by being held in a pair of pincers close over the flame of a lamp till covered with soot, which is then burnt off by means of a blow pipe. Then the article should be immersed in a hot solution of from one to five parts of sulphuric acid and twenty parts of water —the quantity of acid depends on the quality of the silver the articles are made of, the coarser the silver the more acidulated. The time for the articles to remain in the solution also depends on the quality of the silver, whilst good sterling silver will be whitened in almost an instant, commoner silver will take a minute, or even longer ; care is, however, to be taken not to allow the articles to remain too long in the solution, which would turn the surface into an unsightly greyish colour, and the manipulation will have to be commenced afresh. As soon as the desired whiteness of the articles whilst in the acid is observed, they are removed and quickly thrown into lukewarm water ; it is advisable to have an additional vessel with warm water at hand, to place the articles in after having been removed from the first. The articles are then immersed in boxwood sawdust, kept in an iron vessel near the stove, or any warm place, when, after thoroughly drying in the sawdust, the articles will be found to look like new. Any places on the articles desired to look bright, are burnished with a steel burnisher. If the articles are soft-soldered or very fragile the first process of annealing must be dispensed with.

Single-beat Escapement.—[*Echappement à coup perdu.—Die Hemmung mit einem verlorenen Schlage.*]—An escapement in which the escape wheel moves only at every alternate vibration of the balance or pendulum. The chronometer and duplex are the best known examples of single-beat escapement.

Skive.—[*Scie en diamant.—Die Diamantscheibe zum Stein-schneiden.*]—A circular saw used for slitting stones. It consists of a disc of iron fixed on a spindle between two collars or nuts. The free part is slightly dished to secure rigidity. Its edge is charged with diamond powder by pressing a hard stone against it, and gently pouring a little powder between the edge of the skive and the stone.

Slide Gauge.—[*Calibre à coulisse.—Die Schublehre.*]—A measuring instrument consisting of one fixed and one sliding jaw. It is generally provided with a vernier, and used for obtaining exact sizes. Its construction may be gathered from the engraving (Fig. 1, p. 368), which originally appeared in a useful little work on watchmaking by the late Charles Frodsham. To his son, Mr. H. M. Frodsham, I am indebted for the use of this block. The gauge there shown consists of a brass stock, to which one of the tempered steel chops is screwed. A brass slide, which carries a corresponding steel chop, works freely in a dovetailed groove in the stock, the upper surface of the slide being level with the upper surface of the stock. A set screw at one end serves to tighten the dovetail and so fix the gauge, when it is desired to do so. The stock is divided into inches, as denoted by the large figures at the outer edge, and the inches into tenths, the even tenths being marked by figures. Each tenth is again subdivided into five equal parts, representing a fiftieth or ·02 of an inch. The vernier is engraved on the slide.

A length equal to nineteen-fiftieths of an inch is divided into twenty equal parts ; each part is therefore a twentieth of nine-teen-fiftieths, or ·019 of an inch. Then, as the divisions on the stock are each ·02 of an inch, while those of the vernier are only ·019, it follows that their difference is ·001, or one-thousandth of an inch. When the chops of the instrument are closed, the zero points of the stock and the vernier should exactly coincide. Then, when the chops are opened, the distance between them

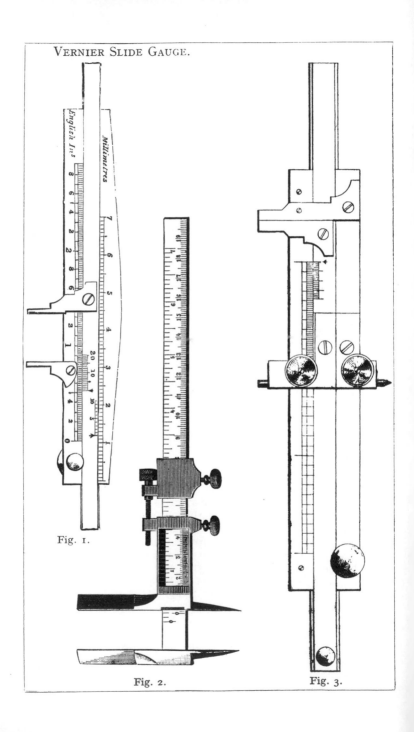

VERNIER SLIDE GAUGE.

Fig. 1.

Fig. 2.

Fig. 3.

will be indicated by the position of the zero point of the vernier with relation to the scale on the stock, plus as many thousandths of an inch as there are divisions of the vernier from zero before one coincides with a division on the stock. For instance, if the reading showed five-tenths, 1 division of another tenth and 4 divisions of the vernier, the measurement would be—

$$\cdot5 + \cdot02 + \cdot004 = \cdot524 \text{ of an inch.}$$

The scale at the back of the stock is divided into millimetres, the vernier being nine millimetres divided into ten equal parts.

Fig. 2 is a light useful instrument by Boley, with points for taking and marking depths. There is also a sliding clamp which is fixed when the position of the jaws is somewhere near what is required. Great exactness may then be obtained by means of the adjusting screw. Fig. 3 shows a most useful vernier gauge by Mr. R. Gardner, with sliding points which may be shifted to unequal heights for accurately marking depths, &c. (In the cut the centre is passed through the hole in the stock and the hole in the slide, merely as a proof that the two holes are coincident when the jaws are closed and the scale at zero. It will be understood that in use one centre would be fixed in the stock and another in the slide.)

Slide Rest.—[*Burin-fixe.*—*Der Support.*]—A tool holder for a lathe or mandril mounted on a bed with dovetailed side, which in its turn rests in another dovetailed bed at right angles to the first. By means of screws each slide can be caused to traverse its bed. (See Lathe.)

Slide Rule.—For rapid calculation, Routledge's Engineers' Slide Rule is very useful to watchmakers, as Mr. Charles Frodsham has pointed out. Mr. T. Hewitt gives the following as a sample of its capabilities. It will be observed there are four scales, A, B, C, D ; A and D being on the rule, and B and C on the slide. When the dagger is reversed (thus ⤙) it signifies that the slide should be taken out and turned end for end.

To resolve varying trial times into 24 hours.
Example. Watch—1·3 sec. in 5 h. 34 min. = 334m.

	A ⤙ Set 334	Below 1·3 sec.
	B ⤙ To 1440	Find 5·6 sec. time in 24h.

1,440 is a constant, it is the number of minutes in 24 hours.
To find moment of inertia of balance from bending moment of balance spring.

	14,400	A ⤙ Set 158	Below bending moment
	Train	B ⤙ To 1	Find moment of inertia

N.B.—All bending moments under 15·8 will have moments of inertia, ·0,

	16,200	A ⤙ To 1988	Below bending moment	[&c.
	Train	B ⤙ Set 1	Find moment of inertia	
	18,000	A ⤙ To 2465	Below bending moment	
	Train	B ⤙ Set 1	Find moments of inertia	

N.B.—Bending moment under 1·990 will have moments of inertia, ·00,
To find cubical contents of barrel. [&c.

Y

C ← Set the height	Find cubical contents
D ← To 1·128	Above diameter

To find radius of gyration of balance,

C ← Set balance weight	Below moment of inertia
D ← To 19·66	Find radius of gyration

To find length of cylindrical balance spring,

A ← To 3·1416	Below diameter
B ← Set No. of turns Find inches, to which add pitch, × turns for length

To find length of flat balance spring,

A ← Set 3·1416	Below radius
B ← To No. of turns Find inches, to which add pitch, × turns for length

To find bending moments of balance spring of varying lengths.

A ← Set length	Below length
B ← To bending moment Find bending moment

To increase the balance arc by reducing the moment of inertia

B → To moment of inertia	Find moment of inertia.
D ← Set balance arc	Above balance arc wanted

To lighten a balance for mean time,

B → To moment of inertia	Find moment of inertia
D ← Set No. of vibrations	Above No. of vibrations required.

To apply balance spring to time,

B ← To bending moment	Find bending moment
D ← Set one of vibrations.	Above No. of vibrations wanted.

To find impulse angle of chronometer escapement, balance arc of lever escapement, &c.

A ← To radius of impulse	Below radius of wheel
B ← Set angle between wheel teeth.	Find impulse angle

Example. To find impulse angle of chronometer.

Radius of wheel   = ·25.
Radius of impulse = ·13.
Wheel tooth angle = 24°.

A ← To ·13	Below ·25
B ← Set 24°	Find 46·2 = 46° 12.

46° 12′—·3° for clearance = 43° 12′ balance angle.

If moments of inertia and radii of gyration are not known, an approximation may be arrived at by substituting weight and diameter of balance.

Diameters and radii of gyration vary in inverse ratios as the square roots of the weights.

B → Set weight	Find weights
D ← To radius of diameter.	Above radii or diameters.

Sliding Tongs.—[*Pince à bouche.—Die Schiebzange.*]—A pair of tongs, with a thimble to slide to and fro the handles, by

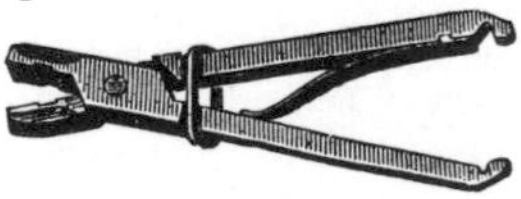

Fig. 1. 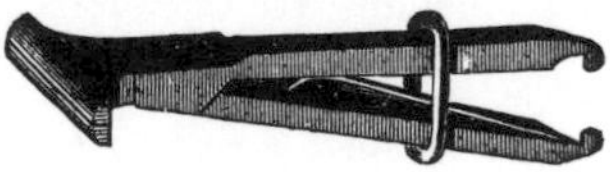   Fig. 2.

means of which an object placed in the jaws may be held fast. The most common form is shown in Fig. 1. For many purposes sliding tongs with vice-shaped jaws, as in Fig. 2, are preferable.

Slitting File.—[*Lime à fendre.—Die Schraubenkopffeile.*]—A very thin file which cuts on the edge. A screw-slitting file used principally for cutting the slits on screw-heads, cuts both on the edge and sides, and makes a V-shaped groove. A file cutting only on the edge is more generally called a safe-sided lever notch file, or a pinion bottoming file.

Snail.—[*Limaçon.—Der Excenter.*]—A cam shaped like a snail, used generally for gradually lifting and suddenly discharging a lever.

Snailing.—[*Adoucissage en colimaçon.—Der excentrische Zierschliff. Der Sonnenschliff.*]—Finishing a surface by means of curved eccentric lines.

Fusee caps, steel keyless wheels, &c., are snailed with a copper mill like the sketch. The face of the mill is hollowed out, leaving only a thin projecting rim. The work to be operated upon is placed so as to work just freely between the centres of a pair of turns, the rest of which has been removed from the holder. There is projecting from the snailing mill a foot which fits the holder of the turns containing the work, so that the mill and the turns may be attached to each other, and by means of the screw in the rest holder adjusted at any required distance. The edge of the mill is then brought into contact with the face of the work, the mill being slightly angled so as to touch on one side only. The snailing mill is fixed in nearly the right position in relation to the work, and the final angling made by moving round the runner of the turns, the arbor of the work being centred near the top of the runner. As the mill is rotated the desired curves are produced on the surface of the 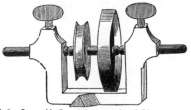 work, the mill being charged with fine " double-washed " emery at first, and sharp red stuff for finishing off. For brass, a bone or ivory mill is used, with oilstone dust first, and red-stuff for finishing off. Fine red-stuff is not suitable. Snailing requires a sharp polishing material, or else instead of distinct curves the surface will be come quite smooth. A long bow should be used to rotate the snailing mill. At first the curves will be rather undecided, but if the mill is set at the proper angle it will not slip on the surface of the work, and will soon produce a nice curve. The last stroke or two should be given downwards, the bow being bent and relieved as it comes to the end of the stroke. (See also Lathe.)

Snap.—[*Cran de fermeture.—Einsprengen.*]—A method of attachment by springing one edge over another, one of the edges being slightly undercut and the other dovetailed to correspond.

Solar Time.—[*Heure solaire.—Die Sonnenzeit.*]—The time

recorded by sundials. The solar day is the time between two successive passages of the sun over the meridian. (See Time.)

Soldering.—[*Soudage.*—*Löthen.*]—The operation of joining two pieces of metal by heating, and interposing a film of metal that fuses at a lower temperature than the pieces to be joined.

Hard Solder requires the application of a red heat for melting it, and can only be used for metals that require a high temperature for fusing them, such as gold, silver, and brass.

Soft Solder, melting with but little heat, is used for metals fusing at a low temperature, or when repairing articles previously joined with hard solder which it is desired not to melt.

It should be remembered that the thinner the layer of solder the better, and the nearer the fusing point of the solder approaches the fusing points of the joined surfaces the stronger will be the joints. The surfaces to be joined should be heated to the melting point of the solder.

Gold Soldering.—The article to be soldered is placed upon a bunch of old binding wire, hammered flat, or on a piece of charcoal. If a breach or crack has to be filled, a small thin plate of the same quality of gold as the article under repair should be used. Rub borax and water to a thin paste on a piece of slate, brush *one side* of the plate with this, and run small pallions of suitable solder evenly over it. The plate is then boiled in diluted sulphuric acid, and hammered or rolled very thin. A bit of this gold plate of a shape to fill the breach is cut off. Any old solderings near the breach should be coated with a paste of red-stuff mixed with water, and to preserve the polish and colour of the article, it should be covered with equal parts of borax and charcoal pounded up together and mixed into a paste with water. This " black stuff," which must be carefully excluded from the part to be soldered, is dried. Any stones or settings in the article should be covered with a thick paste of whiting and water ; or some bury the part in a piece of raw potato. If it is a ring that is being soldered on the opposite side to the settings, a piece of charcoal may also with advantage be placed through the ring. When all necessary precautions have been taken, the breach is boraxed, and the piece of plate laid in, and heat directed to it by means of a blowpipe. Care must be taken not to apply too much heat. When the solder begins to flow, the plate will drop slightly, and the solder round the edges of it glisten. By following this method a strong job is made, the colour of the article preserved, and very little cleaning is required afterwards. Perhaps the greatest mistake made by novices in soldering is that in their anxiety to see the solder flow, they direct the flame too suddenly to it, and in consequence the dampness of the borax causes the solder, if used loose, to corn, and it will not run at all. The heat should be applied to the surrounding parts first, gradually approached to the solder, and stopped the moment the solder glistens.

Great care is needed in dealing with very low quality gold rings when broken. File the edges flat so that no light is seen through when brought together. Cut a VERY thin piece of silver solder, a trifle larger than the section of the ends. Cover the ends with borax, and place the piece of silver solder between them. Apply heat with the blowpipe till the solder begins to glisten.

The Gripper, as shown below, is a great help in many soldering jobs. Its arms have universal joints to allow of adjustments in any position. In the first view it is arranged for holding in juxtaposition two pieces of wire, but as seen in the second view a dial or other dissimilar article is grasped quite as readily.

Gold Solder.—For 18-carat gold—18-carat gold, 12 parts ; fine silver, 2 parts ; brass wire, 1 part. For lower qualities of gold substitute for the 18-carat gold the same standard as the article to be soldered, and add the same proportion of silver and brass wire as given above. For the brass wire, pins are generally used, as they contain a little tin, which is an excellent ingredient for causing the solder to flow. Some jewellers use copper instead of the brass pins, and add a little zinc. Ordinary silver solder is quite unsuitable for gold work which has to be coloured.

Silver Solder.—To 1 ounce of standard silver add 6½ dwt. of white pins ; melt the silver first with a good piece of borax.

Jewellers' Solder.—Fine silver, 19 parts ; copper, 1 part ; brass pins, 10 parts.

White Silver Solder.—Equal parts of silver and tin ; melt the silver first.

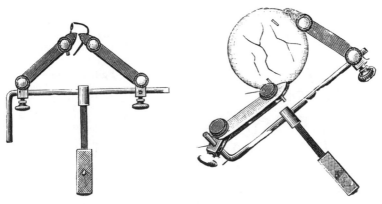

EXCELSIOR SOLDERING GRIPPER.

Soft Solder.—Tin, 2 parts ; lead, 1 part ; melt the lead first.

Solder for Pewter and Britannia Metal.—Tin, 10 parts; lead 5 parts ; bismuth, 1 to 3 parts, according to the work ; melt the lead first.

Soldering Fluxes.—With hard solder use borax ; with soft solder, chloride of zinc prepared by dissolving small pieces of zinc in spirits of salts till no more can be dissolved. A little spirits of ammonia, added to the chloride of zinc, will prevent it from rusting iron or steel.

Dissolving Soft Solder.—Nitric acid is the quickest solvent, and may be used safely, for not lower than 12-carat gold. The safest solvent suitable for all grades of gold and silver goods, which is recommended by Mr. George E. Gee, is prepared as follows :—Reduce to fine powder 2 ozs. of green copperas and 1 oz. of saltpetre, add 10 ozs. of water, and boil for some time in a cast iron saucepan. On cooling, it will become crystallized either wholly or partially. Pour off any remaining uncrystallized, and boil again, when it also will crystallize in cooling. Dissolve the crystals by placing them in a pipkin, and adding to 1 part of crystal 8 parts of spirits of salts. Pour on 4 parts of boiling water, keep the mixture hot, and immerse the work to be operated upon. In a short time the whole of the solder will be removed without changing the colour of the work.

Solder for Aluminium. — Zinc 70 parts, copper 15, aluminium 15. As a flux 2 parts balsam copaiba, 1 Venice turpentine, and add a few drops of lemon juice. Use an aluminum soldering bit.

Split Seconds.—*Second es rattrappantes —Der Doppel Chronograph. Das Doppel Chronoscop.*]—A form of double chronograph in which there are two centre-seconds hands, one under the other, usually of different metal for contrast. When the

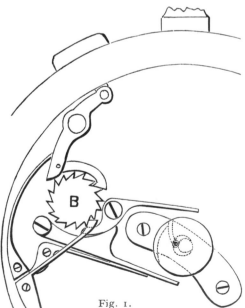

Fig. 1.

chronograph is started, the two hands travel together until a button in the band of the case is pressed, when the under one remains stationary, while the other continues to travel till stopped by the chronograph push-piece.

The split seconds mechanism is shown in the engraving, Fig. 1, p. 374. Attached to the pipe of the lower hand is a brake disc which may be clasped by two stop levers. It will be observed that at present the tail of one of the levers is supported on the tip of one of the teeth of the ratchet wheel B, and when the button in the band of the case is pressed the click pulls round the ratchet wheel a little, the tail of the lever then sinks into the space between two of the teeth, and allows the levers to clasp the brake disc.

The connection between the two hands is made in the following way. A very light curved spring arm is fixed at one extremity to the brake disc on the pipe of the lower hand. A small roller, preferably of jewel, carried by the free end of the arm, bears on the edge of a heart-shaped cam fixed to the pipe of the upper hand ; so that when the lower hand is released, the pressure of the spring causes the roller to fly to the point of the heart piece nearest the centre, and the two hands are then coincident.

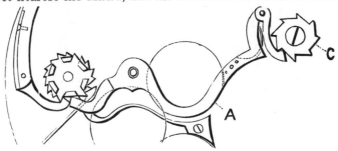

Fig. 2.

Formerly when the chronograph hand was returned to zero by means of the push piece at the pendant, while the lower or split hand was locked, the subsidiary button had also to be pressed, in order to release the lower hand. Messrs. Baume & Co. made a little addition to obviate this second push. The ratchet wheel C, Fig. 2, is fixed to the arbor of the split seconds ratchet B, Fig. 1. On the left of Fig. 2 is the usual castle and ratchet for actuating the chronograph, and in connection therewith is an additional lever, A, which moves to and fro according as a projection or space of the castle ratchet is presented to its left-hand extremity. At the other end of the lever A is a jointed spring click, to give motion to the ratchet wheel C. The motion of the lever A is so proportioned that when the chronograph is being used, and the seconds hands are travelling together, the jointed click simply slides to and fro on the same tooth of the ratchet wheel c, no matter how often the hands are returned to zero by the depression of the principal push piece. But should the split seconds be put into action, the ratchet wheel c is slightly advanced, and its new position allows the pawl to engage with a fresh tooth of the ratchet wheel c; then, when the principal push piece is depressed the ratchet wheel c, and with it the ratchet wheel B, is carried round sufficiently far to take the stop levers off the brake disc. The remainder of the actions have already been described under the head of Chronograph.

Spotting.—[*Adoucissage en taches.—Der rundliche Zierschliff Marmorieren.*]—The process of finishing chronometer, and occasionally watch plates by polishing thereon equidistant circular patches.

The plate that is to be spotted is fixed to the top of a slide rest, and the marks are made with a small bone or ivory tube, which screws into the bottom of the upright spindle. The material used to produce the pattern is a mixture of oilstone dust and sharp red-stuff. The plate, when fixed in position on the platform of the tool, is dabbed all over with the end of the finger dipped in this composition, which must not be at all dry or thick. This upright spindle carrying the spotter is kept constantly rotating by a band from a foot wheel. A spiral spring round the arbor of the spotter keeps it off the work, and a little pressure on a knob at the top brings the spotter into action. The pattern is made by turning the handle of the slide rest equal amounts after each spot till a row is finished, and then moving the transverse

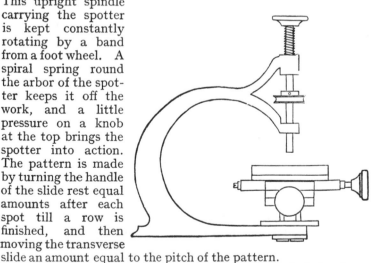

slide an amount equal to the pitch of the pattern.

Sprung Over. Sprung Above.—[*Spirale sur le balancier.— Die Spiralfeder oberhalb der Unruh.*]—A watch in which the balance spring is attached to the staff above the balance, the index or regulator being fixed to the balance cock.

Sprung Under.—[*Spirale sous le balancier.—Die Spiralfeder* *unterhalb der Unruh.*]—A watch in which the balance spring is attached to the staff underneath the balance. This arrangement is adopted in full-plate watches for the convenience of getting the index on the top plate. The stud is usually screwed into the plate, a most inconvenient arrangement, for the balance spring has then to be unpinned every time the balance staff is removed. Mr. John Jones avoided this by fixing the stud to the plate by means of a screw and a steady pin, a much better plan.

Stackfreed.—An eccentric wheel or cam attached to the barrels of watches before the invention of the fusee, in order to equalize the force transmitted. A spring pressed upon the edge of the eccentric wheel.

Staff.—[*Axe.—Die Welle.*]—Arbor, axis.

Staff Punch.—This is a useful punch for driving pallet and other staffs and colleted wheels to correct position. It may be made from a piece of polished steel. A large hole is first drilled transversely near one end, and a smaller one of a size to allow of the passage of ordinary pivots is then drilled from the end to meet it. The mouth of the smaller hole is chamfered, and rests on the shoulder of the staff to be driven ; the pivot passing into the large hole is secure from damage during the operation.

Staff Gauge.—This gauge is for obtaining the extreme length of balance and pallet staffs, and is useful for many similar purposes. There are two arbors, each capable of being tightened by a binding screw, and the upper one is threaded to receive an adjusting nut. The tool is held in the vice, and the endstones having been removed from the holes of the staff pivots, the movement is held in the left hand, so that the face of the bottom arbor takes the place of the foot endstone, while the right hand brings down the top arbor till it rests on the cock jewel. The top arbor is then bound, and the adjusting nut turned till it rests on the face of the gauge. The top arbor may then be unbound, the movement removed,

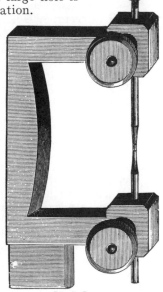

Staff Gauge.

and the arbor brought down again till stopped by the adjusting nut.

Stake.—[*Enclume.—Der Amboss.*]—An anvil or rest. A hollowed stake is used for " bumping " wheels, *i.e.* altering the plane of the teeth with relation to the hole. A stake with a beak to it, like the sketch, a form much used by jewellers, will be found useful by jobbers for shaping rings and the like. (See Polishing Stake and Riveting Stake.)

Star Finger.—[*Doigt d'arrêtage.—Der Stellungs zahn.*]—The part of the going barrel stop work which is fixed to the barrel arbor.

Star Wheel.—[*Etoile.—Der Stern.*]—(1) A wheel with pointed teeth.

(2) The wheel with twelve divisions to which the snail of a striking clock is fixed.

(3) A toothed wheel with a snail on one face as in the sketch, used in musical boxes to adjust the cylinder endways, so that the different tunes may be played in rotation.

(4) The wheel of the going-barrel stop work which is pivoted to the barrel is called indifferently a star wheel or a Maltese cross.

Steady Pin.—[*Pied.—Der Stellstift.*]—A pin used to secure the relative positions of two pieces of metal. Where great accuracy is needed, the holding screws alone cannot be depended on to ensure this. One or two holes are therefore drilled and broached through the pieces while they are together, and a pin or pins with nicely rounded ends, carefully fitted.

For rounding up the ends of steady pins a tool may be made by filing two half-round grooves at right angles to each other across the end of a piece of round steel, which is then hardened and tempered.

Steel.—[*Acier.—Der Stahl.*]—See Iron.

Steel Watch Cases.—The dark blue-grey approaching a black on steel watch cases is usually obtained by exposing the case to a dull red heat in a muffle filled with super-heated steam for several hours, which produces a coating of magnetic oxide of iron, which is intensely hard and also resists the action of air and water. Under the name of " Ferrodyne " is an excellent preparation for restoring black cases. A black surface may be given to steel or iron by adding a slight coating of copper from a cyanide of copper bath by electro-deposit ; on this a film of silver is deposited, which can be blackened with chloride of platinum dissolved in hot water and applied with a brush.

Stem Winding.—[*Remontoir au pendant. Der Bügelaufzug Remontoir.*]—Winding by means of a stem running through the pendant of a watch. The ordinary method of keyless winding. (See Keyless.)

Stirrup.—[*Etrier pour pendule mercuriel.*—*Der Rahmen (Queck-silberpendel.*)]—In a mercurial pendulum, the bottom of the rod on which the glass jar of mercury rests.

Stogden, Matthew.—Inventor of half-quarter repeating mechanism. He died in abject poverty, about 1770, at an advanced age.

Stop Work.—[*Arrêtage.*—*Die Stellung.*]—An arrangement for preventing the overwinding of a mainspring or clock weight. Fusee stop work consists usually of an arm pivoted to a stud fixed on the watch or clock plate, against which a weak spring presses, though sometimes the arm is fixed at one end, and is sufficiently yielding to act as a spring. The chain or gut as it coils on the last groove of the fusee pushes this arm in the path of the fusee snail, a projecting nose on the end of the fusee, and stops it. By the time the fusee has made one rotation the

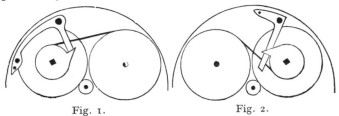

Fig. 1.   Fig. 2.

reverse way, the altered position of the chain or gut has permitted the arm to be pressed out of the way of the projecting nose of the fusee.

Fig. 1 shows a solid stop work for the usual relative positions of fusee and barrel, in which the stop finger is subjected to a thrust, the stress being taken by a screw of adequate size passing freely through a hole in the stop and tapped into the plate. Fig. 2 is the stop work for a reversed fusee, in which the stop finger is subjected to a tensile strain. It will be seen that for one rotation of the fusee in the first arrangement the chain would raise the end of the stop finger more than its pitch on the fusee, but in the second figure the finger would be raised less than the pitch, and, consequently, if the spring of the stop is so stiff as to bend the chain a little, the stop sometimes fails to obtain a sufficient hold. Mr. Schoof says that if the face of the stop where the fusee snail butts is undercut instead of being left square, and the snail correspondingly bevelled, the stop finger is drawn safely home, and then if the spring is made sufficiently weak, a sound action may always be secured with a reversed fusee.

As a stop for the going-barrel, a star wheel or Maltese cross revolves with the barrel four times. Each time it passes a finger fixed to the barrel arbor, the finger passes into and out of a slit, but at the beginning of the fifth revolution the finger, owing to the altered contour of the star wheel at the part with which it then comes into contact, stops it. It will be observed that in the fusee stop work any extra pressure on the key after the winding is completed is transferred to the plate, but in the going barrel the resisting base is really the escapement of the watch or clock, and pressure after winding would injuriously affect the vibration of the balance. Many different kinds of stop work have been arranged for the going-barrel in which the resisting base is transferred to the plate, but no one has come into

Fig. 3.

general use, either because they are wanting in simplicity or take up too much room.

The going-barrel stop work usually stops the watch after the

Fig. 4.

barrel has made four rotations. The finger should be of somewhat longer radius than the wheel, for then, with a given pressure on the winding work, there is less strain transferred to the star wheel. The circular portion of the star wheel that acts as a stop should be of sensibly larger radius than the rest of the wheel, and should make contact with a correspondingly shaped part of the finger piece when the finger points exactly to the centre of the star wheel as shown in Fig. 3. There are five spaces in the star wheel and therefore the angular distance between one space and the next = 72° (360° ÷ 5). Taking the finger to be in the line of centres as shown it has already moved the star wheel 72° ÷ 2 = 36°, and moving on it will pass out of the space when the star wheel has travelled 36° more. It is easy to see that if you start by drawing the centre of the finger in this position and make it such a length that it will just be leaving the space, the star wheel will have been moved one-fifth of a turn, and the next space will be in the proper position to receive the finger as it comes round again. As to the width of the finger at the tip the proportion shown is very good ; if wider the roots of the teeth of the star wheel would be too weak, if much narrower the neck of the finger when reduced to clear the corners of the teeth would be the weak points. Where a sharp corner on the finger piece is left to dig into the stop, trouble is sure to arise. The finger tip should be of adequate width, so that the neck may be left of sufficient strength with out rubbing on the corners of the spaces as the finger passes in and out. The screw should be cut with a fine thread, and have a head as large as the star will allow. The thickness of metal it screws into is so small that a coarse thread is almost sure to overturn, and many of the failures may be traced to such a cause.

The arrangement shown in Fig. 4, though, I believe, old in principle, has recently been introduced by Mr. Hammersley. The stop ring is sprung into a groove in the barrel, and contains a sufficient number of notches for the barrel to rotate till the spring has run down. In winding, when the finger enters the last notch, a projection on the ring is drawn to the ball of the finger, thus forming a very solid and strong stop.

Straight Line Lever.—[*Ancre ligne droite.*—*Die Ankerhemmung in gerader Linie.*]—A form of the lever escapement chiefly used in foreign watches, in which the escape wheel arbor, the pallet staff, and the balance staff are planted in a straight line.

Striking Work.—[*Sonnerie.*—*Das Schlagwerk.*]—The part of a clock devoted to striking.

Rack Striking Work.—Fig. 1 is a view of the front plate of an English striking clock on the rack principle, the invention of Barlow, which is the most reliable and the most generally used now except by turret clockmakers, the majority of whom still prefer the locking plate. The going train occupies the right and centre, and the striking train the left hand on the other side of the plate. The wheels of the striking train are indicated by dotted circles. The connection between the going train and the striking work is by means of the motion * wheel on the centre

* Clockmakers call this, which corresponds to the cannon pinion in watches, the minute wheel. I use the term motion wheel to distinguish it from the wheel gearing with it, to which the name " minute wheel " is also applied.

arbor, and connection is made between the striking train and the striking work by the gathering pallet, which is fixed to the arbor of the last wheel but one of the striking train, and also by the warning piece, which projects from the boss of the lifting piece.

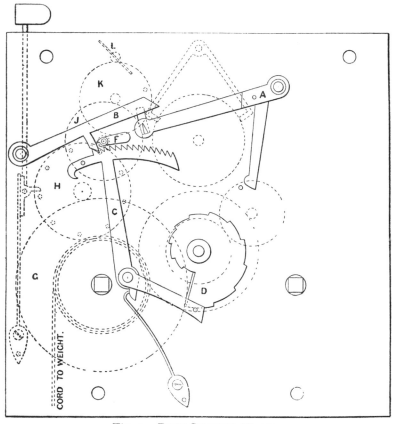

Fig. 1.—RACK STRIKING WORK.

A, Lifting piece.	D, Hour snail.	H, Pin wheel.
B, Rack hook.	F, Tail of gather-	J, Third wheel.
C, Rack.	ing pallet.	K, Warning wheel.
	G, Great wheel.	L, Fly.

This warning piece goes through a slotted hole in the plate, and during the interval between warning and striking stands in the path of a pin in the last wheel of the striking train, called the warning wheel. The motion wheel on the centre arbor, turning once in an hour, gears with the minute wheel, which has an equal number of teeth. These two wheels are indicated by dotted circles. There is, projecting from the face of the minute wheel, a pin which in passing raises the lifting piece every hour. Except for a few minutes before the clock strikes, the striking train is kept from running by the tail of the gathering pallet resting on a pin in the rack. Just before the hour, as the boss of the lifting piece lifts the rack hook, the rack, impelled by a spring at its tail, falls back until the pin in the lower arm of the rack is stopped by the snail. This occurs before the lifting piece is released by the pin in the minute wheel, and in this position the warning piece stops the train. Exactly at the hour the pin in the minute wheel gets past the lifting piece, which then falls, and the train is free. For every hour struck the gathering pallet, which is really a one-toothed pinion, gathers up one tooth of the rack. After it has gathered up the last tooth, its tail is caught up by the pin in the rack, and the striking ceases.

In the drawing the gathering pallet is just about to gather up the last tooth of the rack. This operation will bring the locking pin, which projects from the face of the rack, into the path of the longer end of the gathering pallet and stop its motion.

It is essential that the moment the rack hook is lifted the rack should be free to fall, and that the form of gathering pallet tail where it rests on the pin in the rack should be such as to accelerate rather than retard the movement of the rack. Within reasonable limits the position of the resting pin is unimportant, provided the surface of the tail where it rests, the pin, and the centre of motion of the rack form an angle of less than 90°, so that the tail passes the pin with a sort of wedge-like action. And with the half-round termination of the gathering-pallet tail this condition is secured. To insist that the end of the tail should then be filed to form a slanting plane is quite unnecessary and would spoil the symmetry of the tail.

The steps of the snail are arranged so that at one o'clock it permits only sufficient motion of the rack for one tooth to be gathered up, and at every succeeding hour additional motion equal to one extra tooth.

The centre of motion of the rack hook, the point of its tooth, and the centre of motion of the rack form a right angle ; but in planting the work it is better to put the centre of the rack hook higher on the plate rather than lower, for if lower the pressure of the rack teeth may cause the hook to jump out occasionally, a very tiresome fault. The extreme point of the lower arm of the lifting piece should point to the centre of the motion wheel which carries the pin lifting it, or more to the centre of the plate than the other way, to avoid the possibility of engaging friction in their action.

The spot where the spring which causes the rack tail to fall To the snail should be screwed to the plate is not definite. A generally accepted clockmakers' rule is that it should be nearly in a line with the centres of motion of the rack and of the gathering pallet. I would prefer to plant it rather to the side in

HOUR AND HALF-HOUR STRIKING MECHANISM.
Fig. 2.

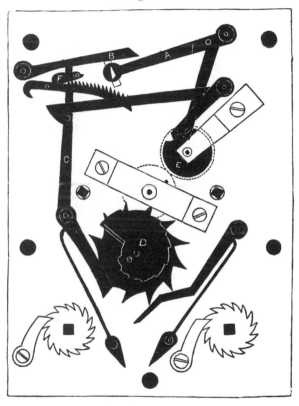

the direction of the pressure as in the drawing, because in the movement of the free tip of the spring on the rack tail, which induces alternately pushing and dragging friction, the sputtering tendency of pushing friction is less pronounced than when the spring is screwed on the other side.

When putting the striking train together observe that the pin in the wheel K is about in the position shown, so as to allow but little run to the projecting catch of the lifting piece, and that a pin in the pin wheel is not close to the hammer tail, but placed so that the hammer tail drops off just before the train is locked ; this allows all the run possible on to the pin.

The lower arm of the rack and the lower arm of the lifting piece are made of brass, and thin, so as to yield when the hands of the clock are turned back ; the lower extremity of the lifting piece is a little wider, and bent to a slight angle with the plane of the arm, so as not to butt as it comes into contact with the pin when this is being done. If the clock is required to repeat, the snail, instead of being placed upon the centre arbor, may, with advantage, be mounted on a stud with a star wheel, as shown in Fig. 2. Indeed, many good clockmakers always mount it so, because the position of the snail is then more definite owing to the backlash of the motion wheels.

Half-Hour Striking.—The usual way of getting the clock to strike one at the half-hour, is by making the first tooth of the rack lower than the rest, and placing the second pin in the minute wheel a little nearer the centre than the hour pin, so that the rack hook is lifted free of the first tooth only at the half-hour. This adjustment is rather delicate, and the action is liable to occasionally fail altogether or to strike the full hour, from the pin getting bent or from uneven wear of the parts. The arrangement shown in Fig. 2 appears to be safer. One arm of a bell crank lever rests on a cam fixed to the minute wheel. The cam is shaped so that just before the half-hour the other extremity of the bell crank lever catches a pin placed in the rack, and permits it to move the distance of but one tooth. This is the position shown in the drawing. After the half-hour has struck, the cam carries the catch free of the pin.

Quarters.—The quarter-hours may also be sounded on the short-toothed rack principle without a special train. In this case there are three extra teeth in the rack, all lower than the hour teeth, and arranged in steps, the outermost tooth being the lowest of all. There are then three lifting pins in the motion wheel besides the one for the hour, and they are placed at such a distance from the centre of the wheel that the first one after the hour lifts the rack sufficiently to allow the lowest tooth only to escape when one stroke is given on the bell. The half-hour pin being a little further from the centre allows two teeth to escape, and so two blows are given. The third pin after the hour is still a little further from the centre, and allows the three short teeth to pass, so that of course three blows are given. This arrangement answers well if the gradations in the height of the teeth are sufficiently pronounced to ensure uniform action and the pins in the wheel are of adequate size so that they do not get bent. Using but one bell is, of course, apt to cause uncertainty as to the

z

meaning of the sounds between the hours of 1 and 3. By providing a second bell ting-tang quarters can be struck. Pins to lift the second hammer are placed on the other side of the pin wheel and that hammer arbor is slid longitudinally out of the way as the hour approaches and allowed to spring back to its normal position when the hour is struck.

The snail is mounted on a star wheel placed so that a pin in the motion wheel on the centre arbor moves it one tooth for each revolution of the motion wheel. The pin, in moving the star wheel, presses back the click or "jumper," which not only keeps the star wheel steady, but also completes its forward motion after the pin has pushed the tooth past the projecting centre of the click.

Division of the Hour Snail.—The length of the lower arm of the rack, from the centre of the stud hole to the centre of the pin, should be equal to the distance between the centre of the stud hole and the centre of the snail. The difference between the radius of the top and the radius of the bottom step of the snail may be obtained by multiplying the distance of twelve teeth of the rack by the length of the lower arm, and dividing the product by the length of the upper arm. Divide the circumference of a circular piece of stout, well-hammered brass plate into 12 parts, and draw radial lines as shown in Fig. 3. Each of these spaces is devoted to a step of the snail. Draw circles representing the top and bottom step. Divide the distance between these two circles into eleven equal parts, and at each division draw a circle which will represent a step of the snail. The rise from one step to another should be sloped as shown, so as to throw off the pin in the rack arm if the striking train has been allowed to run down, and it should be resting on the snail when it is desired to turn the hands back. The rise from the bottom to the top step is bevelled off, as shown by the double line, so as to push the pin in the rack arm on one side, and allow it to ride over the snail if it is in the way when the clock is going. Clockmakers generally

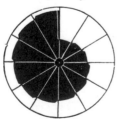

mark off the snail on the clock itself after the rest of the striking work is planted. A steel pointer is fixed in the hole of the lower rack arm, and the star wheel jumped forward twelve teeth by means of the pin in the motion wheel. After each jump, a line is marked on the blank snail with the pointer in the rack arm by moving the rack arm. These twelve lines correspond to the twelve radial lines in Fig. 3. The

Fig. 3.

motion wheel is then turned sufficiently to carry the pin in it free of the star wheel, and the star wheel click fastened back, so as to leave the star wheel and blank snail quite free on their stud. The rack hook is placed in the first tooth of the rack, and while the pointer in the rack is pressed on the blank snail, the latter is rotated a little, so that a curve is traced on it. The rack hook is then placed in the second, and afterwards in the succeeding teeth consecutively, and the operation repeated till the twelve curves are marked. There is one advantage in marking off the snail in this way. Should there be any inaccuracy in the division of the teeth of the rack, the steps of the snail are thus varied to suit it.

Clocks striking one at the hour have the hammer tail lifted by a snail on the minute wheel. Although a snail of this kind is not divided into steps, it may be set out on the plan of Fig. 3, the contour of the snail being represented by a curve cutting through the intersection of the circles with the radial lines. If formed in this manner, it is evident the hammer tail will be lifted uniformly throughout the revolution of the minute wheel. Sometimes these snails are wrongly made so as to do nearly the whole of the lifting as the hour approaches, and the extra strain on the minute wheel stops the clock, or causes the motion wheel to slip on the centre arbor.

Three Train Quarter Chime Clock.—The engraving on p. 389 is a front elevation of the mechanism of a quarter clock. The going train occupies the centre of the plate ; the striking train is planted on the left, and the chiming on the right hand. All the train wheels are represented by circles, except the fusee wheel of the going train.

Going Train.

Fusee Wheel 96	Centre Wheel 84	Third Wheel 78
Pinion .. 8	Pinion .. 7	Pinion .. 7

Striking Train.

Fusee Wheel 84	Pin Wheel 64	Pallet Wheel 70	Warning Wheel 60
Pinion .. 8	Pinion .. 8	Pinion .. 7	Fly Wheel .. 7
	8 pins in pin wheel		

As the gathering pallet makes one complete revolution for every blow struck, the pin wheel must contain as many times more teeth than the pinion on the gathering pallet arbor, as there are pins in the pin wheel. The number of teeth in the pallet wheel must also be a multiple of the teeth in the pinion on the warning wheel arbor.

Chiming Train.

Fusee Wheel 100	Second Wheel 80	Pallet Wheel 64	Warning Wheel 50
Pinion .. 8	Pinion .. 8	Pinion .. 8	Fly Pinion .. 8
	Chime Wheel 40		

The barrels for the going and striking parts are each $2\frac{3}{8}$ inches in diameter, and the barrels for the chiming part $2\frac{3}{4}$ inches, and the rough rule for the size of the fusee wheel is that it should just freely go into the barrel.

There are four pins in the minute wheel for raising the quarter lifting piece, and, therefore, the quarter rack hook every quarter of an hour. One, two, three, or four quarters are chimed according to the position of the quarter snail, which turns with the minute wheel. At the hour when the quarter rack is allowed to fall its greatest distance, it falls against the bent arm of the hour rack hook, and releases the hour rack. As the last tooth of the quarter rack is gathered up, the pin in the rack pulls over the hour warning lever, and lets off the striking train. The position of the pieces in the drawing is as they would be directly after the hour was struck.

This is clearly a much better arrangement than the flirt, which absorbs more power and is less certain in its action.

Marking the Chime Barrel.—The chime barrel is of brass, and should be as large in diameter as can be conveniently got in. To mark off the positions of the pins for the Cambridge chimes, as noted under the head of " Chimes," first trace four circles round the barrel while it is in the lathe, at distances apart corresponding to the positions of the four hammer tails. There are five chimes of four bells each for every rotation of the barrel, and a rest, equal to two or three notes between each chime. Assuming the rest to be equal to three notes, divide the circumference of the barrel into thirty-five equal parts, and draw lines at these points across the barrel. Call the hammer for the highest note 1, and that for the lowest note 4. Then the first pin is to be inserted where one of the lines across the barrel crosses the first circle ; the second pin where the next line crosses the second circle ; the third pin where the third line crosses the third circle ; and the fourth pin where the fourth line crosses the fourth circle, because the notes of the first chime are in the order 1, 2, 3, 4. Then miss three lines for the rest. The first note of the second chime is 3, and the pin for it will consequently be inserted where the first line after the rest crosses the third circle, and so on.

Tuning the Bells.—Bells only very slightly out of tune offend a musical ear, and they may easily be corrected to the extent of half a tone. To sharpen the tone make the bell shorter by turning away the edge of it ; to flatten the tone, thin the back basin-shaped part of the bell by turning some off the outside.

Difficulties with quarter clocks sometimes arise from the method usually adopted for knocking out the hour rack hook, which, as just explained, is effected by the falling of the quarter rack, aided by a strong rack spring. Failure to strike the hour occasionally occurs through too weak a quarter rack spring, or one that has been bent in cleaning. There is also a liability of the clock striking the hour at the third quarter, through the

389

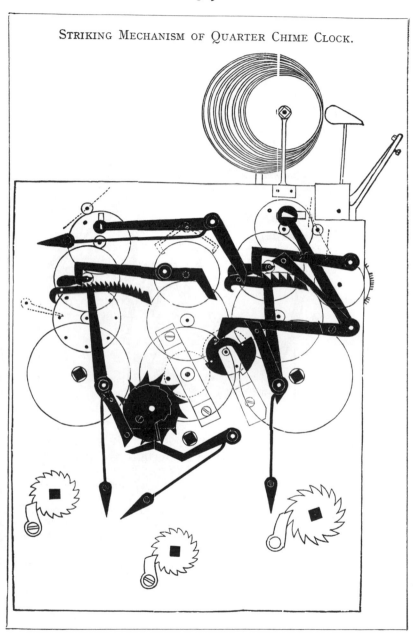

STRIKING MECHANISM OF QUARTER CHIME CLOCK.

quarter rack teeth being too fine, which admits of the quarter rack lifting the hour rack hook.

A suggested improvement will be understood from the sub-joined figure. If, instead of the dotted arm A, which, as on page 389, is attached to the hour rack hook, an arm like B is brought down behind the minute wheel that carries the quarter snail, and a pin placed in the wheel so as to lift the hour rack hook by pressing on the arm B and drop it just before the hour, the

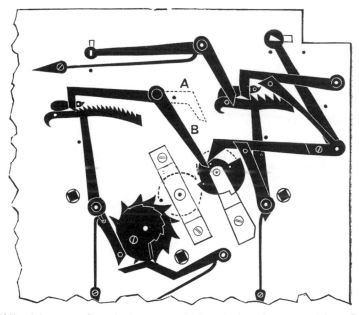

difficulties mentioned above would be obviated. By raising the hour rack hook in this manner coarse teeth to the quarter rack and the abnormally strong quarter rack spring are avoided. The quarter rack must, of course, be made to fall before the hour rack hook is lifted.

Carriage Clock Striking Work.—Fig. 4 is an example of hour and half-hour striking work for small clocks. Though the principle is the same as in Figs. 1 and 3, the parts are differently arranged. The snail A is mounted on the hour wheel, and the arm C, which takes the place of the lifting piece, is pivoted to a lever which is impelled to the position shown by a weak spring. In the cannon pinion F are two pins, placed nearer the centre than the other. At the half-hour the pin nearest the centre has pushed back the lever till the first notch D in the arm C has fallen over the pin in the rack hook I and the pin then releases the end of the lever, and the rack hook is pushed far enough to open the striking, though not far enough to allow the rack to fall. At the hour the notch E engages the pin, and pushes the hook clear of the rack. The striking is kept locked by an upright arm on the arbor of I, catching a pin in the last wheel of the striking train. It will be observed that there is no tail to the gathering pallet, consequently the striking is let off at

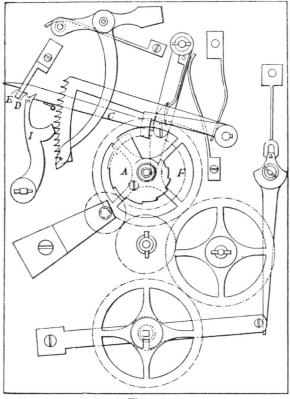

Fig. 4.

once, and there is no preliminary warning. The horizontal lever at the top of the frame and the curved arm coming down in front of the rack are for pushing out the hook when it is desired to repeat the striking of the previous hour. The small wheel below the hour wheel is connected with the cannon pinion through the intermediate wheel on the left, and is for setting hands at the back.

ALARUM.—The wheel at the bottom of the plate in Fig. 4 is connected with the hour wheel by means of the intermediate wheel on the right, and is for letting off the alarum. The notched collar is fixed to the arbor, which may be turned round till the notch corresponds to any desired hour. The wheel rides loose on the arbor, and is pressed forward by the spring at the back. There is a projection on the boss of the wheel facing the collar, and when this projection comes opposite to the notch, it enters, and the spring is thereby allowed to come forward sufficiently to disengage itself from the vertical lever on the extreme right of the figure. To the axis of this lever on the right side of the plate is fixed a pair of pallets, and also a hammer to strike on the bell. On the arbor of the wheel, which engages with the pallets, is a small mainspring barrel, so that if the mainspring is released the hammer rapidly vibrates and strikes the bell. The short end of the lever carries a banking pin which works between a spring fork as shown. One side of the notch in the collar is sloped off at the back, so that as the wheel continues its rotation the projection is pushed out of the notch and the spring again locks the alarm lever.

Cams for Turret Clock Striking.—For lifting the hammer tails of small clocks, pins in the wheel, as shown in Fig. 1, do very well ; but in turret clocks, where the hammers are heavy and it is a consideration to economize the power, steel cams, formed so that at the beginning and end of the lift the end of the lever bears on the cam, should be used. The power should also

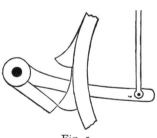

Fig. 5.

be taken from the lever on the same side of the centre of motion as the cams act, as shown in Fig. 5 ; in fact the hammer rods should be as near the cams as is practicable. In the figure the hammer rod is attached to the end of the longer lever, and the shorter lever, pressed downwards by the wheel, is just leaving the face of the cam. I fancy Sir Edmund Beckett was the first to use cams of this form for lifting the hammer. They are so used in the Westminster clock (see Turret Clock), and the arrangement is described by him in detail in his book on clocks. For the convenience of readily fixing them in their relative positions, Messrs. Gillett & Co. make the cams loose with a screwed shank and nut. The wheel then consists of two narrow rims side by side, with a space between them sufficiently wide for the shanks of the cams to pass freely into.

Locking Plate Striking.—Fig. 6 is a drawing of the

quarter train of the Knightsbridge Barracks clock, the whole movement of which is depicted under the head of " Turret." The barrel is on the arbor of the great wheel, and the great wheel making one rotation in three hours has thirty cams, equal to the number of quarters struck during that period. Except for a few minutes before every quarter of an hour, a square pin projecting from the extremity of an arm (A) on the arbor of the fly pinion, bears against the upper of two square blocks on the locking lever (B). This lever is gradually raised by a four-armed snail on the hour wheel, and a few minutes before the quarter it is high enough to allow the pin to escape from the upper to the lower block. A separate view, to show the position of these blocks and the pin in the arm, is given on the left of the principal figure. Exactly at the quarter the snail allows the lever to fall, and the train is free. On the end of the arbor of the second wheel is a cam with three circular hollows, and when the lever is in its lowest position, a roller on it just fits into one of these hollows. As the train begins to move, the cam pressing against the roller lifts the lever, so that as the arm comes round the pin in it passes under the locking blocks. When the great wheel has made a thirtieth of a rotation, a lever for each of the two bells has dropped from the cams, and a " ting-tang " quarter has struck ; the second

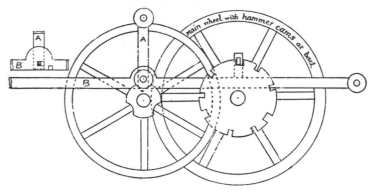

Fig. 6.

wheel has then made one-third of a rotation, and another hollow of the cam is accordingly presented to the roller. The notch in the locking plate for the first quarter is wider than is sufficient for freedom of the finger that drops into it by one-thirtieth of the circumference of the plate, so that the finger again drops into it after the quarter has struck, and the train is again locked by the pin in the arm catching against the upper of the two blocks in the lever. At the second quarter, though, after the first " ting-tang," a hollow of the cam is presented to the roller as before, the locking plate finger meets with the plain part of the periphery of the locking plate, and another quarter is struck before a notch in the locking plate is coincident with the finger. The space between the second and third notches of the locking plate allows three quarters to be struck, and between the third and fourth, four quarters, when the finger again falls into a wide notch ready for the succeeding first quarter.

The hour striking is similar in action to the quarter part. As, however, it would be inconvenient to have 78 cams on the great wheel, which is the number of blows struck in twelve hours, the locking plate is carried on a stud. Attached to it is a wheel of 78 teeth driven by a pinion on the great wheel arbor, having the same number of teeth as there are cams in the great wheel, in this instance 24. The second wheel turns once for every two blows struck, so the cam has two hollows only. There are eleven notches in the locking plate, the first wide enough to allow the finger to return to it after one blow is struck, the others divided round the periphery at distances corresponding to the number of blows to be struck at each succeeding hour.

Stud.—[*Piton.*—*Das Spiralklötzchen.*]—A small piece of metal pierced to receive the outer or upper coil of a balance spring. In watches it is usually of steel, and in marine chronometers generally of brass.

(2) The holder of the fusee stop work.

(3) Any fixed holder used in watch or clock work not otherwise distinguished is called a stud.

Adjustable Stud.—As a rule, the balance spring stud, when screwed to the plate or cock, admits of no alteration in its position, and if removed may be replaced with the certainty that it is un-

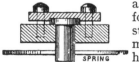

altered with relation to the spring. But for convenience in manufacturing, several studs have been devised to allow of adjustment, either in circle to bring the stud hole right for the spring, or to and from the centre of balance, to suit the diameter of the spring. The adjustable stud shown in the sketch is one of the best of these, and is found mostly in Swiss watches. The wing of the cock is formed into a slot, into which the body of the stud fits. The head of the stud rests on the top of the wing, and is kept in position by a cover plate and two screws.

Stud Remover.—Mr. Daldorph has devised a useful table with four slits as shown in the sketch, for supporting a Swiss cock while pushing out the stud. The holes in the table, and also the slits, allow the tool to be used as a support when pressing a balance spring collet to position.

The engraving shows a pair of tweezers for removing studs such as are pivoted to the balance cock. By placing the tweezers so that the slit embraces the stud and the projecting finger is pressed on the pivot, a tight fitting stud may be readily removed without fear of damaging the balance spring.

Minute Wheel Studs, Repeating Rack Studs, and the like cannot be threaded right up to the root of the shoulder, and chamfering the hole diminishes the holding length of the thread. This may be avoided in the following way, which is recommended by Mr. John Meek. From a piece of wire larger than the size of the shoulder, turn the part which goes into the frame, and form the thread on it ; hollow out behind the shoulder, as shown in Fig. 1; then lay the collar on a stake and hammer all round the shoulder, when it will assume a shape shown in Fig. 2 ; true the face of the undercut.

Fig. 1.

Fig. 2.

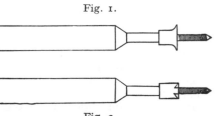

Sully, Henry.—An eminent English watchmaker, born 1680, who settled in France, where he died and was buried in 1728. Author of the *Règle Artificielle du Temps.* Berthoud says, " To Sully and Le Roy the horological art owes its first perfection."

Sun Dial.—[*Cadran solaire.—Die Sonnenuhr.*]—A dial for indicating solar time by means of shadows. The most usual kinds are vertical dials which face the south, and horizontal dials.

The art of dialling is somewhat complex. A glance at the annexed

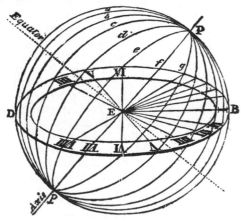

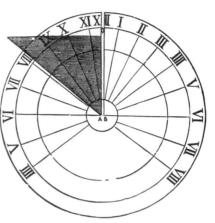

figure will show why, except for places on the equator, the hour spaces are not all equal. A sun dial may be regarded as a circle round the earth, or as the edge of a disc which passes through the centre of the earth from the spot where the dial is fixed, *a, b, c, d, e, f, g,* etc., are longitudinal circles, representing the hours, B the spot where the dial is situated, D the corresponding latitude, P *p* the poles, and E the centre of the earth.

A dial prepared for any particular place, is useless for

Horizontal Sun Dial.

another place in a different latitude, with the exception that a *horizontal* dial for a certain latitude will be a vertical dial for a latitude which is the complement of the first, or what it wants of 90°. That is, a horizontal dial for our latitude of 51½°, would have to be placed in a vertical position facing south in latitude 38½°.

Horizontal Sun Dial.—To set out a horizontal dial first draw two lines, parallel to each other, at a distance equal to the thickness of the gnomon which is to cast the shadow. Next, draw a line at right angles to these, the extremities of which will indicate respectively the hours of 6 in the morning and 6 in the evening. Then, with A and B as centres (see engraving), draw quadrants of circles, and divide each into 90 degrees. Now assuming the dial to be for the latitude of London, lay a rule over B, and draw the first line through 11⅔ degrees, and second through 24¼, third 38 1/12, fourth 53½, and fifth 71 1/15. Proceed the same with the other side. Extend the afternoon hour lines of 4 and 5 across the dial, and these will form the morning hours, while 8 and 7 of the morning hours prolonged will give the same evening hours. To form the style or gnomon, draw a radial line through that degree of the quadrant which corresponds to the latitude = 51½°. This will show the elevation of the style, which is here represented as if lying on the surface of the dial. The thickness of the style must be equal to the distance between A and B. Place the style truly upright on the dial, and it is finished.

Sunk Seconds.—[*Secondes rapportées.*—*Das eingesetzte Senkundenzifferblatt.*]—A watch in which the portion of the watch dial

397

traversed by the seconds hand is sunk below the level of the rest of the dial. With a sunk seconds the hour hand may be closer to the dial than it otherwise could.

Surprise Piece.—[*Surprise.—Der Vorfall.*]—A loose plate under the quarter snail of a repeating watch, which prevents the quarter rack reaching the snail if the mechanism is set going at the hour.

Swing Tool.—[*Equilibre.—Die Schleifvorrichtung am Drehstuhle.*]—A tool in which watchwork is fixed to be polished. It swings on centres so that the work yields to unequal pressure from the polisher. (See Polishing.)

Table Roller.—[*Plateau simple.—Die einfache Rolle (Ankergang).*]—The roller of a lever escapement that carries the impulse pin. The term table was originally applied to distinguish this kind of roller from the part carrying the impulse pin in Massey's escapement, which was in the form of a crank.

Table Tool.—[*Perce-droit.—Die Geradbohrmaschine.*]—A tool for uprighting and drilling. (See Uprighting Tool.)

Tell-Tale Clock.—[*Pendule (ou montre) de contrôle.—Die Kontrolluhr.*]—A clock by which a record is left of the periodical

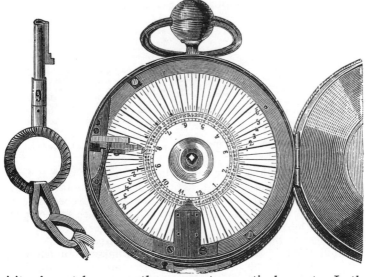

visits of a watchman or other person to a particular spot. In the earliest form invented by John Whitehurst, of Derby, in the last century and still met with, a ring studded with projecting pins rotates, so that the pins successively coincide with an aperture at the desired intervals. As the watchman passes he pushes the pin in ; but should he be late, it has passed the aperture, and cannot be got at. Recent improvements provide for a permanent record of the visits of the watchman. Hahn's Time Detector, one of the best examples, is here drawn to one half of the real size. The detector is carried by the watchman, who at every station he has to visit finds a key (as shown on the left of the engraving) which he inserts into a keyhole in the band of the case opposite the pendant, and gives it one turn. This lifts a steel type, and impresses a figure corresponding to the number of the key on to the paper dial, so that in the morning when the instrument is unlocked there is recorded on the dial the time at which each station was visited. The dial is then removed and a fresh one attached ready for the succeeding night. There is a lever by the catch on the left of the picture. which punctuates the dial when the detector is unlocked, to indicate when this was done.

Tempering. — [*Revenir. — Anlassen.*] — The operation of softening steel to the degree suited for the particular purpose required, the steel having been previously hardened. The following table gives the characteristics of a wide range of temperature.

Tempering Steel.

Colour.	Purpose.	Temperature.	Alloy, whose fusing point is of the same temperature.	Effect on Tallow.
Pale straw	Lancets and Tools for Cutting Iron	420° Fah.	7 lead 4 tin	Vapourizes.
Straw	Watchmakers' Tools	450° ,,	8 ,, 4 ,,	Smokes.
Straw yellow	Pen Knives and Razors	480° ,,	8½ ,, 4 ,,	More smoke.
Nut brown	Small Pinions and Arbors	500° ,,	14 ,, 4 ,,	Dense smoke.
Purple	Large Pinions and Arbors	530° ,,	19 ,, 4 ,,	Black smoke.
Bright blue	Swords and Watch Springs	580° ,,	48 ,, 4 ,,	Flashes if light is applied.
Deep blue	Watch Balance Springs	590° ,,	50 ,, 2 ,,	Continuous burning.
Blackish blue	Chronometer Balance Springs	640° ,,	{ All lead or boiling linseed oil. }	All burns away.

Steel is less oxidised by tempering in an alloy than if tempered in the air, and the required temperature is obtained with much greater certainty. Instead of linseed oil some chronometer makers use olive oil, which boils at 600°, as a bath for tempering balance springs.

Thimble for holding punches.—(See Riveting Stake.)

Thiout l'aine.—A clever French watchmaker. Inventor of many ingenious forms of repeating clocks, curious clocks, &c., described in his *Traité d'Horlogerie*, published 1741.

Third Wheel.—[*Roue petite moyenne.—Das Kleinbodenrad.*]— In a watch, a wheel of the train between the centre and fourth wheels.

Three Part Clock.—Three Train Clock.—[*Horloge à* 3

rouages.—Die Schlaguhr mit drei Federhäusern. Die Vierteluhr.]—
A clock with three trains : the going train, the striking train
and the quarter or chiming train. (See Striking Work.)

Three-Quarter Plate.—[*Trois quarts de platinē.—Das Dreiviert-
elplatinen Werk.*]—A watch in which sufficient of the upper plate
is omitted to allow of the balance vibrating in the same plane as
the plate. (See Plate.)

Throw.—[*Tour à renvoi à corde.—Das Englische Handschwungrad für Grossuhrmacher.*]—A clockmaker's " dead centre " lathe.
At the back of the lathe centre is fixed a large grooved wheel
or pulley, having a handle attached. The workman turns this
wheel with his left hand while holding his tool or cutter with
the right. A gut connects the large throw wheel with a small
pulley rotating freely on the lathe centre.

Tidal Clock.—[*Horloge de marée.—Die Ebbe und Flutuhr.*]—
Ferguson designed a curious and useful clock for showing the
time of high and low water, the state of the tides at any time of

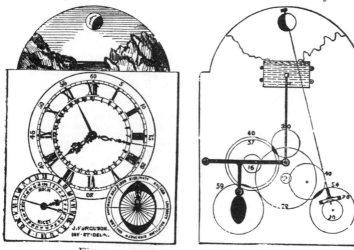

Fig. 1. Fig. 2.

the day and the phases of the moon. The outer circle of the dial
in the left-hand corner of Fig. 1 is divided into twice 12 hours,
with halves and quarters and the inner circle into 29·5 equal
parts for showing the age of the moon ; each day standing
under the time of the moon coming to the meridian on that day.
There are two hands on the end of the arbor coming through
this dial, which go round in 29d. 12h. 45m., and these hands
are set as far apart as the time of high water at the place the
clock is to serve differs from the time the moon comes to

meridian. So that by looking at this dial, one may see at what time the moon will be on the meridian, and at what time it will be high water. On the dial in the right-hand corner, all the different states of the tide are marked. The highest points on the shaded ellipse represent high, and the lowest low water. The index travels round this dial in the time that the moon revolves from the meridian to the meridian again. In the arch above the dials, a blue plate to represent the sea rises and falls as the tides do, and over this a ball, half black and half white, shows the phases of the moon.

The mechanism as it would appear at the back of the dial is shown in Fig. 2. A wheel of 30 fixed to the hour wheel on the centre arbor goes round once in 12 hours, and gears with a wheel of 60, on whose arbor a wheel of 57 drives a wheel of 59, the arbor of which carries the hand for the right-hand dial. On this arbor is an elliptical cam which carries and lets down the tide plate twice in 24 h. 50·5 m. On the arbor of the wheel of 57 is a pinion of 16, driving a wheel of 70, on whose arbor is a pinion of 8 driving an idle wheel of 40. This idle wheel is merely to reverse the direction of the wheel of 54 with which it gears, and which carries the hands of the left-hand dial. The moon is driven from this last arbor by means of a pair of mitre wheels.

Ferguson afterwards devised a cheaper arrangement by making the dials all concentric, but it certainly was not so good as the one here illustrated.

Time.—[*Temps.—Die Zeit.*]

Solar Time.—Solar time is marked by the diurnal revolutions of the earth with regard to the sun, so that the instant the sun is seen at its greatest height above the horizon it is true mid-day, which sometimes takes place 16 m. 18 s. sooner, and at other 14 m. 28 s. later than twelve o'clock mean time. The diurnal rotation of the earth on its axis might naturally be supposed to bring each place to the meridian at regular intervals ; this would be nearly the case if the earth had no other movement ; but it advances at the same time in its orbit, and as the meridians are not perpendicular to the ecliptic, the days are not of equal duration. This may be easily perceived by placing a mark at every 15 degrees of the equator and ecliptic on a terrestrial globe, as by turning it to the westward the marks on the ecliptic, from Aries to Cancer, will come to the brazen meridian sooner than the corresponding ones on the equator, those from Cancer to Libra later, from Libra to Capricorn sooner, and from Capricorn to Aries later ; the marks on the ecliptic and equator only coming to the meridian together at Aries, Cancer, Libra, and Capricorn. True and mean time do not agree, though, on the days in which the sun enters these signs in March, June,

September and December, for the earth moves with greater rapidity in December, when it is nearer the sun, than it does in July, when it is farther from it. The regularity of the earth's motion is also further disturbed by the attraction of the Moon, Venus, and Jupiter. True and mean agree about the 25th December, 15th April, 14th June, and 31st August ; these coincidences vary slightly in different years, because the earth takes about a quarter of a day more than a year to complete a revolution in its orbit, and this error accumulates from leap year till the fourth year after, when the extra day is taken in.

Sundials mark apparent time, while clocks measure *equal* or *mean* time ; if, therefore, a timekeeper, perfectly regular in its motion, were set to apparent solar time, it would be found to agree with it only on four days in the year.

Solar noon at any particular place may be obtained by a transit instrument (page 417), by a sextant (page 363), or by a meridian dial (page 396). The equation table, on page 402, shows the amount to be added to or subtracted from true solar time to give mean solar time, that is, the time shown by a clock. But if it is desired that the clock should show the mean time or Greenwich, or some other standard meridian, a further correction for longitude must be made. When it is noon at Greenwich the succeeding day is just beginning at the Antipodes, so that all places to the East of Greenwich are fast and those to the West slow. The East is divided into 360 degrees and the day into 24 hours of 60 minutes each, therefore every degree of deviation is equal to four minutes of time and every 15 degrees to one hour. As the day advances and the morning hours are recorded at Australia the date there is one day later than at Greenwich ; and in the later hours of the American day the date in America will be one day behind that at Greenwich.

To facilitate international communications and railway travelling over wide continents, standards of time have been arranged with Greenwich as the initial meridian. At the end of this book is a table showing the difference between mean time at Greenwich and other places.

Greenwich time is the standard adopted throughout Great Britain, Belgium, Luxemburg and Spain. In Ireland, Dublin time is kept.

Amsterdam time instead of Greenwich is now adopted as the standard for Holland.

" Mid-European time," one hour fast of Greenwich, has been adopted as the standard throughout Germany, Austria, Hungary, Sweden, Norway, Denmark, Switzerland, Italy, Bosnia and Servia.

" Eastern European time," two hours fast of Greenwich, serves

A A

for European Turkey, Bulgaria and Roumania. In Russia most of the railways adopt St. Petersburg time, which is 1 min. 13 secs. fast of East European time.

Paris time is the standard for France, Algeria and Tunis ; Lisbon time for Portugal.

Five different standards are now established in America. A central meridian, 90° west and 6 hours slow of Greenwich, which serves for the Mississippi Valley, Missouri Valley, Upper Lakes, and Texas, is called "Central Time," or "Valley Time." A meridian, 75° west and 5 hours slow of Greenwich, called "Eastern Time," serves for the district from Maine to Florida, from Ohio to Alabama, and the Lower Lakes, and northward

EQUATION TABLE COMPUTED TO MINUTES OF TIME.							
Clock faster than the Sun.	Min.	Clock slower than the Sun.	Min.	Clock faster than the Sun.	Min.	Clock slower than the Sun.	Min.
December 26	1	April 17	1	June 17	1	September 2	1
,, 28	2	,, 22	2	,, 22	2	,, 6	2
,, 30	3	,, 28	3	,, 26	3	,, 8	3
January 1	4	May 6	4	July 1	4	,, 11	4
,, 3	5	,, 23	3	,, 7	5	,, 14	5
,, 5	6	June 1	2	,, 14	6	,, 17	6
,, 7	7	,, 7	1	August 8	5	,, 20	7
,, 9	8	,, 14	0	,, 14	4	,, 23	8
,, 12	9			,, 19	3	,, 26	9
,, 15	10			,, 23	2	,, 29	10
,, 18	11			,, 27	1	October 2	11
,, 21	12			,, 31	0	,, 5	12
,, 25	13					,, 9	13
,, 30	14					,, 12	14
February 24	13					,, 17	15
March 1	12					,, 23	16
,, 6	11					November 14	15
,, 10	10					,, 19	14
,, 14	9					,, 23	13
,, 17	8					,, 26	12
,, 21	7					,, 29	11
,, 24	6					December 2	10
,, 27	5					,, 4	9
,, 30	4					,, 7	8
April 3	3					,, 9	7
,, 6	2					,, 11	6
,, 10	1					,, 13	5
,, 15	0					,, 15	4
						,, 18	3
						,, 20	2
						,, 22	1
						,, 25	0

NOTE.—In an equation table for use in any year of the four from leap year to leap year, absolute exactness is impossible, on account of the error in the computation of the year, which is referred to on p. 401. Seconds and intermediate days are therefore omitted.

through Canada. A meridian, 60° west and 4 hours slow of Greenwich, known as " Atlantic Time," serves for Newfoundland, New Brunswick, and Nova Scotia. A meridian, 105° west and 7 hours slow of Greenwich, known as " Mountain Time," serves for the Rocky Mountain region. A meridian, 120° west and 8 hours slow of Greenwich, known as " Pacific Time," serves for the Pacific States and British Columbia.

The standard for Western Australia is 8 hours, South Australia 10 hours, Victoria, Queensland and Tasmania 10 hours, and New Zealand 11½ hours fast of Greenwich.

A standard two hours fast of Greenwich serves now for the whole of South Africa.

Madras time is the standard for India.

In Japan the standard is 9 hours fast of Greenwich.

Unlike the civil day, which starts at midnight, the *A tronomical Mean Solar Day* is reckoned from noon to noon, and the hours are counted continuously from 1 to 24, instead of being divided into two equal spaces of 12 hours each. The day begins at the noon which succeeds the opening of the civil day; thus half-past six o'clock in the morning of, say, the 2nd day of January, would be expressed by astronomers as January 1st, 18 hours 30 minutes. In an Astronomical Regulator the hour circle is accordingly divided into 24, and the hour hand goes round once in 24 hours.

Cycle of the Sun.—A cycle of the sun is a period of 28 years, after which the days of the week again fall on the same days of the month as during the first year of the former cycle. The cycle of the sun has no relation to the sun's course, but was invented for the purpose of finding the dominical letter which points out the days of the month on which the Sundays fall during each year of the cycle. Cycles of the sun date nine years before the Christian era. If it be required to know the year of the cycle in 1892, nine added will make 1901, which, divided by 28, gives the quotient 67, the number of cycles that have passed, and the remainder 25 will be the year of the cycle answering to 1892.

Sidereal Time, the standard used by astronomers, is measured by the diurnal rotation of the earth, which turns on its axis in 23 hours 56 min. 4·1 sec. The sidereal day is therefore 3 min. 56 sec. less than the mean solar day, and a clock to show sidereal time must have its pendulum a trifle shorter than a mean time clock with the same train. About the 15th of April the sidereal clock and the mean time clock would agree, but from that time the divergence between the two would be increased each day by 3 min. 56 sec.

Mean time clocks, though, can be regulated by the stars with greater facility than by the sun, for the motion of the earth with regard to the fixed stars is uniform, and a star will always appear at the meridian 3 min. 56 sec. sooner than it did on the preceding day. In the absence of a transit instrument and a table giving the right ascension of particular stars, choose a window having a southern aspect, from which the steeple of a church, a chimney, or any other fixed point may be seen. To the side of the window attach a thin plate of brass having a small hole in it in such a manner that by looking through the hole towards the edge of the elevated object some of the fixed stars may be seen ; the progress of one of these being watched, the instant it vanishes behind the fixed point a signal is made to a person observing the clock, who then notes the exact time at which the star disappeared, and on the following night the same star will vanish behind the same object 3 m. 56 s. sooner.

Days.	STARS GAIN.		
	hrs.	min.	sec
1	0	3	56
2	0	7	52
3	0	11	48
4	0	15	44
5	0	19	34
6	0	23	35
7	0	27	31
8	0	31	27
9	0	35	23
10	0	39	19
11	0	43	15
12	0	47	11
13	0	51	7
14	0	55	3
15	0	58	58
16	1	2	54
17	1	6	50
18	1	10	46
19	1	14	42
20	1	18	38

If a clock mark 10 h. when the observation is made, when the star vanishes the following night it should indicate 3 m. 56 s. less than 10 h. If several cloudy nights have rendered it impossible to compare the clock with the star, it will be necessary to multiply 3 m. 56 s. by the number of days that have elapsed since the observation, and the product deducted from the hour the clock then indicated gives the time the clock ought to show. The same star can only be observed during a few weeks, for as it gains nearly one hour in a fortnight it will, in a short time, come to the meridian in broad daylight and become invisible ; to continue the observation, another star must be selected. In making the observation, care must be taken that a planet is not observed instead of a star; Mars, Jupiter, and Saturn are those most likely to occasion this error, more especially Saturn, which from being the most distant of the three resembles a star of the first magnitude. The planets may, however, be easily distinguished, for being comparatively near the earth, they appear larger than the stars ; their light also is steady because reflected, while the fixed stars scintillate and have a twinkling light. A sure means of distinguishing between them is to watch a star attentively for a few nights; if it change its place with regard to the other stars it is a planet.

The Nautical Day commences when the sun is on the meridian ; 8 blows are then struck on the ship's bell, and the afternoon watch is begun. At 12.30 one blow is struck, and time is spoken of as " one bell " ; at one o'clock, 2 bells ; at 1.30, 3 bells ; at 2, 4 bells ; 2.30, 5 bells ; 3, 6 bells ; 3.30, 7 bells ; at 4 o'clock, 8 bells again. At 4 o'clock begins the first dog watch, which lasts two hours, the periods being struck as before, ending at 6 o'clock with four bells. Then begins the second dog watch, also of two hours' duration, ending at 8 o'clock, half-hour intervals being struck 1, 2, 3, 8 ; 8 bells marking the completion of the second dog watch. Next comes the middle watch, lasting four hours, and struck like the afternoon watch. The night watch, the morning watch, the forenoon watch, each of four hours similarly marked, follow in succession ; the forenoon watch ending at noon with 8 bells completes the day.

There is a distinction to be noticed between the sign Aries and the constellation of that name. The first point of the sign Aries or the equinoctial point ♈ is the zero from which the right ascension, or longitude, of celestial bodies is measured, just as Greenwich is an initial meridian for measuring the longitude of terrestrial places. Ancient astronomers called it the first point of Aries because in their time the phrase correctly described its position, but the vernal equinox retrogrades $50\frac{1}{4}$ seconds of a degree each year, and so the first point of Aries is now really in the constellation Pisces. The moment the point ♈ passes the meridian it is sidereal noon, and sidereal time would then be 0 hour 0 min. 0 sec.

The civil year began on 25th March before 1752, when the present reckoning for the year to commence on 1st January was adopted.

Duration of a Year.—The sidereal year starts with the spring equinox, when the sun enters the sign Aries, that is, when the sun crosses from the south to the north of the equator. The earth in its revolution round the sun makes rather over 366 rotations or 366 sidereal days, which are equal to 365 solar days. The sidereal year is equal to 365 days 6 hours 9 min. 11 sec., nearly, of mean solar time. The earth, on the completion of its revolution, returns to the same place among the stars, but not exactly at the spring equinox, owing to the precession of the equinoxes, so in order that the year may accord with the seasons the sidereal year is disregarded in favour of the equinoctial, tropical or solar year, taken as 365 days 5 hours 48 min. 48 sec. Among the Romans no regular account was taken of the difference between the year and 365 days till B.C. 45. Then the surplus was reckoned as six hours, making one day in four years ; and one day was accordingly added to every fourth year. There still remained the apparently trifling difference of 11 min. 11 sec. between the civil and the tropical year ; this, however, produced an error of about seven days in 900 years. In 1582, Pope Gregory XII. struck out ten days, which represented the accumulated error, from the calendar, and it was decided that three leap years should be omitted every 400 years ; thus, as 1600 was leap year, the years 1700, 1800, and 1900 were not, but 2000 will be leap year. This rectification was not adopted in England till 1752, when eleven days were omitted from the calendar. As our year still exceeds the true year, although by an extremely small fraction, another leap year in addition to those should be omitted once in 4,000 years.

The Golden Number.—Meton, an Athenian astronomer, B.C. 432, discovered that after a period of 19 years the new and full moons returned on the same days of the month as they had done before ; this period is called the cycle of the moon. The Greeks thought so highly of this calculation that they had it written in letters of gold, hence the name golden number ; and at the Council of Nice, A.D. 325, it was determined that Meton's cycle should be used to regulate the movable feasts of the Church. Jesus Christ was born in the second year of the lunar cycle. Therefore if we add one to the present year and divide the sum by 19, the remainder will give the year of Meton's cycle. 1892 + 1 divided by 19 leaves a remainder of 12, which is, therefore, the golden number for 1892.

The Epact serves to find the moon's age by showing the number of days which must be added to each lunar year, in order to complete a solar year. A lunar month is composed of 29 days 12 hours 44 min. 3 sec., or rather more than 29·5 days ; twelve lunar months are, therefore, nearly 11 days short of the solar year—thus, the new moons in one year will fall 11 days earlier than they did in the preceding year, so that were it new moon on January 1st, it would be nearly 11 days old on the 1st of January of the ensuing year, and 22 days on the third year ; on the fourth year it would be 33 ; but 30 days are taken off as an intercalary month (the moon having made a revolution in that time), and the three remaining would be the Epact ; the Epact thus continues to vary, until the expiration of 19 years, the new moons again return in the same order as before. If the solar year were exactly 11 days longer than 12 lunar months, it would only be necessary to multiply the golden number by 11, divide the product by 30, and the remainder would be the Epact : but as the difference is not quite 11 days, one must be taken from the golden number, the remainder multiplied by 11, and the product, if less than 30, shows the Epact; but if more it must be divided by 30, and the remainder is the Epact for that year. The golden number for 1892 being 12, 12 multiplied by 11 = 121, and 121 divided by 30 leaves a remainder of 1, which is the Epact for 1892.

To find the moon's age upon any particular day, add the number placed against the month in the following table to the Epact and the day of the month ; the product, if under 30, will be the moon's age.   Should it exceed this number, divide by 30, and the remainder will show it :—

January ..	2	April ..	2	July ..	5	October ..	8
February..	3	May ..	3	August	7	November	10
March ..	1	June ..	4	September	7	December	10

From the irregularity of the number of days in the calendar months and other causes, it is difficult to make an exact calculation, but the error resulting from this rule does not exceed one day.

The Number of Direction.—The Council of Nice decided, A.D., 325, that Easter Day is always the first Sunday after the full moon which happens upon or next after the 21st of March. Easter Day, therefore, cannot take place earlier than the 22nd of March, or later then the 25th of April.   The number of Direction is that day of the thirty-five on which Easter Sunday falls.

The Roman Indiction.—The Roman Indiction was a period of 15 years, appointed A.D. 312 by the Emperor Constantine for the payment of certain taxes.

The Julian Period.—The Julian Period of 7,980 years is the product obtained by multiplying together 28, 19, and 15, which numbers represent the cycles of the Sun, the Moon, and the Roman Indiction.   The beginning of the Julian period is reckoned from 709 before the Creation of the World, so that its completion will occur A.D. 3267.

Time Ball.—[*Der Zeitball.*]—A ball suffered to fall a short distance to indicate time.   The time ball at Greenwich Observatory is of very thin copper, sliding on a square wooden mast. There is a groove down one side of the mast, to allow room for an iron rod attached to the bottom of the ball which terminates in the piston *a*.   When the ball is in its highest position two clips support the piston, as shown in the figure.   At the instant the ball is desired to fall an electric current attracts the armature *e*, the detent *d* is released, the levers *c* and *b* fall, thus opening the clips that hold the piston.   The ball falls freely at first, but as the piston enters the cylinder *f*, which it fits loosely, the resistance of the imprisoned air moderates its velocity.   The ball may be raised by means of a rack on the end of the piston taking into a winding pinion, which is thrown out of gear when the winding is accomplished.   A large ball may be cheaply made with a wire skeleton frame covered with canvas and painted.   It should be loaded inside with lead, to ensure its dropping with celerity.   A large spiral spring, padded

at the top, would
afford a sufficient
cushion for the ball
to drop on to, and be
less costly than the
piston and cylinder.
The trigger arrange-
ment might also be let
off by hand, to save
expense.   It may be
taken as a very good
rule that the diameter
of a time ball should
be one foot for every
ten feet it is above the
ground, and that its fall
should be at least half
as much again as its
diameter.

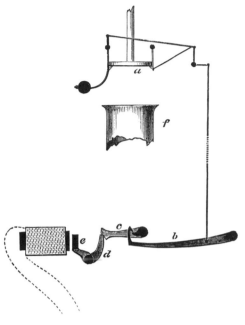

Timepiece. — [*Pen-
dule sans sonnerie.–Die
Uhr ohne Schlagwerk.*]
—Among watch and
clockmakers any time-
keeper above the size
of a watch, which does not strike at the hours, is called a timepiece.   A clock is understood to mean a timekeeper that strikes.

Timing Box.—[*Der Werkbehälter.*]—A brass box for the reception of an uncased watch movement while it is being timed.   The
edge of the pillar plate rests in a rebate, and a cover with a glass
let into the top screws on to the box, and keeps the movement in
position.

Timing.—[*Réglage.—Das Regulieren.*]—The art of proportion-
ing the balance spring and balance.   Getting a watch or clock to
time.   (See Balance Spring.)

Timing French Pendules.—Escape wheels of French pendules rotate twice in a minute, so that the pendulum makes four
times as many vibrations per minute as there are teeth in the
escape wheel.   A pendule may therefore be quickly brought to
time by counting if the beats of the pendulum per minute equal
four times the number of teeth in the escape wheel.

Timing Repeating Carriage Clocks.—The quickest method
is to listen to the first blow of the hammer on the bell (or gong)
at each hour and half-hour, noting at the same time the position
of the seconds hand on the regulator.   Thus : supposing the
blow is given exactly at 12 h. o m. o s. make a note of it, and

compare again at 12 h. 30 m. 0 s., when the half-hour blow is given at, say, 12 h. 30 m. 5 s., make a note of this also, but do not alter the index, as this difference may rise from the imperfect position of the half-hour pin in the cannon wheel ; at 1 h. 0 m. 0 s., compare as before, and as it is the same pin coming into action as at 12 h. 0 m. 0 s., any difference in the time of the first blow will be indicative of a gain or loss, as the case may be ; thus : it may be given at 1 h. 0 m. 4 s., which will show a loss of 4 s. in the last hour, and the clock may now be regulated accordingly. The blow may next be given at 1 h. 30 m. 8 s., showing the effect the moving of the index has had in making it gain 1 s. in the preceding half-hour ; regulate and compare again at 2 h. 0 m. 0 s., and if the first blow is given at 2 h. 0 m. 3 s. it will show that the clock has been keeping correct time for the last half-hour.

Timing in Positions.—[*Réglage dans les positions.*—*Das Regulieren in den Lagen.*]—The art of adjusting the balance springs and balance so that the watch keeps time in different positions. In ordinary work two positions are taken—vertical and horizontal, called hanging and lying, or pendant up and dial up. In the finest work the watch is also timed in the quarters, that is, in an open-faced watch with the 3 on the dial upwards and with the 9 upwards. (See Balance Spring.)

Timing in Reverse.—Getting a watch to time in positions by throwing the balance out of poise when the short vibrations are over a turn and a quarter in extent. (See Balance Spring.)

Timing Screws. Quarter Screws. Timing Nuts. Mean Time Screws.—[*Die Regulierschrauben* (*Unruh*).]—Four screws or nuts placed at equal distances round the rim of a watch compensation balance, which are used for getting the watch to mean time. In a marine chronometer there are two timing nuts, one at each end of the arm. (See Compensation Balance.)

Ting Tang Clock.—[*Horloge à 2 timbres.*—*Die Vierteluhr mit zwei Glocken.*]—A clock that sounds the half-hours or quarters on two bells only, as distinguished from a chime clock which runs through a series of notes.

Tipsy Key.—(See Bréguet Key.)

Tompion, Thomas.—" The father of English watchmaking," born 1638, died 1713, and was buried in Westminster Abbey. He patented the cylinder escapement in 1695 (No. 344) ; he applied to watches the balance spring, under the direction of Hooke, the inventor, and to clocks the rack-striking work, as devised by Barlow.

Torsion Pendulum.—[*Pendule à torsion.*—*Das Drehpendel.*]— A pendulum in which the bob rotates by the twisting of the suspending rod or spring. Rotating pendulums of this kind will not bear comparison with vibrating pendulums for timekeeping. They are only used when a long duration of the motion of the pendulum is required. Small clocks to go a twelvemonth without winding are made with torsion pendulums about six inches long, which make fifteen excursions a minute. The time occupied in the excursion of such a pendulum depends on the power of the suspending rod to resist torsion, and the mass and distance from its centre of motion of the bob. In fact, the action of the bob and suspending rod is very analogous to that of a balance and balance spring.

Touch Stone. Touch Needles.—[*Pierre de touche.*—*Der Probierstein.*]—A touch stone is a piece of fine grained dark silicious stone, used in conjunction with touch needles for estimating the quantity of gold. Touch needles are small bars of gold, one each of all the different standards likely to be tested. The gold which it is desired to test is rubbed on the stone, and beside it is rubbed one of the needles which in the judgment of the operator is nearest in quality to it. The two streaks are then subjected to the action of nitric acid, and the experience of the operator enables him afterwards to judge of the difference in quality by the difference in the colour of the two streaks.

Tourbillion.—[*Tourbillon.*—*Der Tourbillon.*]—A carriage in which the escapement of a watch is fitted so that it revolves round the fourth wheel. The idea of the tourbillion, one of the numerous inventions of Bréguet, is to get rid of position errors.

"*Tourbillion*" *Escapements.*—In Figs. 1 and 2 the chrono-

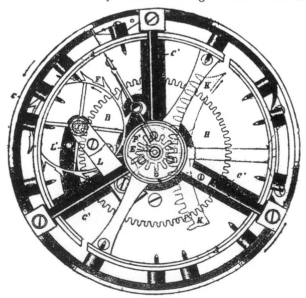

Fig. 1.

meter escapement is shown, but the lever is equally applicable to the tourbillion. It will be observed that the fourth wheel, H, is screwed to the plate, and is stationary, and that the tourbillion cage or carriage, C C, is caused to revolve round it. The third wheel, K, gears with the pinion, G, which is fixed to the carriage, and as the escape pinion, I, gears with the fixed fourth wheel, the motion transmitted by the third wheel causes the escape pinion to turn on its axis, and also to roll around the fourth wheel, which, as far as its connection with the escape pinion is concerned, may be regarded as a circular rack. The lower pivot D of the carriage runs in the main plate of the movement, and carries the seconds hand ; the upper pivot E rotates in a high and long bridge which spans the carriage. L L are the cocks for escape pinion pivots.

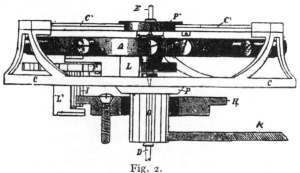

Fig. 2.

B. Bonniksen has invented a more compact arrangement which he calls a karrusel, in which the carriage driven by the

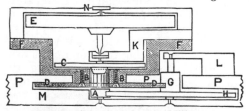

Fig. 3. Bonniksen's Tourbillion.

A, fourth pinion ; B, pivot of carriage ; C, fourth wheel ; D, wheel attached to carriage driven by third pinion G ; H, third wheel ; E, balance ; F, body of carriage ; K, L, N, cocks ; M, bar screwed to pillar plate P.

third pinion rotates once in $52\frac{1}{2}$ minutes, which gives sufficiently quick change of position for all practical purposes. Fig. 3 shows Bonniksen's karrusel in which the section of the rotating carriage is shaded.

Train.—[*Rouage.*—*Das Laufwerk.*]—The toothed wheels in a watch or clock that connect the barrel or fusee with the escapement. In a going barrel watch the teeth round the barrel drive the centre pinion, to which is attached the centre wheel; the centre wheel drives the third wheel pinion, which carries the third wheel; the third wheel drives the fourth wheel pinion, on which the fourth wheel is mounted; the fourth wheel drives the escape pinion, to which the escape wheel is fixed. The number of teeth in the various wheels and pinions is determined by the following considerations:—The centre arbor to which the minute hand is fixed always turns once in an hour, the fourth wheel to the arbor of which the seconds hand is fixed turns once in a minute, so that the product obtained by multiplying together the number of teeth in the centre and third wheels must be 60 times the product obtained by multiplying together the numbers of third and fourth pinions. Two other points may be settled before deciding the rest of the train; 1st, the number of turns the barrel makes in 30 hours, which is the time allowed from winding to winding. Four turns would be a suitable number, and in that case the barrel would contain $7\frac{1}{2}$ times the number of teeth in the centre pinion. 2nd, the number of vibrations made by the balance in an hour. If 18,000 be decided on, then, assuming the escape wheel to have the usual number of 15 teeth, the escape pinion must make 10 rotations a minute, and the fourth wheel must have ten times as many teeth as the escape pinion. The barrel teeth and centre pinion, which have considerable pressure to bear, must be of adequate strength, but the pitch of the teeth and the size of the wheels are gradually diminished as the train nears the escapement. In the last wheels of a train small and light wheels are especially needed, so that they get quickly into motion directly the escapement is unlocked, and are stopped with but little shock when the escapement is locked again. The remarks on the train of a going-barrel watch apply equally to the going train of a clock. The considerations which guide in deciding the numbers for the striking train of a clock are the number of blows to be struck from winding to winding, the fall of the weight or turns of the barrel or fusee, as the case may be, and the number of pins in the pin wheel. English lever watches usually have either a 16,200 or an 18,000 train. Foreign watches, both lever and horizontal, have 18,000 trains as a rule.

To Calculate Clock Trains.—Divide the number of pendulum vibrations per hour by twice the number of escape wheel teeth, the quotient will be the number of turns of escape wheel per hour. Multiply this quotient by the number of escape pinion teeth, and divide the product by the number of third wheel. This quotient will be the number of times the teeth of third wheel pinion must be contained in centre wheel.

For Example.—Take a pendulum vibrating 5,400 times an hour, escape wheel of 30, pinions of 8, and third wheel of 72. Then $5,400 \div 60 = 90$. And $\dfrac{90 \times 8}{72} = 10$. That is, the centre wheel must have ten times as many teeth as third wheel pinion, or ten times $8 = 80$.

The centre pinion and great wheel need not be considered in connection with the rest of the train, but only in relation to the fall of the weight, or turns of mainspring, as the case may be. Divide the fall of the weight (or twice the fall, if double cord and pulley are used) by the circumference of the barrel (taken at the centre of the cord) ; the quotient will be the number of turns the barrel must make. Take this number as a divisor, and the number of turns made by the centre wheel during the period from winding to winding as the dividend ; the quotient will be the number of times the centre pinion must be contained in great wheel. Or if the numbers of great wheel and centre pinion and the fall of weight are fixed, to find the circumference of barrel, divide the number of turns of centre wheel by the proportion between centre pinion and great wheel ; take the quotient obtained as a divisor and the fall of weight as a dividend (or twice the fall if pulley is used), and the quotient will be the circumference of barrel. To take an ordinary regulator or 8-day clock as an example—192 (number of turns of centre pinion in 8 days) ÷ 12 (proportion between centre pinion and great wheel) = 16 (number of turns of barrel). Then if fall of cord = 40 inches, $\dfrac{40 \times 2}{16} = 5$, which would be circumference of barrel at the centre of the cord.

If the numbers of the wheels are given, the vibrations per hour of the pendulum may be obtained by dividing the product of the wheel teeth multiplied together by the product of the pinions multiplied together, and multiplying the quotient by twice the number of escape wheel teeth.

The numbers generally used by London clockmakers for clocks with less than half-seconds pendulum are, centre wheel 84, gearing with a pinion of 7 ; third wheel 78, gearing with a pinion of 7.

The product obtained by multiplying together the centre and third wheels = $84 \times 78 = 6,552$. The two pinions multiplied together = $7 \times 7 = 49$. Then $6,552 \div 49 = 133\cdot7$. So that for every turn of the centre wheel the escape pinion turns $133\cdot7$ times. And supposing the escape wheel to have 34 teeth, $34 \times 2 = 68$, and $133\cdot7 \times 68 = 9,091\cdot6$, which is the number of vibrations made by the pendulum in an hour.

The length of the pendulum, and therefore the number of escape wheel teeth, in clocks of this class, is generally decided with reference to the room to be had in the clock case, with this restriction, the escape wheel should not have less than 20, nor more than 40 teeth, or the performance will not be satisfactory. The length of the pendulum for all escape wheels within this limit is given in the table at the end of the book. The length there stated is of course the theoretical length, and the ready rule adopted by clockmakers is to measure from the CENTRE arbor to the bottom of the inside of the case, in order to ascertain the greatest length of pendulum which can be used. For instance, if from the centre arbor to the bottom of the case is 10 inches, they would decide to use a 10-inch pendulum, and cut the escape wheel accordingly with the number of teeth required as shown in the table. But they would make the pendulum rod of such a length as just to clear the bottom of the case when the pendulum was fixed in the clock.

It occasionally happens that a clock gains even when the pendulum is lowered as far as the case will allow. If then a piece be cut off the bottom of the bob the clock may generally be brought to time.

NOTE.—A Table of Diameters and Circumferences of Circles is given at the end of the book.

In the clocks just referred to, the great wheel has 96 teeth and gears with a pinion of eight.

Month Clocks have an intermediate wheel and pinion between the great and centre wheels. This extra wheel and pinion must have a proportion to each other of 4 to 1 to enable the 8-day clock to go 32 days from winding to winding. The weight will have to be four times as heavy, plus the extra friction, or if the same weight is used, there must be a proportionately longer fall.

Six Month Clocks have two extra wheels and pinions between the great and centre wheels, one pair having a proportion of $4\frac{1}{2}$ to 1 and the other of 6 to 1. But there is an enormous amount of extra friction generated in these clocks, and they are not to be recommended.

Lunation Train.—The following numbers are suitable for a lunation train, *i.e.* a train the first pinion of which (placed on the centre wheel arbor) revolves once in an hour, and the last wheel (which carries the moon on its axis) once in 29 days 12 hours 44 minutes 3 seconds.

Pinions of 6, 9 and 37. Wheels of 91, 91 and 171.

These numbers give the duration of a lunation within a fraction of a second. It is immaterial in what order the pinions gear with the wheels. Brocot's lunation train is given under the head of " Calendar."

Motion of the Earth Train.—As stated at page 405 the

earth takes 365 days 5 hours 48 minutes 49·7 seconds to perform its annual revolution round the sun. The following numbers will represent nearly the motion of the earth, supposing the first pinion to turn in twelve hours.

Pinions, 8, 7, 7.   Wheels, 50, 69, 83.

Ferguson devised the simple arrangement shown in Fig. 1 for indicating the age of the moon. The thick wheel of 57 turns once in 24 hours, and gears with the two wheels of 57 and 59,

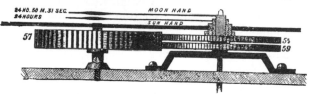

Fig. 1.

which consequently are a little out of pitch. By adding a pair of mitre wheels, as shown in Fig. 2, the rotation of the moon might also be indicated.

Solar and Sidereal Train.—The following remarkably simple

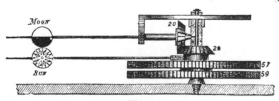

Fig. 2.

arrangement for showing the motion of the sun and the motion of the stars is by Ferguson. The two wheels are of the same

diameter as each other, and the two pinions also. The wheel of 61, turning once in 24 hours, carries the sun hand and drives the pinions, which are fixed together. The upper pinion drives the wheel of 73, which carries the star plate.

To Calculate a Lever Watch Train.—The fourth wheel turning 60 times for one turn of the centre wheel, the numbers of teeth in centre and third wheels multiplied together must be 60 times the product obtained by multiplying together the teeth of third and fourth pinions. *For Example,* to take the seconds train most in use for lever watches having third and fourth pinions of 8, we should have $8 \times 8 = 64$, and $64 \times 60 = 3,840$. Any two numbers which when multiplied together make 3,840 would be suitable for the centre and third wheels. But, unless some special numbers are desired, the calculation need not be carried further, because it is evident the two numbers we already have (64 and 60) will answer the condition. The escape wheel having 15 teeth turns once for every 30 vibrations of the balance, and the train of 16,200, which is the most usual for lever watches, we have 16,200 ÷ 30 = 540 turns per hour for the escape pinion. As the fourth wheel turns 60 times an hour, the numbers for fourth wheel and escape pinion must be in the same ratio as 540

NOTE.—A full table of ordinary trains, with length of pendulum, is given at the end of the book, and good examples of striking and quarter trains under the head of Quarter Clock, see page 387.

TABLE OF LEVER TRAINS.

Centre Wheel	Third Wheel	Fourth Wheel	Third Pinion	Fourth Pinion	Scape Pinion		Centre Wheel	Third Wheel	Fourth Wheel	Third Pinion	Fourth Pinion	Scape Pinion	
64	60	64	8	8	8	⎫	80	75	80	10	10	8	⎫
64	60	56	8	8	7		64	60	80	8	8	8	
60	56	56	8	7	7		64	60	70	8	8	7	
60	49	56	7	7	7		64	56	70	8	7	7	
64	60	48	8	8	6		60	49	70	7	7	7	
60	56	48	8	7	6	Trains of 14400	*64	60	60	8	8	6	Trains of 18000
60	49	48	7	7	6	4 vibrations per sec.	60	56	60	8	7	6	5 vibrations per sec.
60	42	48	7	6	6	Equidivisional seconds.	60	49	60	7	7	6	Equidivisional seconds.
56	45	48	7	6	6		60	42	60	7	6	6	
48	45	48	6	6	6		56	45	60	7	6	6	
60	48	48	8	6	6		48	45	60	6	6	6	
64	45	48	8	6	6	⎭	60	48	60	8	6	6	
64	60	72	8	8	8	⎫	64	45	60	8	6	6	⎭
64	60	63	8	8	7								
64	56	63	8	7	7								
60	49	63	7	7	7								
64	60	54	8	8	6	Trains of 16200							
60	56	54	8	7	6	4½ vibrations per sec.	* This Train is also the one						
60	49	54	7	7	6	Imperfect seconds.	most generally used in Hori-						
60	42	54	7	6	6		zontal Watches.						
56	45	54	7	6	6								
48	45	54	6	6	6								
60	48	54	8	6	6								
64	42	54	8	6	6								
80	75	72	10	10	8	⎭							

and 60, that is (540 ÷ 60 = 9) as 9 to 1. And if we decide on 7 for the escape pinion, we have 7 × 9 = 63 for the fourth wheel.

NOTE.—With a train of 14,400, the escape wheel turns 8 times; with 16,200, 9 times; and with 18,000, 10 times a minute.

Marine Chronometer Trains.

	Great Wheel.	Centre Wheel.	Third Wheel.	Fourth Wheel.	Centre Pinion	Third Pinion	Fourth Pinion	Scape Pinion	Vibrations per Hour	Turns of Fusee.
2 Day...	90	90	80	80	14	12	10	10	14400	8¼
8 Day...	144	90	80	80	12	12	10	10	14400	16

NOTE.—The two-day Marine Chronometer Train beating 14,400 is changed to an 18,000 train, and used for full-plate pocket chronometers, by substituting for the escape pinion of 10 an escape pinion of 8.

Independent Seconds Train.—See page 195.

Transit Instrument.—[*Lunette méridienne.*—*Das Meridian Instrument.*]—A telescope placed in the meridian and mounted in bearings so that it may be turned from north to south. One horizontal and three or more vertical wires, or " spider lines,"

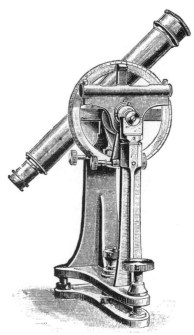

are stretched across the focus, and by looking through the telescope the time of the passage of the sun or a star across the wires may be accurately noted. The engraving shows a very handy instrument of small size, by Mr. John Short, which is exactly suited to the needs of the watchmaker, and contains special points of excellence. The foundation or base plate is made to be screwed down permanently to a window sill or other solid stone or brickwork. The entire instrument may then be lifted on and off the foundation plate with the assurance that when replaced it always occupies the same position without readjustment.

In setting up a transit instrument it is necessary before attempting to take observations to (1) level it. There are two spirit levels attached to a transit instrument, one striding across the **V**-shaped bearings, and the other at right angles to it. The levels themselves should be tested by turning each one end for end when the bubble is in the centre. If the level is correct, the bubble will also be in the centre after turning. (2) Adjust the instrument for collimation, that is, ascertain if the line of sight

B B

is exactly at right angles to the axis on which the instrument turns. Look through the telescope and observe some distant mark that just coincides with the centre vertical wire—a watch dial makes a convenient mark for the purpose if it can be fixed in a suitable spot ; then reverse the telescope in its bearings, and if the wire and the mark are not now coincident the wire must be shifted by means of the collimation screws at the side of the telescope near the eye-piece and the test repeated. (3) Adjust the instrument in azimuth, that is, ascertain if it points truly north and south. Observations of the polar star are the most usual method of testing the azimuth adjustment. The polar star passes the meridian twice in twenty-four hours, once above and once below the pole, and if the instrument is correctly in azimuth these transits will take place exactly twelve sidereal hours apart. Correction for azimuth may be made by means of the screws provided for shifting one of the V bearings to and fro. The time at which certain stars pass in the meridian of Greenwich is given in the Nautical Almanac and other publications. A suitable star having been selected, the instrument is pointed to the known altitude of the star and clamped there. The altitudes given are those for places in the latitude of Greenwich, and would require correction for other latitudes. Mr. Latimer Clark suggests a very simple method. Set the instrument so that it points exactly right to a star, the latitude of which for Greenwich is given ; then turn round the graduated circle at the side of the instrument till the given altitude is indicated, and fix the circle there. No further correction for latitude need then be made as long as the instrument remains in the same place. The instant the star passes the centre wire will be the time given in the Nautical Almanac as the right ascension* of the star, plus or minus the correction for longitude of the particular place where the instrument is fixed. This, of course, is sidereal time. (See Time.)

The passage of the sun across the vertical wire of the telescope gives solar noon. As the sun is a large body, the usual course is to observe the instant when the edge of the sun enters and leaves the centre wire, and take the mean of the two. This may be converted into mean solar noon by correcting for longitude and adding or subtracting the amount given in the equation table under the head of " Time."

Trial Number.—[*Somaire d'essai.*—*Die Gangnummer.*]—A symbol used to express the relative excellence of watches and chronometers in competitive trials. The Greenwich method of

* The right ascension of a star is really its angular distance reckoned not in degrees, but in sidereal time and measured eastwards along the Equator from the point of intersection of the Equator and Ecliptic, generally called the first point of Aries.

arriving at the trial number, which is also used at several foreign observatories, is to add to the difference between the greatest and least (in seconds) twice the difference between one week and the next.

Tripoli.—[*Terrepourrie.—Der Trippel.*]—An infusorial earth used by watch jewellers for polishing the softer stones.

Tripping.—[*Manquement.—Das Vorbeilaufen eines Gangrad-zahnes.*]—In an escapement, a tooth of the escape wheel running past the locking face.

Trunk Dial.—A spring timepiece with a large dial below which the case is continued for a few inches to admit of a longer pendulum than could otherwise be used.

Tumbler.—(See Gathering Pallet.)

Turns.—[*Tour.—Der Drehstuhl.*]—A small dead centre lathe in which watchmakers turn and pivot arbors, &c. The work is centred in the turns and has a ferrule fixed to it, so that it may be rotated either with a bow or by a cord connected with a wheel to which motion is given by the hand or foot.

Every one who knows anything of watchmaking is tolerably conversant with the common turns shown in the engraving. They are used for turning arbors, pivoting generally, and other purposes. To make the best use of the tool a good stock of runners is a necessity. Steel runners with male centres are handy when turning bouchons or other hollow pieces. Brass

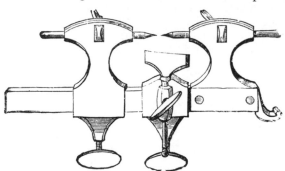

Fig. 1.

runners like Fig. 2 are used for polishing pivots on, and for repolishing pivots that have become marked ; some watch-makers use an ivory bed. Thin edged brass runners with small

Fig. 2.   Fig. 3.   Fig. 4.

holes through which the pivot projects, as in Fig. 3, are used for rounding pivot ends. A runner with a hole in the edge large enough to allow the pivot to enter up to the cone is used to support the pivot while burnishing on an escape wheel. Sometimes a broad-headed screw is used on the pivoting runners, as in Fig. 4, to support the polisher when polishing straight pivots. (See also Lathe.)

Turning Arbor.—[*Arbre.—Der Drehstift.*]—An arbor on which objects that require to be operated on in the lathe may be mounted. It is centred at each end, and has a ferrule fixed to it.

Turning Pendant.—[*Pendant tournant.—Der drehbare Bügelknopf.*]—A device to prevent the bow of a watch case from being twisted off. The bow is attached to a collar which turns stiffly on the pendant.

Turret Clock.—[*Horloge de clocher.—Die Turmuhr.*]—A large clock in which the dials are distinct from the movement. The movement of a turret clock should, if possible, be arranged so as to get a direct fall for the weight, and on the same side of the arbor as the wheel for the leading off rod is, if this latter is placed horizontally with the barrel. Where guide pulleys are used to conduct the wire rope away from the barrel, not only is a heavier weight required, but there will be more or less trouble from friction of the rope. The barrel should be of iron, cast hollow; its diameter and length would be determined by the amount of fall to be obtained. Steel wire rope should be used for the attachment of the driving weights, and the larger the barrel is in reason the better, for with a very small barrel there is danger of crippling the wire ropes. If a pulley is used with the weight, it should also be large in diameter. When the leading off rods of a turret clock are long, they should be made large in diameter. If often happens that turret clocks with large dials and long and slender leading off rods vary considerably in their timekeeping from the twisting of the rods. It is not the weight of the hands alone, but the action of the wind on the hands, which has to be taken into consideration. Hence also the desirability of counter-balancing the hands as far as possible on the *outside* of the dial.

Among the causes that occasionally lead to the failure of turret clocks, a current of air acting on the pendulum is one of the most frequent, especially where the pendulum is a long one. Pinions should be rather too small than too large. Large pinions, added to want of truth in some parts of the train, are responsible for many stoppages. With heavy hands, want of lubrication, or of freedom in the dial work is also a source of trouble.

Westminster Clock.—The Westminster clock tower is 40 feet square. There are four dials, 180 feet above the ground level; each of them is 22½ feet in diameter, or nearly 400 square feet in area. Cast-iron framework forms the divisions and figures, the spaces being filled in with opalescent glass. The hour figures are 2 feet long and the minute spaces 1 foot square. The hour hands are solid, and cast of gun-metal. For lightness the minute hands are tubular, they are of copper, the shells being thin, but strengthened by diaphragms at intervals. The copper tubes are tapered and closed at the tips, their open ends being fitted to gun-metal centres, which also form the outside counter-poises. Each minute hand measures 11 feet from its centre of motion to the point, besides the counterpoise of 3 feet, so that the load on the clock when the hands are subjected to a high wind or covered with snow can be appreciated.

The movement is contained in a frame made up of two cast iron girders fifteen and a half feet long placed side by side four feet apart, and braced together. There are three trains or sets of wheels, each one driven by a separate weight ; the " going " or " watch " train that drives the hands, and is controlled by the escapement and pendulum, occupies the centre of the frame ; on the left hand in the drawing on page 424 is the hour striking train, which only moves once an hour, when it is released by the going train, and locks itself after it has struck the number of blows corresponding to the hour of the day ; on the right is the quarter train, which is released by the going train and allows a chime of either one, two, three, or four quarters as required,* and again locks itself. The first quarter denotes fifteen minutes past the hour, and the interval between the first and second and between the second and third quarters is fifteen minutes, but the fourth quarter is let off 20 seconds before the hour so that it shall be completed before the first blow of the hour.

As it has less to do, the going train is lighter than either of the striking trains, and in all three the strength of the wheels and other parts is greater near the weight barrels, and is gradually diminished as the velocity of the parts increases. In the going train the parts near the escapement can hardly be too light, for it is necessary that they should get into action quickly, directly they are unlocked, and give as light a blow or shock as possible when they are locked again. The four pairs of hands are driven by four horizontal minute arbors placed high above the move-ment, and leading each one to the centre of one of the dials. Each dial has separate motion wheels for reducing the rate of travelling of the hour hand, the motion work being carried on the walls of the clock room. It will be seen from the drawing on page 424 that connection between the movement and the minute arbors is made by means of the oblique shaft, *a*, and the mitre wheels, *b, c, d,* and *e.*

* The musical notation is given under the head of " Chimes."

The numbers of the watch train are :—great wheel 180, driving a pinion of 48 on the right for the hand work, and one of 12 on the left ; on the arbor of the latter is the second wheel of 120 driving the pinion of 16, which carries the third wheel of 90, driving the escape pinion of 9. The great wheel is 2 feet 3 inches in diameter and the escape wheel 12 inches. The pendulum beats once in two seconds and weighs 685 lbs. The construction of double three-legged gravity escapement is given under the head of " Gravity " and the method of compensating the pendulum under the head of " Pendulum."

For striking the quarters the four-armed cam or snail, g, turns once in an hour. It is gradually pressing down the lever h. The quarter train is held by the locking lever j, which rests on the upper one of two blocks on the lever k. The lever h acts on the lever k, and as the quarter hour approaches the lever k rises and allows the locking lever j to escape from the first locking block to the second one, which is rather lower on the lever ; this allows the train to move a little, and causes the noise generally known as warning. When the quarter is to be sounded the lever k falls free of the locking lever j and the train of wheels begins to run, the lever k being lifted sufficiently high by the cam l to disengage the tongue m from the notch of the locking plate or count wheel n, in which it is resting. If one chime only is to be struck the tongue m descends into the same notch of the locking plate, for that notch is wide enough to receive the tongue again after the small angular movement made by the plate, and the upper block on the lever k catches the locking lever j as it comes round. But at the next quarter, after one chime has been sounded, the tip of the tongue rests on the periphery of the locking plate till another chime is struck, when it falls into the next notch. The locking plate makes one rotation in three hours, and it will be observed that it is spaced out to allow of three sets of quarters. The interval between the hammer blows is kept constant by the resistance of the air against the revolving fly o, which is composed of two large blades of sheet iron.

The action for letting off the hour striking is very similar to that for discharging the quarters, except that there is a double warning. The hour striking train is held by a stop on the locking lever, resting against the upper of the two blocks on the lever t. A few minutes before the hour the locking lever falls on the lower block and is released thirty seconds before the hour by the snail r, which revolves once in an hour.* The locking

* The four-armed snail attached to the hour snail is for actuating a lever which stops the winding of the quarter part when the time for striking the quarters approaches.

lever is then held by a small independent lever till two seconds before the hour, when a snail on the second wheel arbor which rotates once every fifteen minutes allows one extremity of a rocking lever to fall, and the other extremity then hits up the independent lever and releases the locking lever. By the time the two seconds have elapsed the first stroke is sounded on the bell. While one o'clock is striking, the lever t is held clear of the locking lever by the cam w ; the tongue on the lever then descends into the wild notch of the locking plate s ; at two o'clock it is retained on the edge of the plate till two blows have been struck, and so on, the locking plate which turns once in twelve hours being divided so as to allow all the hours to be struck in rotation.

Around the side of the great wheel (x) of the hour part are ten cams * for pressing down the lever, which through the intervention of the wire rope shown on the drawing raises the hammer of the great bell in the chamber above. This wheel is 3 feet in diameter, has 140 teeth, and gears with a pinion of 21 ; the second wheel has 90 teeth, and gears with a pinion of 15 on the arbor of the locking lever. The great wheel of the quarter past (y) is 3 feet in diameter, and the side of it is spaced out for 60 cams. This wheel has 150 teeth, gearing with a pinion of 20 ; the second wheel of 90 teeth gears with a pinion of 15 on the arbor of the locking lever.

Attached to the clock frame over the hour striking lever is a strong curved spring, as shown in the drawing, to check the upward motion of the lever. The length of the wire rope connecting this lever with the bell hammer lever is so adjusted that the hammer is lifted after the last blow is struck ; so that when the train is again released the lifted arm is disengaged from the cam at once ; and the hammer immediately falls.

To maintain the vibration of the pendulum during the twenty minutes or so that it takes to wind the going part of the clock, Mr. Denison invented a special kind of maintainer. The back bearing of the winding-pinion arbor is carried in a loose link slung from the barrel arbor. To obtain a resisting base so that the winding pinion should not run round the wheel with which it gears, a click presses against the ratchet teeth on the side of the great wheel, and so drives the clock. But as the great wheel travels on, the back end of the winding arbor in following it is taken out of the horizontal line and soon becomes so oblique that the winder has to stop and let it down to its normal position again. Though this maintaining work is ingenious, it is not in my judgment so good as the continuous sun and planet

* Under the head of " Striking Work " the action of these cams is shown more clearly.

424

maintainer. For clocks of moderate size that take but a few minutes to wind, I would prefer a spring maintainer.

To obtain a sufficient purchase in winding the hour and quarter parts there is an intermediate wheel and pinion to each, and the bearing of the arbor of the intermediate pinion is formed of an eccentric bush, so that the pinion may be readily disengaged from the wheel when the time for striking approaches,

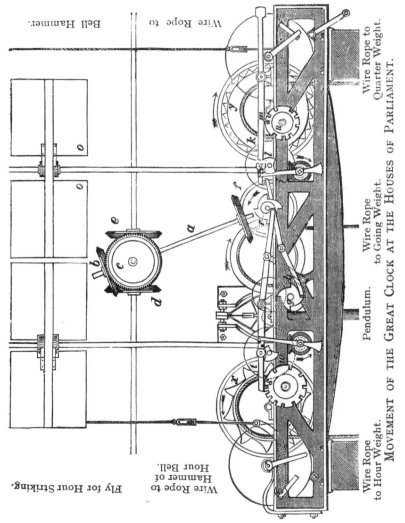

MOVEMENT OF THE GREAT CLOCK AT THE HOUSES OF PARLIAMENT.

or when the winding is completed. The hour pinion is shown out of gear, the lever attached to the eccentric being pushed away from the spring catch, while the one for the quarter winding is shown in gear ready for winding.

The clock frame is not in the centre of the room, but placed so as to allow a space of about two feet clear from one of the walls, to which a very strong cast iron bracket is fixed, and from this bracket the pendulum is hung.

The bells are arranged in a chamber above the dials, and hung from massive wrought iron framing. The hour bell is 9 feet in diameter ; 8¾ inches thick at the sound bow, and weighs 13 tons 11 cwt. It is struck by a hammer with a cast iron head weighing 4 cwt., which is lifted 9 inches vertically and 13 inches altogether from the bell before it falls. There are four quarter bells weighing, respectively, 78 cwt., 33½ cwt., 26 cwt., and 21 cwt.

The hammers for the quarters are each about one-fortieth of the weight of the bell it strikes. To prevent the hammers jarring on the bells they are kept from contact by india-rubber buffers on which the shanks fall.

Knightsbridge Barracks Clocks.—On page 426 is a sketch of a turret movement by Messrs. Thwaites and Reed, which is fixed at Knightsbridge barracks. The clock has a zinc and iron pendulum, double three-legged gravity escapement, shows the time on four six-feet dials, and strikes the hours and ting-tang quarters.

The wheels and pinions are of the following numbers :—

Going Train.	Wheels	120	96	90
	Pinions	10	9	8
	Minute Pinion (gearing with great wheel) 40.			
Striking Train.	Wheels	120	96	
	Pinions	10	8	
	20 cams on main wheel to lift hour hammer.			
Quarter Train.	Wheels	120	96	
	Pinions	12	8	
	30 cams on main wheel to lift quarter hammers.			

The great wheel and barrel for the going part occupy the centre of the frame, and the remainder of the train is placed horizontally on the right. The striking train is arranged on the left, and the quarter train on the right of going part. From the arbor of the set hands dial, the bevels for the hands are driven. On this arbor, behind the set hands dial which conceals it, is a four-armed snail or cam which every quarter of an hour depresses one end of a lever, the other end of which then unlocks the quarter locking plate, fixed to the end of the quarter barrel. Three pins may be observed sticking out from the face of this locking plate. As the plate turns once in three hours, and these pins are equidistant, one of them at every hour depresses one end of the long lever, seen in front of the frame, causing the other end to lift the stop from the hour locking plate, and so release the striking train. The cams attached to the further

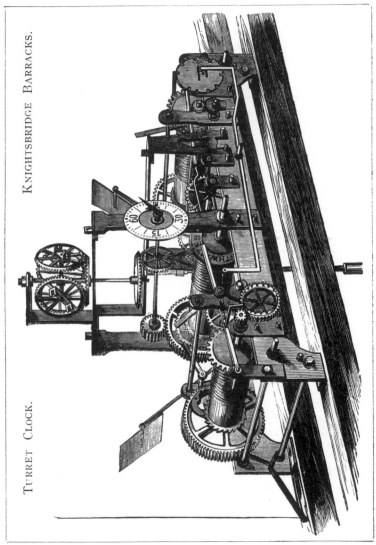

KNIGHTSBRIDGE BARRACKS.

TURRET CLOCK.

faces of the striking and quarter great wheels, for lifting the hammers, are of the kind shown on page 392, under the head of

Striking Work, where also the action of the locking plate striking is fully explained.

I have selected this movement as an example of a good and economical arrangement. When it is desired that the striking should begin exactly at the hour the striking is let off by a snail on the set hands arbor. But by the method of letting off the striking by pins in the quarter locking plate, as here employed, a lesser interval between the last quarter and the first stroke of the hour suffices, and the duration of the interval is constant. With good work the error either way is small, and may practically be disregarded.

The following table gives the safe load and the breaking strain for different sizes of steel and iron ropes, from Messrs. Newall and Co.'s tests :—

Diameter in decimal parts of an inch.	·33	·29	·25	·22	·16	·13	·1
Iron.							
Breaking Strain lbs.	2520	1920	1380	960	600	360	250
Working Load ...	672	448	336	224	150	90	60
Steel.							
Breaking Strain lbs.	5040	3540	2760	1920	1200	720	500
Working Load ...	1344	896	672	448	300	180	120

Its strength is not sensibly affected whether the rope is galvanized or not.

Tweezers.—[*Brucelles.*—*Die Spiralzange.*]—Spring fingers for holding minute pieces of watch work and for manipulating balance springs, &c. Drawings of tweezers for special purposes are given under the heads of Balance Spring, Collet, Stud.

Two-Pin Escapement.—[*Echappement à goupilles.*—*Die Anker-hemmung von Savage.*]—A variety of the lever escapements in which the unlocking and impulse actions are divided between two small gold pins in the roller and one in the lever. Latterly the two small pins in the roller are discarded in favour of a broad one of jewel. (See Lever Escapement.)

Unlocking Resistance. — [*Résistance au dégagement.* — *Der Auslösungswiderstand.*]—The resistance opposed to unlocking by the adhesion between the locking faces of the pallets and the tips of the escape wheel teeth, and in the case of lever pallets by the draw of the locking faces.

Up and Down Indicator.—[*Indicateur de haut et bas.*—*Das Auf und Ab-Werk.*]—Mechanism for indicating when a watch or chronometer requires winding. When a fusee is used this is easily accomplished. A pinion on the fusee arbor drives a wheel, the two so proportioned that the wheel makes something less than a turn during the period from winding to winding. For instance, if the fusee makes 4 turns, 12 teeth in the pinion and 60 in the wheel would cause a hand fixed to the arbor,

of the latter to make ·8 of a turn for 4 × 12 = 48 and 48 ÷ 60
= 0·8.

The up and down hand usually travels through $\frac{5}{8}$ of a circle,
and a pinion of 10 is generally used. The number of teeth in
the wheel can then be readily ascertained by multiplying 10 times
the number of fusee turns by 6 and dividing the product by 5.

Up and down wheels for $\frac{5}{8}$ of circle :—

4	turns of fusee	10 and	48	
$4\frac{1}{3}$	,,	10 ,,	52	
$8\frac{1}{2}$	,,	10 ,,	102	(2-day marine)
16	,,	10 ,,	192	(8-day ,,)

Two-day marine chronometers have sometimes a pinion of 12
and a wheel of 120, using a trifle less than $\frac{6}{7}$ of the circle.

With a going barrel the application of up and down work is
not so simple. In the first place, anything fixed to the barrel
arbor or the face of the barrel diminishes the width of the
mainspring. Then, although a wheel fixed to the barrel arbor
would cause the indicating hand to travel one way when the
watch is wound, the hand would not return with the running
down of the mainspring, for the arbor is then stationary and
the barrel moves, and the fact that both the arbor and the
barrel move in the same direction renders it still more difficult
to devise a satisfactory up and down work. Appended are two
of the best examples.

Samuel Stanley's Patent Up and Down Indicator.—Here
the wheel A gears in the barrel teeth and carries a small wheel
planted on its face. This small wheel gears in and revolves
around two wheels of equal diameters, but of unequal or different
numbers of teeth. The arbor of one of these two wheels carries
the up and down hand : the cannon of the other wheel passes

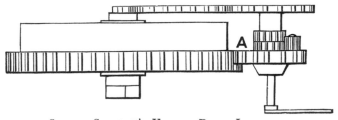

SAMUEL STANLEY'S UP AND DOWN INDICATOR.

through the top plate and has a wheel attached to it, which
is connected to the barrel arbor by an idle wheel and a
pinion.

While the watch is being wound the mechanism is actuated
by the barrel arbor, the two equal diameter wheels turn together,

and the up and down hand moves. As the watch goes, the mechanism is actuated by the barrel; one of the equal diameter wheels is dead and the other turns by reason of its different number of teeth, and the hand moves the reverse way. This indicator has the great advantage of being easily made and fitted to any existing rocking-bar keyless watch.

Donne's Patent Up and Down Indicator.—This is here shown in conjunction with Donne's resting barrel as already described under the head of " Mainspring," but it is equally applicable to ordinary going barrel watches and clocks.

Upon the arbor *a* is loosely mounted a spur wheel, which gears with the main wheel, and a similar wheel, which through an idle wheel mounted upon the pillar plate, gears with the winding wheel.

These wheels are formed upon their inner faces with contrate or bevelled teeth, which engage the corresponding teeth of an intermediate bevelled wheel, mounted loosely upon an axis fixed to the arbor *a*. Rotary motion may be given by either of the

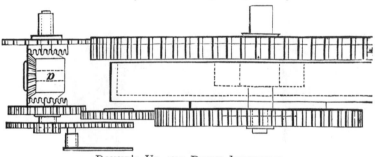

DONNE'S UP AND DOWN INDICATOR,

contrate wheels to the arbor *a*, whilst the other contrate wheel remains stationary, so that when the winding wheel is rotated the up and down hand will be turned in one direction, and during the running down of the mainspring the main wheel will give motion to the opposite loose wheel and the up and down hand will be moved in the opposite direction.

Uprighting.—[*Plantage.*—*Geradestellen.*]—A method of ensuring the correct position of a pivot hole in a watch or clock plate with relation to the corresponding hole in the opposite plate, so that the arbor which is supported in the two holes shall be perfectly upright, *i.e.* perpendicular to the plates.

Uprighting Tool. Table Tool. Drilling Tool.—[*Outil à planter.*—*Der Geradesteller.*]—A tool for uprighting and drilling. The engraving shows a very good example of a tool of this kind by Boley. A pillar at the back of a table on which the work is laid supports an arm. This arm terminates in a long hole exactly over the centre of, and perpendicular to the table. A perfectly cylindrical arbor or runner passing through the hole carries at its lower end a pointed centre or a drill as may be required. At the top of the arm is a ferrule so that the drill holder may be rotated for drilling. Two centring runners are shown with the tool, one for passing through the

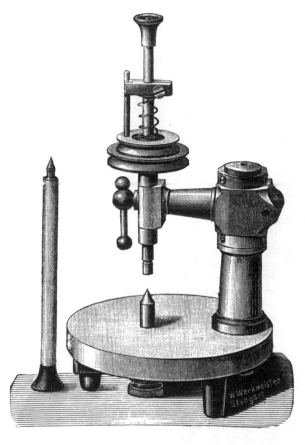

hole carries at its lower end a pointed centre or a drill as may be required ; the other is for passing through a corresponding hole in the table. The latter is handy in cases where it would be inconvenient or impracticable to upright from the top. Chucks of different sizes for holding the drills are provided.

The chief requirements in an uprighting tool are that the holes in the arm and in the table shall be exactly opposite and straight with each other, and also perfectly perpendicular with the table. If the holes are in line, a true runner fitting them should pass from one to the other without binding. The readiest way of testing if the runners are perpendicular to the table is first to ascertain that the runners are true in themselves, and then to fasten to each of the runners in turn a piece of wire extending horizontally to nearly the edge of the table. The point of the runner is pressed on a plate, and on rotating the runner the wire will clear the face of the table exactly the same distance all round if the tool is correct.

Uprighting tools are not used so much now as formerly, for as accuracy of drilling is more absolutely ensured if the work rotates, the mandril is now generally preferred where extreme exactness is required. However, from the readiness with which the work may be adjusted in a table tool, it is not without its advantages.

Velocity.—[*Vélocité.—Die Geschwindigkeit.*]—Speed ; rate of travelling.

Velocity of Sound.—In noting the time of the striking of a clock or the firing of a gun at a distance, a correction should be made for the velocity of sound, which at a temperature of 50° Fahr. is 1,110 feet a second. The direction of the wind, though it may intensify or deaden the sound, would not affect its velocity.

Verge.—[*Verge Die Spindel.*]—The pallet axis of the verge escapement.

(2) The pallet axis of a clock.

Verge, or " Crown Wheel " Escapement.—[*Echappement à verge.—Die Spindelhemmung.*]—A recoil escapement in which the pallet axis is set at right angles to the axis of the escape wheel.

The verge, the earliest probably of all the escapements, is shown in the engraving. It has no pretension to accuracy in presence of such escapements as the lever and chronometer.

The balance in this escapement has no free arc, and its vibration is limited to about 110° each way. The escape wheel, or " crown wheel " as it is called, has generally either 11 or 13 teeth, and in the plan of the watch its arbor lies horizontally. The balance staff, or verge, is made as small as proper strength will allow, and planted close to the wheel so that the tips of the teeth just clear it. The pallets, which form part of the verge, are placed at an angle of 95° or 100° with each other. The latter angle is generally preferred.

The drawing is a plan of the escape wheel and verge as they lie in the watch. The width of the pallets apart, from centre to centre, is equal to the diameter of the wheel. A tooth of the escape wheel is just leaving the upper pallet (*c*) ; as it drops off, the under tooth will reach the root of the lower pallet (*d*), but the motion of the verge will not be at once reversed. The escape wheel will recoil until the impetus of the balance is exhausted. The teeth of the wheel are undercut to free the face of the pallet during the recoil.

Generally in French, and occasionally in English watches, the pallets are even more open. An increased vibration of the balance and less recoil can be obtained with a larger angle, but to get sufficient impulse the verge must be planted closer to the wheel. This necessitates cutting away a part of the body of the verge to free the wheel teeth. Then, as the wheel tooth impinges on the pallet almost close to the centre of the verge, there is more friction on the pivots, and the wheel tooth gets so small a leverage that the escapement often sets, unless the balance is very light. On the other hand, with the opening between the pallets only 90°, as it is in many English watches, the vibration

VERGE ESCAPEMENT.

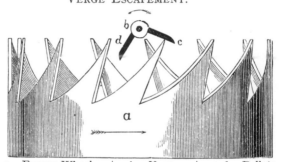

a. Escape Wheel. | *b*. Verge. | *c d*. Pallets.

of the balance is too small and the recoil too great. An opening of about 100° avoids the drawbacks incidental to the two extremes, and may therefore be adopted with advantage.

To ensure good performance the body or arbor of the verge should be upright, and when in the frame and viewed through the follower potance hole should be seen crossing the balance wheel hole of the dovetail, the view taken being of course in a line with the arbor of the balance wheel pinion when in the follower. The drops off the pallets should be equal, and the balance wheel teeth true.

Vernier.—[*Vernier.—Der Nonius.*]—A secondary scale for obtaining very exact measurements. If a scale already exists in which inch s are divided each into 10 equal parts, and another scale having $\frac{9}{10}$ of an inch divided into 10 equal parts is applied to it so that the two scales coincide at zero, they will also coincide at $\frac{9}{10}$ of an inch, but nowhere else. Starting from zero the difference between each of the first divisions would be $\frac{1}{100}$ of an inch. The second division on secondary scale, would be $\frac{2}{100}$ short of the second division on the primary scale, and so on. In use, if the zero point of the secondary scale were coincident with a division of the primary scale the secondary would not be needed, but when the zero point did not coincide, as many $\frac{1}{100}$ an inch would be added to the register of the primary scale as divisions of the secondary scale were passed before one coincided with a division of the primary scale. (See Slide Gauge and Micrometer.)

Vertical Escapement.—An escapement in which the pallet axis or the balance staff is set at right angles to the axis of the escape wheel. (See Verge Escapement.)

Vibrator.—[*Régler.—Die Reguliermaschine.*]—A standard balance and balance spring used by watch-timers for comparison (see p. 26).

Vice.—[*Etau.—Der Schraubstock.*]—A tool fixed to the work

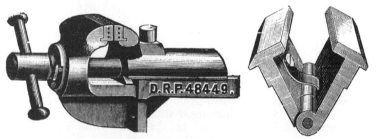

bench, consisting of one fixed and one moving jaw, the distance between which is generally regulated by a screw. For nearly all operations connected with watchmaking either the work or the tool is gripped in the vice. The old staple in which the jaws open at an angle is rapidly giving way to paralleled vices, of which one is shown on the left hand. For some purposes a vice mounted on a post so as to be adjustable at an angle is useful. Such a vice is shown in use under the head of Handwheel.

Vice Clams.—Shields of wood, usually beech or of soft metal, are interposed between the jaws of a vice and work that has been polished. The hinged brass clams shown in the sketch are very useful. They are kept open by means of a spring. Besides affording protection to the article gripped, they are serviceable as a brass stake, and with holes drilled in them answer readily for many purposes in which a riveting stake is used.

Vienna Lime.—[*Chaux de Vienne.—Der Wiener Kalk.*]—A polishing medium much favoured in American factories. (See Polishing, page 317.)

C C

Virgule.—[*Virgule.—Die Virgülhemmung.*]—An escapement having points of resemblance to the verge and to the horizontal. Obsolete.

Vulliamy, Benjamin Lewis (born 1780, died 1855).—Succeeded his father as a clock and watch maker in Pall Mall, and established a high reputation for the excellence of his work. He was the maker of the large clock at the General Post Office, and published a pamphlet on the construction of the dead-beat escapement for clocks, besides other small technical works.

Warning.—[*Toto.—Anlauf.*]—The partial unlocking of a striking train. (See Striking Work.)

Warning Piece.—[*Détent.—Die Anlaufpalette.*]—In a striking clock, a finger on the lifting piece that holds the striking train during the interval between warning and striking. (See Striking Work.)

Warning Wheel.—[*Roue de toto.—Das Anlaufrad.*]—The last wheel but one in a striking train. It carries a pin which butts on the warning piece during the interval between warning and striking. (See Striking Work.)

Watch.—[*Montre.—Die Taschenuhr.*]—A pocket timekeeper. *Cleaning and Repairing Watches.*—Before starting to clean a watch note any faults that can be observed, especially such as would be likely to interfere with its going. The construction of watches is so varied that it is difficult to give instructions that shall be applicable to all alike ; but, after describing methods of cleaning, I will add Mr. Gray's procedure of examination and repair of the type of Swiss watch most commonly met with. Many of the remarks will be found to refer just as pertinently to an English watch, and the special features of the latter are dealt with fully under the various headings of the book. Presuming this preliminary examination to have been made, and the mainspring let down, the movement is taken apart, the various pieces with their screws being placed in a movement tray with partitions (see Movement Tray) or a movement box similarly arranged.

Formerly the plates, wheels, and pinions were cleaned by brushing alone, and this practice is still followed by some ; but immersing the pieces in a bath of some detergent is the method now more generally adopted. Benzine, paraffin, spirits of wine, " essences," composed of petroleum, and sulphuric ether, soap, powdered hartshorn in spirits of wine, and other preparations are used for this purpose ; some of these are good in particular cases, and not so suitable in others. The most difficult watches to clean are those that have been lying by for years, or in stock for a long time, or used by any one whose body or avocation tarnishes the gilding or thickens the oil.

Dry Brushing.—The plates are first wiped with a clean chamois leather, the work is then held in tissue paper and brushed with a circular motion. The brushes are cleaned by rubbing them on chalk, a dry crust, or a burnt bone.* Common chalk is too coarse to be used with safety. French prepared chalk answers better. There is, however, a new composition called " Calcinas," that cleans the brush more thoroughly, and has a renovating effect on the gilding, and should be preferred. The leaves of all pinions should be thoroughly cleaned with pointed peg-wood.

Benzine and Paraffin may be taken to be as good detergents as any for watchwork. If the oil is apparently in fair condition it will be sufficient to immerse the parts in benzine and then brush them with a goat-hair brush. Perfectly clean benzine should be used for the balance and balance spring. When the oil is very thick dip the pieces in paraffin first and benzine afterwards

Should the plates and bars be tarnished, treat them with cyanide, as recommended under the head of " Gilding." All screws should be removed before subjecting the parts to the cyanide.

Holes.—Brass holes should be thoroughly cleaned by twirling in them a pointed piece of peg-wood, the soiled point being cut away as often as is necessary—that is, until it is clean when removed from the hole. Jewel holes are also pegged, but will be found to require but little attention in this direction after being well brushed. Endstones should always be removed and cleaned.

Small and delicate pieces require great care in handling. Slender levers are brushed best in one direction by drawing the brush along them while they are held in brass tweezers of tissue paper on the board. Dust may be removed from the balance spring with a camel-hair brush used circularly, and unless the spring is sticky it need not be taken off. A sticky spring should be placed in benzine and afterwards between tissue paper, which is then gently dabbed with a watch brush.

Mainsprings.—If the oil appears to be good and the watch has been in hand before, it will not be necessary to take the spring out. With a strange watch it is better to remove the spring for examination. If it comes out almost closed together, and not more than twice the size of the barrel, its elasticity is insufficient, especially if for a going barrel, and a new one is needed. If it distends sufficiently, and the oil is sticky, immerse the spring in paraffin for a few minutes, and then, without disturbing the coils more than is necessary, rub them with a bit of clean rag lapped over the sides, starting from the outside and working gradually to the eye. See that the spring was free in the barrel ; though the spring was the right height it may have

* One of the large bones from a cow-heel is said to be the best. This, after being well boiled, is placed is a rather slow fire for an hour or two.

been bound close to the arbor, and the barrel may require turning out a little there ; or the hook in the arbor may be a little too high, or too low, or too thick.

If a fusee watch, the chain may be soaked in paraffin and then drawn gently through blotting paper and afterwards through tissue paper. Put it then in the bluing pan for a few seconds over gentle heat, and if any oil should exude wipe it with tissue again. Observe that nothing adheres to any of the links.

Any wide holes may be bushed (see Bouchon) and other trifling faults made good. If substantial repairs are required, and the watch has to be put by, a paper with the faults noted should be placed in the movement tray or box with the job.

Repairing Swiss Watches.—Before removing the movement from the case, wind the watch a little, if down, and try it by the ear in each of the following positions, viz. with the 6 up, 12 up, dial and cock up. You can then usually detect the following faults : Not in beat, wheel rubbing in cylinder passage, cylinder pivots acting on shoulders instead of their end, incorrect fourth depth with escape pinion, balance spring rubbing on balance cock, or centre wheel, &c. Next ascertain that the centre pinion or set square is free of the glass, also of the bottom of the case ; see that the band of the case is *well* free of the teeth of the barrel *when shut* ; it is often free in thin gold cases when open, but shutting the case pinches the band in and fouls the barrel : to try it put a piece of paper between the teeth and barrel and shut the case ; if foul it will mark or cut the paper. See that dirt cups on winding and set squares are free of the dome ; frequently the dome presses on the centre bar and binds the centre pinion, causing, if not instant stoppage, the oil to disappear, and the pivots to cut. See that the balance is free of the case ; if it is much out of flat it will probably be foul of the case or centre wheel. See if the fly spring, when the cover is shut, is foul of the balance. Put a key on the set square, and turn the hands to see that they are free of themselves, the dial, and the glass ; if they do not turn truly it will proceed from either the centre holes being out of the upright, a bent set square, or a badly-fitted pinion.

Take the movement from the case *in paper* and lock the train, by putting a bristle through either fourth, or escape wheel ; remove cock and balance, being particularly careful not to strain the balance spring ; put the balance and cock in a tray, and remove the hands. (See Watch Hands.) Remove dial and motion work, using brass pliers to take hold of the cannon pinion to avoid marking it.

At this stage, if there is reason to suspect that the escapement is faulty, examine it as recommended under the head of Cylinder Escapement, page 120.

After examining the escapement it will be necessary to look over all jewel holes, noting cracked ones, and in brass, those that are too wide ; trying end shakes, &c. ; also to see that the passage in the escape cock, for the wheel teeth, is not too close, so as to draw off the oil, as when this is the case it is impossible to get the piece to go for any length of time.

Now take the movement completely down—some workmen use a brass block, with a series of holes drilled in it, to place the screws in ; it is a good plan, as if left in their respective bars or cocks they are apt to get lost. Having the piece down, examine all pivots to see that none are cut or bent. The barrel and its arbor and stop work should also receive attention ; it should turn with freedom, and perfectly true ; any want of truth in these particulars being fatal to good going.

In the event of the cylinder pivots being bent some workmen have recourse to a pair of smooth pliers, made just hot enough to turn the colour of the pivot to be straightened to a blue ; but in this class of work it is rare to meet with a pivot so hard as to require this treatment. It may be straightened by placing a bouchon over it as advised on page 121, but it will generally be sufficient after filling the body of the cylinder with shellac, and at the same time fixing either a bone or brass ferrule, to use a bell-metal polisher on the Jacot tool ; taking care to select a notch slightly larger than the pivot, which you have previously measured with the gauge that accompanies the tool for that purpose. Then use a smaller notch, finishing with a burnisher made expressly for this tool, and sharpened on a No. 1 emery stone, or emery of similar coarseness on a zinc or lead block ; the latter being the better material, the most convenient size being a square block about seven inches long and an inch and a quarter wide, got up true on each of its four sides. The burnisher should be put in a Swiss handle, similar to a pen-holder, and nearly as long, fastened in with shellac or sealing wax ; it can thus be set perfectly straight with the handle. In sharpening, the block should rest against the front of the work-board, pointing from you, and be plentifully supplied with emery and oil, mixed not too thickly ; the handle being held lightly in the right hand, and the first finger of the left applied on the top of burnisher ; the stroke should be point to heel, lifting it from the block for the return stroke. For reducing a pivot the burnisher should be cut on a No. 2 stone, or emery of a similar grade.

Should a pivot be broken, a new plug may be fitted as described on page 123.

Pivoting Cylinder.– To centre the new plug use a steel runner similar to the one nsed for rounding up the end of a pivot, but with larger holes ; these should be nicely chamfered out, hardened and polished ; the extreme end of cylinder will work in one of these holes, which should be plentifully supplied with oil. The top pivot being protected by running in a brass runner, having a hole sufficiently large to admit the pivot freely, the shoulder taking the thrust, you can thus turn the extreme end of the plug true with the body of the cylinder. Having centred the plug, it only remains to turn the hollow and pivot, leaving the latter a little larger than it will be ultimately required, burnishing it down this amount with first the rough and then the fine burnisher.

If the upper pivot is the one broken, it will sometimes be possible with a high cylinder to do without a new plug, by knocking out the old one sufficiently to turn another pivot on it this is not so good as replacing the plug with a new one, as the plug has a tendency to draw the oil away from the wheel teeth.

There is yet one other way of replacing a pivot that is broken, viz., by drilling through the old plug and inserting a piece of steel somewhat larger than the shoulder of the old one. The cylinder is filled with shellac to withstand the pressure, and the centring runner, just described, may be used, and a recess turned in the end of the plug sufficiently deep to start the drill truly. Having the hollow turned sufficiently deep to bury the angle of the drill, remove the centring runner and replace it with the one described on page 142 : it has a hole in it to take a drill, which, for this purpose, should be short and strong, and not relieved much behind the cutting part. If ground to cut only one way, and tapered in thickness to the point, it will work quickly and well. Use turpentine as a lubricator. The pressure on the drill should be relieved at the return stroke of the bow, and if the drill is sufficiently hard and not driven too rapidly, the drilling will proceed pleasantly. Having drilled the plug through, insert a piece of steel, previously hardened, tempered, and polished down to size, and not too taper, or a piece of a Swiss cutting pivot broach blued may be used. With a light hammer tap the piece for the new pivot into its place, and with the centring runner turn the extreme end of the plug to a centre. Proceed to form the pivot as described when speaking of a new plug. The side shake in cylinder pivot holes should be greater than to ordinary train holes ; if too little, the watch, though performing well when the oil is fresh, fails to do so when it commences to thicken. For directions in replacing a broken cylinder see page 122.

Escape Pinion.—To work in a new pinion when the old

one is present will give no trouble as to height. The pinion being removed from the wheel, all the measurements can be taken from it, and transferred to the new one.

When the old pinion is absent, proceed in the following manner : having put in the cylinder, a brass collet similar to a pivoting ferrule should be roughed out, its thickness being a little in excess of the height you imagine the escape pinion head to be. This is placed in the escape wheel sink, and the wheel on top of it, you can then offer the wheel to the cylinder passage, and see whether it is a correct height. Reduce the thickness of the collet until the wheel is a correct height. Having removed the wheel and cylinder measure over the collet and lower jewel hole, previously removing the end piece. Next measure the thickness of the jewel hole ; this, being deducted from the previous measurement, will give the distance from the lower pivot shoulder to the seat of wheel. The height of pinion head will be less than this, of course ; how much will depend on the position of the fourth wheel. The leaves should come just through the wheel teeth ; any excess is bad, because it attracts the oil from the hole. After screwing on the escape cock, the distance over both jewel holes should be measured, care being taken not to bend down the cock in doing so ; deduct the thickness of both jewel holes, and you have the distance from shoulder to shoulder, or the height. These measurements should be set down on paper as they are taken. The work of running in the pinion may be done in the following order. See that the leaves run true ; next, as one end of the pinion is always polished, that end should be kept for the face ; a very little touching up with a bell-metal polisher and diamantine will be all that is necessary to do to it. Turn down the leaves until the wheel will just pass on a short distance, and gradually let it down into its proper position, finishing with oil-stone dust and steel polisher. The distance from the lower pivot shoulder you will have obtained by the means described just now. The wheel should fit firmly on to the pinion, and the seat be but slightly undercut, almost square. If properly fitted, very little riveting will suffice ; on the other hand, if loosely fitted, no amount of riveting will make it sound, and probably the pinion face will be bulged and spoilt, in endeavouring to get the wheel tight. Having marked just above the boss of wheel with a safe-edged file, remove the wheel and proceed to turn down the leaves to this mark, taking care not to remove any of the diameter of the arbor in doing so. Now reduce the length of the arbors, leaving each about half a pivot longer than they will be required, cutting them off with the graver, and at the same time turning rather long centres. Having turned off one end it should be passed through a hole in a runner, and the extreme point turned true and conical ; the opposite end should then be treated in the same manner.   If this is done, the pinion will be kept centred ; if a file is used for making the centres it will probably get out of truth.   Now proceed to mark the place for the shoulder of the upper pivot, leaving it a shade high ; knowing where this shoulder will be you can see how deep the hollow will have to be cut.   If there is much distance above the seat of the wheel, a slight hollow will suffice ; but if, as is usually the case, there is but little room, the hollow will have to be deep and rather large, with the object of preventing the oil from being drawn away from the pivot.   Cutting a hollow in a foreign pinion is not such a difficult matter as in an English one, as the steel is softer.   Having stoned up your graver to a long point, the sides, instead of being placed quite flat on the stone to remove the burr, should be slightly raised in sharpening, so that a small facet is ground on each side of the belly, thus reducing the otherwise too acute angles, and strengthening the point.   This should be   done on a smooth Arkansas stone ; Turkey is too rough ; then, with a bell-metal polisher and diamantine, polish both sides and top angle ; when finished, the point should be allowed to rest on the nail, and if properly done it will hang in and not slip.   With a tool in this condition and a weak bow, the hollows can be cut both quickly and well ; if a still better finish is desired, nearly dry red-stuff on a peg will give it a nice bright appearance. Having completed the upper hollow of the pinion, and turned down the arbor to meet the root (it should be considerably smaller at the root of the hollow than at the shoulder), the pinion can be reversed in the turns, the lower hollow cut, and the pivot roughed out ;   the height being measured from the shoulder of the top pivot.   If the pinion requires facing, do it at this stage.   Finally, turn down the pivots, leaving them a shade larger than they will be required, to allow for burnishing, being particularly careful to turn the shoulder perfectly square.   The last cut should be down the shoulder of the pivot, so as to leave a *very slight notch* ; when burnished this will disappear, and no lump be perceptible in the corner.   The burnishing will, of course, be done on the Jacot tool and with a flat burnisher.   Having completed the pivots, and rounded the ends off and burnished them, so that they will not scratch the nail if tried on it, it only remains to turn off the corners of pivot shoulders with a polished graver, and to rivet on the wheel, using a hollow steel punch, and revolving the wheel   a   little   between   each   blow   of   the   hammer.   Its truth in flat should be examined from time to time in the callipers.

A word on the ferrule used in working in the escape pinion.

The ordinary screw ferrule is not at all adapted for this purpose, being too heavy and thick. A small brass or bone pivoting ferrule, dished in form, is the most convenient ; first, broached to fit on the arbors of pinion, and after it is shortened a similar one, with a hole sufficiently large to fit on the leaves. There will be no necessity to remove it until the piece is completed. The groove in these ferrules should be deep, to avoid the hair slipping off, and square to prevent the hair jamming. Use a moderately weak bow and horse-hair to turn the leaves down, and afterwards complete the turning, with a very weak bow and human hair.

New Escape Cock.—There is nothing very difficult about the making of a new cock, and if the jewel is sound, it can be reset without much trouble. A piece of good hard brass, of suitable size, is placed in position, the spot for the screw hole marked through, the frame, and drilled to fit the screw with only just freedom. Then screw the cock down firmly, and drill the steady-pin holes through the holes in the frame, afterwards broaching them true and upright, from the cock side, of course. The pins should be made of hard wire, filed with a very smooth old pin-file, and burnished ; the ends also rounded off and burnished ; their size should be such as will not go quite through the cock, when twisting them in while fixed in the pin-vice. The under side of the holes in the cock should be slightly chamfered. Put in the pins and cut them off, file the tops flat, and placing the lower side of the cock on a pinion riveting stake, drive the pins through until they project about one and a half times their diameter ; if they are longer they will be liable to get bent in taking on and off. Some workmen use a little oilstone dust to make the pins hold, but if properly fitted it is unnecessary. Now open the pin-holes in the frame a little, and if the work has been done carefully the cock will go on and off smoothly and without sticking, the pins will each fit independently, and not depend on one another, as is often the case. Put in the balance, and note if the cock is too high, and if so, how much ; the best and quickest way to reduce it is to put the frame in the mandril, centring it by the lower escape hole, and turn down the bar, previously placing a small piece of peg cut taper under the nose of the cock to prevent it from springing from the cutter. Having turned it down sufficiently, allowing a little for stoning up, test the correctness of the centring of the lower cylinder hole by pegging it true in the mandril. Cut a good sound peg to a long and taper point, that will enter freely the hole to be tested. Having brought up the hand-rest to within say half an inch of the hole, a flat is cut on one side of the peg to prevent it turning round, the point is then put into the hole, the flat resting on the tee, which should be adjusted at such a height that the peg is supported horizontally, or if anything, the unsupported end highest ; now on the mandril being rotated, if the hole is not exactly centred the end of the peg will exhibit this error in a magnified degree. Having slightly slackened the dog-screws of the mandril, the work can be minutely shifted either with the fingers or by tapping it with a tool handle or some similar object to avoid bruising the frame. Having corrected the centring and tightened the dog-screws, the cock can be screwed on and centred with a graver similar to the one for cutting a hollow. The cock is now ready to receive the jewel hole ; if it is to be sent to the jeweller, send the frame and cock together with the pinion, and he will give the correct end shake ; if you decide to set the jewel yourself, see instructions under the head of Jewelling. Having the hole set, round off the end of the cock, striking a circle to file to, with the dividers set to mark a circle whose diameter is that of the bar or rather more. Having stoned down the top of the cock, and seen that it is free of the balance, take off the edges square with a very smooth file, cut the wheel passage. Shellac the cock by its top to a flat piece of brass about an inch and a half in diameter, having a hole in the centre so that you can feel the jewel hole with the pump-centre ; coat a tooth of the escape wheel with rouge, place the pivot in its hole, and mark on the cock the position the passage must occupy ; select a cutter about as broad as the wheel tooth and polished on its faces, and place it in the slide-rest, and cut the passage in the following manner : The palm of the left hand being applied to the face plate of the mandril, it can be rocked to and fro about a quarter of a turn, the right hand meanwhile advancing the cutter by means of the slide screw. The passage should be well free of the wheel teeth, both sides and top, to avoid drawing the oil from the teeth ; by putting a layer of rouge on the teeth and turning the wheel this can be tested..

Fourth Wheel.—This is a fruitful source of error in the lower class of foreign work ; it is so often out of truth or round, and its depth with the escape pinion incorrect. If you have reason to suspect the correctness of this depth it will be best to try it first in the frame, and then in the depth-tool, for although you can see the action, and judge of the size of the pinion better in this tool, no allowance can be made for side shake, &c. Try then the running of escape pinion while in the frame, by pressing with a fine peg on the end of the escape pivot with the left hand, while the right turns the fourth wheel with another peg ; if it does not run smoothly, but you can distinctly feel every leaf come into action, be sure that something is wrong. Unfortunately, pinions of six make very bad depths, even when correctly sized and pitched. If you have a sector, the best plan will be to try the wheel and pinion by it, and if the wheel is incorrect change it ; if the depth is too deep the wheel can be re-cut on the engine ; but if it is too small it must be exchanged. Stretch the old wheel with the hammer and re-cut until the depth is right, then exchange the wheel for one of exactly that size, turning a sink in a piece of brass as a gauge to send for the new wheel. The wheel, as bought, is generally much too thick ; it can be cemented by its polished face to a true chuck, and turned down to the required thickness, with a sharp cutter in the slide-rest ; its thickness can be reduced with less trouble and more certainty thus than by filing. Stone out all cutter marks with water of Ayr stone, and finish it with bluestone. Before opening the hole to size try its truth on a *true* arbor, and if the hole is not in the centre, as is sometimes the case, correct it by one of the following ways. First you can put a piece of brass in the mandril and turn a sink to fit the wheel, afterwards with a narrow cutter in the slide-rest, turning the hole out true. Or the wheel may be fastened with wax to a chuck and run true by its teeth with either a peg or the thumb nail. If the watch is a good one the bars and crosses of the wheel should be smoothed with an old fine nicking-file, used lengthways, and finally with box-wood and very fine oilstone dust, or two pieces of water of Ayr stone rubbed together with oil, and the resulting thick paste used in the same way on a box-wood slip. The face of the wheel can be touched up with a tin polisher and red-stuff. The thickness of the wheel will be determined by the depth of seat on the pinion. Before fitting on the wheel remove any burr thrown up in riveting on the old one. While truing up the seat and deepening the hollow a little, the shoulder of the seconds pivot should work in a hollow brass runner, having a hole sufficiently large to clear the seconds pivot. Let the wheel on, using a sharp broach and turning it one way, not backwards and forwards, to keep it true ; it should fit tightly on to the pinion, and before finally putting it on take a slight chamfer off both sides of the hole to remove any burr thrown up by the broach ; rivet on the wheel as directed under the head of escape pinion.

Seconds Pivot.—In the event of a broken pivot, or one so cut as to render a new one necessary, a hollow chuck, somewhat smaller than the fourth wheel, may be used, and the wheel fastened to it by cement. The pinion having been set in the cement as true as possible, a spirit lamp is placed under it, and while the work is rotated a peg is gently pressed against it to get it correct ; the lamp is then removed, but the peg is kept in contact with the rotating work until the cement is set. The pinion can be run perfectly true by the hollow at the root of the old pivot, or the tips of the pinion leaves. It will generally be best to entirely remove the shoulder of the old pivot, previously taking the height or distance from shoulder to shoulder. Having turned away all the shoulder of the old pivot down to the bottom of the hollow, and at the same time cut a centre sufficiently deep to start the drill true, drill the pinion, using for this purpose a drill of a similar shape to that described for replacing a cylinder pivot. The depth of this hole willl depend on its size, about twice its diameter will generally be sufficient. Turn a piece of hardened and tempered steel down nearly to size, shaping the shoulder similar to the old one, and finishing with the steel polisher and fine oilstone dust. It should be almost parallel, and go two-thirds of the way in before being driven. Having warmed the chuck and removed the wheel, it can be rested on a pinion riveting stake, and the piece that is to form the new pivot fixed in position by a few light taps of the hammer. Pass the end through the centring runner, turn the centre true, and proceed to shape the shoulder and pivot, with the graver, in the turns, afterwards removing all roughness with a steel polisher and sharp red-stuff ; finally, finish it with a fine burnisher in the Jacot tool. A special centre adapted for seconds pivots accompanies the best of these tools ; it has a longer bearing for the pivot than the ordinary centres. Having turned off the extreme corner of the shoulder with a polished graver, it only remains to round up the end of the pivot.

In the event of the back pivot being broken, the best method of centring it will be by means of the centring runner described for centring cylinder plugs, turning a recess for the drill with a graver, and using a drill fixed in a runner in the turns, and for the back runner, a hollow one, the *shoulder* of seconds pivot taking the thrust. Care should be taken in making this plug that it is not too taper, or the arbor may be split.

Working in a Fourth Pinion.—Having selected a pinion of the correct size for the third wheel, and fixed to the long arbor an old screw ferrule, cut a thin box-wood slip to a thin edge, and with rather sharp red-stuff and oil proceed to polish out the leaves, resting the pinion on a hard cork, or piece of soft wood. The screw ferrule on the arbor enables you to press the first finger of the left hand against it, and thus the pinion is held while polishing ; the natural elasticity of the cork or wood allows the pinion to give a little to the motion of the polisher, thus keeping it flat. The leaves having been polished out with wet red-stuff, and finished with fine stuff, or diamantine, the truth of the leaves can be tested by running it in the turns. Should the centres of the pinion not be perfect, they must be made so before trying it, by turning it through a runner. Should the leaved portion of pinion on trial prove out of truth, it must be corrected. If, while the pinion is in the turns, a piece of soft lead pencil is held on the rest, so that its point just touches the top of the leaves, those that are furthest from the centre will be marked, thus forming a guide for the correction of the arbor. The *marked side* of the arbor being placed *downwards*; in contact with either a soft steel or brass stake, the upper or hollow side can be stretched by a few light blows from the pane of a small hammer ; the blows should be distributed at equal distances over the arbor, and, as these pinions are usually rather soft, some care is required not to overdo it. Having by this means got the leaves to run true, the arbors can be shortened to little more than the ultimate length of the pinions, and the centres turned true. In some watches the banking, instead of being against the stud in the cock, is against the arbor of the fourth wheel ; in this case the diameter of the arbor is of importance, as if too small and the watch is caused to bank by external agitation, the pin would jam against the arbor of the fourth wheel, and stop the watch. Again, in some callipers of movements, the fourth pinion head comes close to the plane of the balance, and in some positions, if the pinion head is too high, or from excess of end shake the banking pin touches it, a cause of stoppage rather difficult to detect, is created. The old pinion being removed from the wheel, all the measurements can be taken directly from it. The first thing will be to turn down the leaves to form a seat for the wheel, measuring the height from the pinion face. Care must be taken in fitting a pinion to an old wheel, that the leaves fit into the marks made by the old pinion, otherwise a difficulty will be found in securing the wheel. Having fitted the wheel, try its truth in round in the turns, and if untrue, shift its position on the pinion until it runs quite true, then mark the wheel and a leaf of the pinion, so that its position can be found again. Shorten the leaves, leaving just sufficient to rivet soundly. If too much is left to be riveted the pinion face will be bulged and spoilt. If the leaves project the thickness of a sheet of this paper, ·10 mm., it will be sufficient if the wheel fits properly ; there should be but a slight undercut to ensure a sound rivet. Rivet on the wheel, using a steel or bell-metal stake to support the pinion, and a polished steel punch of such a size that it fits just freely over the arbor. A piece of tissue paper between the face of pinion and stake will protect it during the riveting, and if care is taken to shift the wheel a little every few blows, it will be secured true and flat. The face of the rivet can be turned flat, and glossed, and the hollow cut. Turn the arbor to size, leaving a slight shoulder close to the wheel to prevent the polisher coming in contact with it. Then polish, burnish, and mark the position of the upper pivot shoulder, measuring from the pinion face. The pivot being turned down to within a shade of proper size, the pinion can be reversed in the centres, and the seconds pivot turned down, its position being fixed by measuring from the upper pivot shoulder. The pivots, smoothed with red-stuff, are burnished on the Jacot tool to size, leaving only the rounding up, and turning off the extreme corners to complete the work. The size of hollow necessary in the pinion face is regulated by the length of shoulder there is. Where this is extremely short, a hollow of considerable depth and breadth is required ; on the other hand, where the shoulder is of considerable length, a small hollow will suffice.

Third Pinion.—In those callipers of movements, in which, from the reversed position of the centre wheel the power is received and transmitted from opposite ends of the pinion (as in the case of the third wheel in the ordinary foreign or barrel movement), one commonly finds, after the watch has been going for any length of time, the upper pivot considerably worn or cut. This proceeds in many cases from the pivot in the first place being too small to withstand the necessarily great pressure. In addition to this, the leaved portion of the pinion having to be so high to meet the centre wheel with safety, renders it difficult to get a sufficient shoulder to the pivot. If, in addition to this, there is an insufficient hollow, and that but roughly cut, the oil rapidly disappears, and the pivot becomes cut. In repairing these pivots, it will generally not be sufficient to turn out the marks and repolish. If this is done, it often becomes cut again very quickly. It is much better to replace the pivot with one of the original size, and at the same time to see that the extreme corner is cleanly sloped off, and that the hollow is somewhat large, deep, and cleanly cut.

Centre Holes.—In replacing the centre holes great care is required not to get them out of upright or alter the depth of the barrel or third wheel. If both are to be replaced, centre the frame accurately in the mandril and with a narrow cutter and the slide-rest, turn out the hole to receive the stopping (if it is broached it has a tendency to give towards the barrel sink, making the depth deeper), keeping the hole small—not larger than the shoulder of the pivot. A piece of hard wire, drilled and broached out to just fit the pivot tightly, is put on an arbor, and turned taper to fit the hole in frame. Having removed the frame from the mandril, put a broach in the hole, and work it round to roughen, and slightly enlarge the ends of the hole ; drive the stopping in tight from the *inside*, and rivet it with a punch having a slightly rounded face ; a case stake with a piece of paper on it supports the frame while riveting the stopping. Replace the frame in the mandril, and turn off any excess of stopping with the slide-rest ;

and, with a narrow cutter, open the hole to fit the pivot. You can now put in the bar (having previously put in either a hollow or solid stopping) and turn out the hole to size. Where the pivots are taper, it will be necessary to broach the holes slightly at the last ; but in most cases they are parallel, and the holes can be opened with the cutter ; in this way there is no fear of getting the work out of upright.

Centre Pinion.—It is not uncommon to meet with centre pinions having the pivots cut through, or so much as to make it impossible to repair them, rendering a new pinion necessary. This rapid wear proceeds in most cases from the shallowness of the holes and the total absence of any sink for the oil ; in addition to this there frequently is not a sufficient shoulder to the pivot. In selecting a new pinion, difficulty is sometimes found in getting sufficient size in the body, owing to their being too much reduced by the maker ; in addition to this, many are broached from both ends, rendering them untrue when opened out straight. The most convenient way is to turn up a piece of staff steel (hardened and tempered so that you can just cut it) to fit the pinion, and of only just sufficient length to go through the hole ; and fitting the pinion throughout. Having removed the old pinion from the wheel, and measured the height from the face of the pinion to the seat for the wheel, turn down the leaves and let on the wheel tight. The rivet being shortened to the right length and the hollow cut, the wheel can be riveted on, using a polished steel punch that just fits easily over the body of the pinion ; the rivet is then turned true and faced and a brass ferrule opened to fit over the pinion, and secured to the wheel with wax. The position of the lower pivot shoulder is now fixed (measuring from the rivet), and the pivot carefully turned to nearly its correct size. The pinion is now reversed in the centres and the upper pivot turned down. To polish the pivots the pivot polisher, Fig. 24, under the head of " Polishing," may be used.

Barrel and Barrel Arbor.—In foreign watches two varieties of barrel arbor are met with. The first, in which the arbor and ratchet are in one, and secured to the bar by a brass cap covering the ratchet, and forming at the same time the dirt cup. The second, in which the arbor has a large flange or boss bearing against the under side of the bar, is secured to it by the ratchet, which is attached to the boss or flange by three screws, passing through the ratchet into the boss. The latter is a very old plan, and often troublesome to repair, the collet in this case being usually secured by a pin passing right through it and the arbor ; a new collet is necessary if the arbor has been reduced in truing it. The arbor can be secured by wax to a cement chuck, the flange or ratchet, as the case may be, bearing against the face of the chuck, and run true by some part of the arbor which is not worn. The arbor can then be turned true, slightly tapering, and repolished. Where this kind of arbor is met with, the cover is generally in the top of the barrel, and in such cases it will be found better and less trouble to put a new cover, instead of stopping the hole. Where, as is usually the case, the cover is very thin after it has been in and out many times, it gets out of flat, and in such a state it is impossible to get the barrel to rotate truly.

Having roughed out the cover from a piece of hard brass, well hammered, file one side flat, cement it by that side to a chuck (such as a brass-edge turned true), using very little cement, and taking care to press it as close as possible to the chuck while it is cooling. It is now put in the mandril and centred by the hole in the cover, which at this stage should be small. With a cutter in the slide rest, turn the cover slightly thicker than it will ultimately be required, remove it from the chuck, and cut the notch (by which it is removed from the barrel) with a fine square file ; now cement it to a chuck slightly *smaller* than the cover, and turn down the edge to fit the barrel. Doing it this way you can apply the barrel directly to it, and fit with comfort and certainty ; the edge should not be too much bevelled, or it will soon become loose —an angle of about 80° is right for this—and the extreme corner should be slightly rounded off with a piece of water of Ayr stone.

Having broached out the hole in the barrel, and drilled a piece of wire with a hole slightly smaller than the pivot, put it on an arbor and turn it true to fit the hole in the barrel tightly, and slightly longer than the hole, securing it by riveting very carefully; if great care is not taken the barrel will become bulged, and the end shake of arbor altered. Having reduced the stopping with a cutter and the slide rest, to its proper thickness, cement the barrel by its face to a chuck with a hole in the centre on a lathe with a hollow mandril, running it true by the tops of the teeth with a peg, taking care that it runs true in flat, and with a narrow cutter open the hole true to nearly fit the pivot, finishing it with a polished broach (the spindle of the lathe being hollow, allows the broach to pass up it). The cover can now be snapped into its place *without removing the barrel from the chuck*, and its hole opened with the cutter and finished with the broach.

Doing it in this manner, the barrel is sure to be upright and true ; if, on the other hand, both holes are *not* opened together and at the same chucking, it is almost impossible to insure its truth. This is almost as much trouble as making a new barrel ; but many watches are met with, of good quality, having the barrel holes much enlarged, yet showing no signs of wear in the wheel teeth, and in such cases it is better for the watch that the old barrel should be retained. It will probably be harder and with better cut teeth than one obtained at the material dealers.

The barrel cover may be finished in a variety of ways ; it may be polished or spotted, or if it is bright snailed it will have a very nice appearance. Either of these modes of finishing is preferable to gilding if it is thin.

A new collet can be made from an ordinary rough lever roller of large size, the hole being drilled out to nearly fit the arbor ; turn the sides flat, and leave it slightly thicker than it is required ; drill a hole for the pin and one for barrel hook, and rub down both sides flat on an iron block with oilstone dust. Harden it and temper it to a blue colour. Clear out the hole with a broach and let it tightly on to the arbor ; remove it, and polish the sides of the collet underhand on a bell-metal block, perfectly flat ; put the cover on the arbor, also the collet, and broach out the pin hole straight—fitting a *hard* pin of the same taper as the broach. If the cover is bound, free it by polishing down that side of the collet on the polishing block ; great care should be taken that only just sufficient freedom is allowed here. Cut a shallow hollow in the boss of the cover to reduce the friction and retain oil, and a deep hollow in the collet on the under side to supply the lower pivot. Put the hook in, and slope the corners of the collet off with a polished graver.

If teeth are broken out of the ratchet, a new arbor is an absolute necessity. Arbors can be procured from the material dealers, already screwed, hardened, and finished ; but as it is nearly impossible to find one of the correct size in the ratchet and of the right height, the ordinary soft arbors having the teeth already cut are the only ones available.

In selecting a new arbor, one should be chosen having a ratchet nearly as large as the sink in bar will admit ; and the teeth but slightly undercut. See that the ratchet runs true to the centres, turn the top of the ratchet flat, and cut a slight hollow at the root of the square to prevent the oil being drawn up. Reverse the piece in the turns, placing the ferrule on the square, and turn the under side of the ratchet flat, leaving it slightly in excess of the thickness it will be when finished, to allow for polishing. The height for the collet is now marked, and the arbor turned down to size to fit the hole in the screw plate. A barrel arbor plate is the thing to use for this purpose, the ordinary plate having too coarse a thread. In the absence of such a plate or tap, the makeshift mentioned under the head of " Barrel Arbor " on page 54, may serve.

The diameter of the part to be screwed should be such that a barely full thread is formed ; if it is left larger the thread will probably strip off. That part of the arbor which will ultimately be cut off should be turned slightly tapering, so as to form a guide to start the plate, thus avoiding a drunken thread. The arbor is now screwed, holding it by the square in the pin vice. The shoulder which passes through the bar is turned down, leaving very tight in the hole ; the height of this shoulder is measured from the upper side of the ratchet. Next the pivot on which the barrel rotates is turned, leaving it very tight to the hole ; finally the lower pivot is turned, leaving it also very full to the hole. The length of this pivot marked, the square for the stop finger is cut, the finger put on, and the place for the pin hole marked and drilled. The arbor is now ready for hardening. A piece of binding wire is twisted a few times round the body of the arbor, and some common yellow soap plastered over it. It is held in the flame of a spirit lamp until red hot, and then plunged vertically into the water. Care should be taken that it is not overheated, the lowest heat at which it will harden being used. If made too hot it may go out of truth. The object of the soap is to prevent burning ; it also improves the steel which comes out of the water without any scale on it, or, if any, it is easily removed, and the piece is white and clean. The arbor is now placed over the lamp, in a spoon filled with oil, and held there until the oil ignites ; it is now ready for polishing. The tool described on page 328 for polishing centre pivots may be used for this purpose, a pair of laps rather larger than those described being employed. By this tool the whole of the pivots and shoulders are polished perfectly square and flat, using first the steel and oilstone dust and finishing with the bell-metal and red-stuff.

The collet having been roughed out, the centre hook and the two turnscrew holes drilled, it is tapped and turned to thickness on its own arbor, leaving it slightly thick. It is now hardened and tempered in the same manner as the arbor, the sides rubbed down smooth, and polished flat. The arbor is now inserted in the barrel, and the collet screwed on. If the barrel has too much end shake, the shoulder against which the collet bears can be turned back a shade. If care is taken in measuring the heights, no difficulty will be found in this respect. The squares are now rubbed smooth with a steel polisher and oilstone dust, shortened, and the ends polished off either on the cement chuck in a lathe, or in a lantern and the Swiss screw-head tool. Finally the corners of the squares are taken off with the oilstone slip. If the end of the square is required *dead* flat, the fusee-end tool may be used.

Mainspring.—In replacing a broken mainspring it should not be assumed that the old one was the correct strength. It often happens that because it was the nearest to be had, or on account of some fault in the train or escapement, a spring too strong has been applied. With a little experience the workman's judgment will be a fair guide ; but a better plan than sending the old spring as a gauge is to send the barrel, and have a spring that will just drop into it as it is coiled up by the maker ; such a spring will be found to give a sufficient vibration if the train is correct, the balance of proper weight, and the cylinder not too open.

When a spring which is of a proper length is broken close to the eye, it will be sufficient to soften the inner end, punch a fresh hole for the hook, and carefully bend round another eye.

For replacing Balance Spring, see page 25.

Watch Bow.—[*Anneau.*—*Der Uhrbügel.*]—The ring of a watch case to which the guard is attached. (See Bow.)

Watch Movement.—[*Mouvement.*—*Das Taschenuhrwerk.*]—The plates with the wheels and pinions composing the train. A watch without the case.

Sizes of Wycherley's Watch Movements.

Size.	Diameter in inches.	Size.	Diameter in Inches.
No. 4	1·34	No. 12	1·62
No. 6	1·41	No. 14	1·69
No. 8	1·48	No. 16	1·76
No. 10	1·55	No. 18	1·83

Usual Lancashire Size.

No.	Inches.	No.	Inches.	No.	Inches.	No.	Inches.	No.	Inches.
1	1·2	6	1·366	11	1·533	16	1·7	21	1·866
2	1·233	7	1·4	12	1·566	17	1·733	22	1·9
3	1·266	8	1·433	13	1·6	18	1·766	23	1·933
4	1·3	9	1·466	14	1·633	19	1·8	24	1·966
5	1·333	10	1·5	15	1·666	20	1·833	25	2·0

In a modern watch the necessity of economizing space has led to considerable ingenuity in planning the movement. In the

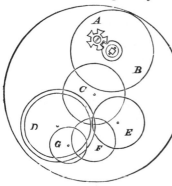

appended drawing A B is the great wheel attached to the barrel containing the mainspring, which forms the motor. On the barrel is planted at A a star wheel, and fixed to the barrel arbor is a finger, which, at each rotation of the arbor, engages with a space of the star wheel and moves it round one division. After four rotations, a swelled portion of circumference butts against the hollow of the finger and stops further movement. This prevents the mainspring being unduly strained by overwinding.

In order to obtain as large a barrel as possible, the great wheel extends from the extreme edge of the movement to the centre pinion, with which it gears. On the centre pinion is mounted the centre wheel C, which drives the third wheel pinion; the third wheel E, attached thereto, drives the fourth wheel pinion. The seconds hand is carried by the fourth wheel F pinion, which is therefore planted so as to obtain a seconds dial of adequate size between the centre and the edge of the main dial. The fourth wheel drives the escape pinion, on which is mounted the escape wheel G. The balance or controller D is as large, or nearly as large, as the barrel, and therefore must also be planted approximately midway between the centre and outside of the movement. The connection between the escape wheel and the balance is through the intervention of a lever or cylinder, or in some other way dependent on the kind of escapement adopted.

The proportion of the various wheels and pinions will be governed by certain data. The centre wheel rotates always once in an hour; the fourth wheel always once in a minute. Then, if the number of vibrations the balance is to make in an hour, and the number of turns the barrel is to make in the interval between the windings of the mainspring are decided, very little latitude will be possible in the relation of the various factors of the train to each other. Four turns are the usual number for the barrel in 30 hours, the period between winding, and either 16,200 or 18,000 the number of vibration per hour of the balance; further, the escape wheel is almost universally made with 15 teeth.

The most favoured trains are—

	Centre.		Third.		Fourth.		Pinions.		
For 18,000 vibrations per hour.	80	...	75	...	80	...	10	10	8
	64	...	60	...	70	...	8	8	7
	64	...	60	...	60	...	8	8	6
For 16,200 vibrations.	64	...	60	...	72	...	8	8	8
	64	...	60	...	63	...	8	8	7

In the old full-plate construction, there are two circular plates which enclose the movement, the balance being outside of the plate furthest from the dial, which is called the top-plate.

In the more modern three-quarter plate movement a portion of the top plate is cut away, and the balance lowered, so that the cock which carries the upper pivot of the balance staff is level with the plate. In this way a much thinner watch is possible, and the escapement can be more readily removed than with the full-plate movement.

Watch Pendant.—[*Pendant.—Der Gehäuseknopf.*]—The little neck of metal connecting the bow to the band of a watch case. (See Bow.)

Wax Lathe.—[*Tour à gommer.—Die Drehbank mit Lackscheibe.*]—A lathe in which the object to be turned is fastened with shellac or sealing wax. (See Lathe.)

Wheel Cutting Engine.—[*Machine à tailler.*—*Die Räder-schneidmaschine.*]—An engine for cutting the teeth in wheels. The wheel to be cut is fixed concentrically to the dividing plate, which is a plate having a number of small holes drilled on its surface. These holes are arranged in circles, each circle contain-

Fig. 1.

ing a different number of holes carefully spaced, so that all the holes in each circle are equidistant from one another. A revolving cutter is brought to bear on the blank wheel and cuts one space, the dividing plate being kept steady the while by a pin pressed by a spring into a hole of the particular circle that has either the same number of holes as the wheel is to have teeth or a multiple of that number. For instance, in cutting a wheel of thirty teeth the 60 or the 90 circle might be used. When the first space is cut, the pin is shifted into the next hole (or next but one or next but two, as the case may be) and the cutter again brought into action, this time cutting the second space, and so on. For steel wheels serrated or milling cutters are used, and for brass generally a single or fly-cutter running at a great speed, or else two or more fly-cutters arranged in a boss so as to follow one another. For feeding the cutter through the work a lever is generally used, but a screw and hand wheel are steadier and leave a smoother surface.

A most essential condition for successfully cutting brass wheels with a fly-cutter is *great speed*, and to obtain this without absorbing unnecessary power the driving cord must be of ample length so as to work without being unduly tight. The cord is often taken once round the pulley on the cutter spindle to get sufficient grip without being tight. A short tight cord will cause the centres of the cutter spindle to become dry and to heat, then the cutter will fail to get through its work, and the point of it is likely to break. It is generally prudent to rough out the spaces with a serrated cutter like a circular saw with very coarse teeth. Fine teeth will clog. This will leave the finishing cutter with less to do, and there will be less fear of distorting the wheel by too heavy a cut. Most cutter spindles are formed with a collar and nut for fixing the circular cutters, and a mortice or a dovetail to receive fly-cutters, which are tightened either with a wedge or a screw. Too much care can hardly be bestowed on getting up the finishing cutters. The sides must be filed up accurately to a gauge ; the slightest want of symmetry between them will be apparent in the wheel teeth, and on the polish of the cutter depends the brilliancy of the work. An expeditious way of making a cutter to match an odd wheel is to turn a piece of steel to a pivot shape just to fit between the teeth of the pattern wheel. File away exactly half of the diameter and well polish the face and point, giving the latter just clearance. A cutter of this kind is not to be recommended for permanent use, because it is too weak and will not bear continued resetting.

Dividing Plate.—On the accuracy of the divisions in the dividing plate depends the truth of the wheels that are cut, and it may be useful to give a method of setting off the holes in a dividing plate. For this purpose as large a disc of mahogany, slate, or other material easily turned and not liable to warp is mounted on the lathe as a chuck. The edge of the disc is carefully turned, and the face of it recessed to receive the blank dividing plate. Divide the circumference of the disc by the number of holes it is desired to have in the largest circle of the dividing plate, and through a piece of steel drill two holes whose centres are apart exactly the distance so obtained.

Through a piece of clock mainspring or other suitable steel band, long enough to encircle the disc, holes are pierced to the number required in the largest circle of the dividing plate, and of exactly the same pitch and size as those already drilled in the piece of steel which now serves as a template for piercing the band ; for after the first hole is pierced through the latter, a pin is passed through it and one hole of the template, while the second hole in the template serves as a guide for the drill or punch in forming the second hole in the band, and so on. When all the holes are formed in the band, the ends of it should be fastened by means of a small cover plate, which is soldered or pinned on, care being taken that the pitch of the two end

holes is kept exactly correct. The band being then placed on the edge of the disc will, it is evident, answer as a guide for drilling off the largest circle of holes, and for any of the other circles having a regular fraction of the number. For the remaining circles the band is cut and disc turned down to suit as required. It is almost unnecessary to add that extreme exactness must be observed in all these operations to ensure a satisfactory result. The steel template is often made of two thicknesses of steel pinned together, leaving a slit between them through which the band is free to slide, and one of the holes forms a holder for a punch. The pin that is fixed to a spring on the lathe, and passes into the holes of the band while the dividing plate is being drilled, should be taper, so as to fill the holes and keep the disc quite steady. In setting off the holes in the template, keep the pitch rather short than long, for if the band is too small for the disc the latter can easily be turned down.

It is obvious that this disc and band method may be used for cutting a wheel or wheels in the lathe if no cutting engine is accessible.

The wheel cutting attachment provided with American lathes is depicted on p. 455. The cutter spindle is fixed to the slide rest and driven from a countershaft at the back of the lathe. Motion is given to the countershaft by a belt which runs from a pulley on the end of it to the treadle wheel. The cutter spindle may be adjusted either horizontally, as shown, or vertically. In the latter case, the idle guide pulleys may be dispensed with.

Wheels and Pinions.—[*Roues et Pignons.*—*Die Räder und Triebe.*]—In the construction of watches and clocks it is necessary to transmit motion from one arbor to another, so that the arbor which is driven rotates more quickly than the one which drives it. If it were practicable to use rollers with smooth edges for transmitting such motion, the diameters of the rollers would be inversely proportionate to the number of rotations made by their arbors in a given time. For instance, the distance apart of two arbors from centre to centre measures 3·7 inches, and it is desired that for every time the arbor from which the power is taken rotates the other shall rotate eight times. The distance between the arbors is divided into nine equal parts, of which eight are taken for the radius of the driver, which rotates only once, and one part for the radius of the follower as it is called, which rotates eight times. Although it is not practicable to drive with small rollers, which would slip unless pressed so tightly together as to cause

Fig. 1.

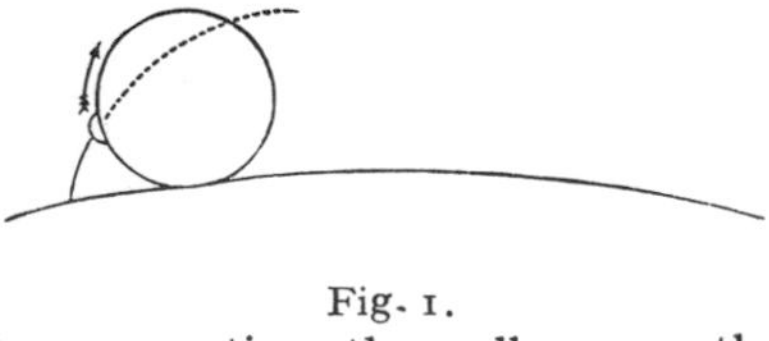

excessive friction, the circles representing the rollers are the basis on which the wheel and pinion are constructed. They are called the pitch circles. The acting part of the teeth of the driver is beyond its pitch circle, and the acting part of the teeth of the follower within its pitch circle. In most of the toothed wheels with which watchmakers are concerned the driver is the wheel and the follower the pinion. The shape for the acting part of the wheel teeth most favoured for horological mechanism is an epicycloid, a curve generated by rolling one circle on another.

In Fig. 1 is shown a portion of a circle representing the pitch-diameter of the wheel, and on it a smaller circle rolling in the direction of the arrow. If these two are made of brass or any thin material, and laid on a sheet of paper, a pencil fixed to the circumference of the small roller will trace a curve as shown. This curve is the acting part of the wheel tooth.

This acting part of the pinion leaves must be produced by the same sized roller as was used for the points of the wheel teeth, but in a different manner. The pinion flank should be hypocy-

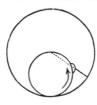

Fig. 2.

cloidal in form. A hypocycloid is obtained by rolling one circle within another instead of upon it. The most convenient size for the generating roller for both wheel and pinion is half the pitch diameter of the pinion. In Fig. 2 is a circle representing the pitch circle of the pinion, with another circle half its size rolling within it, and in this case the point described by the pencil would be a straight radial line, which is a suitable form for the pinion.

Teeth formed in this way will transmit the motion uniformly at the same speed as though the pitch circles rolled on each other without teeth, and will also meet another important requirement. The action between the teeth will take place almost wholly after the line of centres—that is if the pinion has not less than ten leaves. The difference between engaging and disengaging friction is very great, especially if the surfaces in contact are not quite smooth. Wheels which have any considerable portion of their action between the teeth as they are engaging or before the line of centres, not only absorb considerable power thereby but wear out rapidly. With a larger generating circle more of the action between the teeth of the wheel and the leaves of the pinion would take place after the line of centres, which is a consideration with low numbered pinions, but then a larger generating circle traces a pinion leaf too weak at the root.

The pitch circle of the wheel is spaced out so that the teeth and the spaces are equal. To allow of necessary freedom the teeth or leaves of the pinion are less in width than the spaces. The distance between the centre of one leaf and the centre of the next may be divided into ·6 for space and ·4 for leaf.*

The pinion leaves are finished with a semicircular piece projecting beyond the pitch circle as seen in Fig. 3. They would

* The " pitch " of wheels and pinions is the portion of the circumference of the pitch circle between the centre of one tooth and the centre of the next.

work without if properly pitched, but would not be safe as the depth becomes shallow from the wearing of the holes. Some prefer a Gothic-shaped projection like Fig. 4, which is of epicycloidal form the same as the wheel teeth. This is a very suitable form if the pinions are low numbered, for, although with it the action takes place more before the line of centres, a safer depth is ensured.

The teeth of the wheel are extended within the pitch line to allow of clearance for the addendum of the pinion. The root or part of the wheel tooth within the pitch line is generally radial. The corners at the bottom of the tooth may be rounded for strength, but these round corners must not be so full as to engage the points of the pinion leaves. The action should be confined as nearly as possible to the epicycloid on the wheel, and the hy-

Fig. 3.

pocycloid on the pinion. In watches the roots of all the wheels and pinions are left square, except the roots of the barrel or great wheel teeth, and the roots of the centre pinion leaves which should always be rounded for strength. There is then less danger of the teeth stripping if the mainspring breaks.

If the pinion is to be used as the driver and the wheel as the

Fig. 4.

follower, as is the case in the motion work of watches and clocks, the points of the pinion teeth must be epicycloidal, and the roots of the wheel teeth hypocycloidal struck with the same generating circle. For the convenience of using wheels and pinions indiscriminately as drivers and followers, engineers generally use a generating circle whose diameter $=$ the pitch $\times$ 2·22 for the points and roots of all wheels and pinions of the same pitch.

The tip of the addendum is removed in both wheels and pinions.

If more than two wheels gear together, the acting parts of all should be struck from the same sized generating circle. The number of teeth in a wheel bears exactly the same proportion to

TABLE 1.—FOR THE FULL DIAMETER OF WHEELS FROM 10 TO 100 TEETH; THE PITCH DIAMETER BEING = 1.

Number of Teeth	True Diameter	Number of Teeth	True Diameter	Number of Teeth	True Diameter	Number of Teeth	True Diameter
10	1·315	33	1·096	56	1·056	79	1·0400
11	1·286	34	1·093	57	1·055	80	1·0395
12	1·260	35	1·090	58	1·054	81	1·0390
13	1·243	36	1·087	59	1·053	82	1·0385
14	1·225	37	1·085	60	1·052	83	1·0380
15	1·210	38	1·083	61	1·051	84	1·0375
16	1·197	39	1·081	62	1·0508	85	1·0371
17	1·185	40	1·079	63	1·050	86	1·0366
18	1·175	41	1·077	64	1·0493	87	1·0363
19	1·166	42	1·075	65	1·0485	88	1·0360
20	1·157	43	1·073	66	1·0477	89	1·0355
21	1·150	44	1·072	67	1·0470	90	1·0350
22	1·143	45	1·070	68	1·0462	91	1·0346
23	1·137	46	1·068	69	1·0456	92	1·0343
24	1·131	47	1·067	70	1·0450	93	1·0339
25	1·126	48	1·066	71	1·0445	94	1·0335
26	1·121	49	1·064	72	1·0438	95	1·0332
27	1·117	50	1·063	73	1·0431	96	1·0330
28	1·113	51	1·062	74	1·0426	97	1·0325
29	1·109	52	1·061	75	1·0420	98	1·0321
30	1·105	53	1·060	76	1·0415	99	1·0318
31	1·102	54	1·058	77	1·0409	100	1·0315
32	1·099	55	1·057	78	1·0404		

TABLE 2.—FOR PINIONS WITH CIRCULARLY ROUNDED LEAVES LIKE FIG. 3: THE PITCH DIAMETER BEING = 1.

Number of Pinion	Full or True Diameter.	Diameter of Bottom.	Thickness of Teeth.	Diameter for Uneven Teeth.
6	1·209	0·402	0·209	
7	1·180	0·477	0·180	1·121
8	1·157	0·533	0·157	
9	1·140	0·572	0·140	1·106
10	1·126	0·611	0·126	
11	1·114	0·640	0·114	1·092
12	1·105	0·666	0·105	
13	1·097	0·684	0·097	1·081
14	1·090	0·701	0·090	
15	1·084	0·717	0·084	1·072
16	1·078	0·729	0·078	
17	1·074	0·741	0·074	1·063
18	1·070	0·751	0·070	
19	1·066	0·760	0·066	1·059
20	1·063	0·768	0·063	

TABLE 3.—FOR PINIONS WITH EPICYCLOIDAL ADDENDUM LIKE FIG. 4; THE PITCH DIAMETER BEING = 1.

Number of Pinion Leaves	True Diameter.	Diameter of Bottom.	Thickness of Leaves	Diameter for Uneven Numbered Leaves.
6	1·314	0·402	0·209	
7	1·270	0·477	0·180	1·206
8	1·235	0·533	0·157	
9	1·209	0·572	0·140	1·173
10	1·188	0·611	0·126	
11	1·171	0·640	0·114	1·148
12	1·157	0·666	0·105	
13	1·145	0·684	0·097	1·128
14	1·135	0·701	0·090	
15	1·126	0·717	0·084	1·113
16	1·118	0·729	0·078	
17	1·111	0·741	0·074	1·101
18	1·105	0·751	0·070	
19	1·099	0·760	0·066	1·093
20	1·094	0·768	0·063	

the number of teeth in a pinion with which it gears as the diameter of the pitch circles of the wheel and pinion bear to each other. If the pinion whose pitch circle is ·8 of an inch in diameter has 10 teeth, then the wheel with a pitch circle of 6·4 inches in diameter will have 80 teeth, because ·8 is contained 8 times in 6·4, and 10 × 8 = 80. But the outside or full diameter of a wheel or pinion is not proportional to the pitch diameter. The addendum or portion of the tooth beyond the acting part bears reference rather to the size of the generating circle and to the width of the teeth than to the diameter of the wheel or pinion.

The tables on page 459, prepared by Mr. A. Lange, give the amount to be added to the pitch diameters in order to obtain the full diameter of wheels and pinions, in which the teeth and spaces of the wheel are of equal width and the pinion leaves ·4 of the pitch. Mr. Lange has taken the height of the addendum of the wheel to be equal to the width of the tooth, which is sufficiently accurate for all practical purposes. When a slide gauge is used for measuring pinions with an odd number of leaves, the true diameter is not obtained because the points of the teeth are not opposite. To meet this, a separate column is given showing the proper allowance to be made.

These tables will be useful in drawing off the caliper of a watch or clock, and in many other instances when it is desired to find the full diameter from the pitch diameter, or to find the pitch diameter when the full diameter is given.

Example.—The full diameter is required of a pinion having 10 teeth, whose pitch diameter is ·8. In table 2, opposite 10, we find 1·126 as the full diameter of a pinion whose pitch diameter is 1. Then ·8 × 1·126 = ·9008, the full diameter required.

If the full diameter is known and the pitch diameter is required, the full diameter is to be *divided* by the number in the table.

Example.—The full diameter of a pinion with 10 leaves being ·9008, what is its pitch diameter ? Then ·9008 ÷ 1·126 = ·8, which is the pitch diameter as given in the preceding example.

In drawing off the caliper of watches and clocks it often occurs that the numbers of a wheel and pinion are known, and also the distance apart of their centres ; their diameters are then required. Then the pitch diameter of the wheel and pinion together = twice the distance of centres given. This whole distance is then to be divided into two portions. If we take the sum of the wheel and pinion teeth to represent the whole length, then the length of the portion representing the diameter of the wheel will be in proportion to the number of teeth contained in the wheel, and the remaining portion will represent the pitch diameter of the pinion.

Example.—A wheel of 80 and a pinion of 10 are to be planted

3·6 inches apart—centre to centre. Required their respective diameters. Then $3·6 \times 2 = 7·2$, and $90 : 7·2 : : 80 : 6·4$, which is the pitch diameter of the wheel. And $7·2 — 6·4$ gives ·8 for the pitch diameter of the pinion.

Lantern pinions work very smoothly as followers, though they are unsuitable as drivers. The space occupied by the shrouds precludes their use in watches, but in the going parts of clocks they answer well.

For the convenience of ready calculation it may be assumed that the addendum of the wheel teeth increases the size of the wheel by three teeth. For instance, the pitch diameter of a wheel of 80 teeth is 2 inches. Then its pitch diameter would bear the same proportion to its full diameter as 80 does to 83 ; or $80 : 2 : : 83 : 2·07$, which is the full diameter.

In the same way it may be taken that the circular addendum increases the size of the pinion by 1·25 teeth and the epicycloidal addendum by 1·98, or nearly 2 teeth. This is, of course, supposing the width of the pinion leaf to be ·4 of the pitch.

If the pinion is to be used as the driver, it must have the epicycloidal addenda to ensure proper action. I believe an opinion prevails among some watchmakers that the circularly rounded pinions may be used as drivers if they are sectored large, and that they are so used for motion work, but such a practice is altogether wrong.

In the motion work of keyless watches the followers are used as drivers when the hands are being set, and a good form of tooth for motion work generally may be obtained by using for roots and points of both wheels and pinions a generating circle of a diameter equal to twice the pitch. This gives a short tooth which will run smoothly when at full width. The form of gearing suitable for the train permits of too much shake for motion work.

Wheel Stretching Tool.—[*Die Räderstreckmaschine.*]—The

appended engraving represents a new wheel stretching tool in which the wheel after having been stretched remains straight and runs true. The wheel is to be placed between the circular and shaped stake and the adjustable revolving drum, so that it is held between the two with a slight pressure. When the crank of the tool is turned, the revolving drum drags the wheel round and at the same time stretches it perfectly equally.

Whitehurst, John.—A celebrated clockmaker, inventor of " Tell Tale " clocks ; born at Congleton in 1713, died in London in 1788.

Wig Wag.—[*Wig Wag.—Die Einrichtung zum Polieren der Triebe.*]—An engine that imparts a to-and-fro motion to a polisher for pinion leaves and the like.

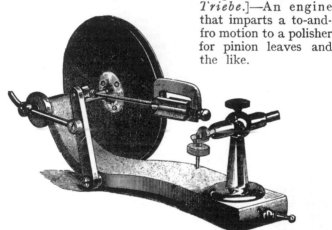

Winder, Bench Winder.—[*Carré à remonter.*—[*Der Uhrschlüssel.*]—A key used for winding watches by a watchmaker. It is about four inches long, and is made generally from a piece of pinion wire. Clock keys are often spoken of as winders.

Birch's Winder.—This is an ingenious key for fitting any size of watch square. The hollow steel square projects from an outer case, and may be projected still further by pressing a knob at the other end of the case. The square is split into halves, the

slits on each side of the body being taper, and widest at the mouth of the case. A pin in the case passes through the wide part of the slits and so, as the square is pressed out of the case, the narrow parts of the slits coming into contact with the pin compel the square to open. The winder is then placed in position, and as the knob is released the key closes on to the winding square. When the winding is completed, the knob is again pressed to release the winder.

Winding Square.—[*Carré.—Der Aufziehzapfen.*]—The squared termination of the barrel or fusee of a watch or clock, to which the key or winder is attached.

Wire Gauge.—[*Filère.—Das Drahtmass.*]

New Legal Standard Wire Gauge.

The following table gives the numbers of the revised and legalized Birmingham Wire Gauge, and their equivalents in decimal parts of an inch :—

B.W.G.	Inches.	B.W.G.	Inches.	B.W.G.	Inches.	B.W.G.	Inches.
7/0	·500	9	·144	24	·022	39	·0052
6/0	·464	10	·128	25	·02	40	·0048
5/0	·432	11	·116	26	·018	41	·0044
4/0	·400	12	·104	27	·0164	42	·004
3/0	·372	13	·092	28	·0148	43	·0036
2/0	·348	14	·080	29	·0136	44	·0022
0	·324	15	·072	30	·0124	45	·0028
1	·300	16	·064	31	·0116	46	·0024
2	·276	17	·056	32	·0108	47	·002
3	·252	18	·048	33	·01	48	·0016
4	·232	19	·040	34	·0092	49	·0012
5	·212	20	·036	35	·0084	50	·0001
6	·192	21	·032	36	·0076		
7	·176	22	·028	37	·0068		
8	·160	23	·024	38	·0060		

Woerd, Charles V., born 1821, died 1888. A clever mechanician who did much to advance the art of machine watchmaking by designing automatic tools for the Waltham Watch Company, of which he was mechanical superintendent from 1875 to 1882.

Wycherley, John, born in 1817, at Prescot, Lancashire, where he founded the machine-made watch movement industry. Died at Southport, 1891.

Zinc.—[*Zinc.—Das Zink.*]—A piece of zinc may be readily severed by making a deep scratch at the place of desired separation and running mercury into the cut. When the mercury has soaked into the zinc, but little force is needed to part it.

464

APPENDIX.

Weight of Lead, Zinc, and Cast-Iron Cylinders One Inch Long.

Diameter in Inches.	Lead.	Zinc.	Iron.	Diameter in Inches.	Lead.	Zinc.	Iron.
·25	·020	·012	·012	3·25	3·400	2·098	2·156
·5	·080	·049	·050	3·5	3·944	2·434	2·491
·75	·180	·111	·114	3·75	4·51	2·783	2·865
1	·321	·198	·204	4	5·149	3·177	3·265
1·25	·503	·310	·319	4·25	5·813	3·587	3·686
1·5	·724	·447	·459	4·5	6·519	3·922	4·134
1·75	·984	·607	·624	4·75	7·265	4·483	4·607
2	1·287	·794	·816	5	8·048	4·966	5·103
2·25	1·630	1·005	1·033	5·25	8·872	5·474	5·626
2·5	2·009	1·239	1·274	5·5	9·737	6·008	6·175
2·75	2·434	1·502	1·544	5·75	10·643	6·567	6·749
3	2·897	1·788	1·837	6	11·590	7·152	7·350

The above table will be found useful for readily ascertaining the weight of cylinders for pendulum bobs and clock weights. Example :—Required the weight of a lead pendulum bob 3 inches diameter, 9 inches long, which has a hole through it ·75 inches in diameter. The weight of a lead cylinder 3 inches diam. in the table is 2·897, which multiplied by 9 (the length given) = 26·07 lbs. Then the weight in the table of a cylinder ·75 inches diameter is ·18, and ·18 × 9 = 1·62. And 26·07 — 1·62 = 24·45, the weight required in lbs.

To Extract Root of Whole Number and Decimal Parts.

If it is desired to extract the root of a whole number and decimal parts, multiply the difference between the root of the whole number and the next higher number in the table on page 465 by the decimal part of the given number, and add the product to the root of the whole number given ; the sum will be the root required, correct at all events to three places of decimals.

EXAMPLES.—Required the square root of 60·2.

$$\sqrt{61} = 7·8102$$
$$\sqrt{60} = 7·7459$$

·0643 × ·2 = ·01286

And 7·7459 + ·01286 = 7·75876 the root required.

Square Root of Numbers from 1 to 200.

To Extract Root of Whole Number and Decimal parts see preceding page.

No.	Square Roots.	No.	Square Roots.	No.	Square Roots.	No.	Square Roots.
1	1·0000	51	7·1414	101	10·0498	151	12·2882
2	1·4142	52	7·2111	102	10·0995	152	12·3288
3	1·7320	53	7·2801	103	10·1488	153	12·3693
4	2·0000	54	7·3484	104	10·1980	154	12·4096
5	2·2360	55	7·4161	105	10·2469	155	12·4498
6	2·4494	56	7·4833	106	10·2956	156	12·4899
7	2·6457	57	7·5498	107	10·3440	157	12·5299
8	2·8284	58	7·6157	108	10·3923	158	12·5698
9	3·0000	59	7·6811	109	10·4403	159	12·6095
10	3·1622	60	7·7459	110	10·4880	160	12·6491
11	3·3166	61	7·8102	111	10·5356	161	12·6885
12	3·4641	62	7·8740	I12	10·5830	162	12·7279
13	3·6055	63	7·9372	113	10·6301	163	12·7671
14	3·7416	64	8·0000	114	10·6770	164	12·8062
15	3·8729	65	8·0622	115	10·7238	165	12·8452
16	4·0000	66	8·1240	116	10·7703	166	12·8840
17	4·1231	67	8·1853	117	10·8166	167	12·9228
18	4·2426	68	8·2462	118	10·8627	168	12·9614
19	4·3588	69	8·3066	119	10·9087	169	13·0000
20	4·4721	70	8·3666	120	10·9544	170	13·0384
21	4·5825	71	8·4261	121	11·0000	171	13·0766
22	4·6904	72	8·4852	122	11·0453	172	13·1148
23	4·7958	73	8·5440	123	11·0905	173	13·1529
24	4·8989	74	8·6023	124	11·1355	174	13·1909
25	5·0000	75	8·6602	125	11·1803	175	13·2287
26	5·0990	76	8·7177	126	11·2249	176	13·2664
27	5·1961	77	8·7749	127	11·2694	177	13·3041
28	5·2915	78	8·8317	128	11·3137	178	13·3416
29	5·3851	79	8·8881	129	11·3578	179	13·3790
30	5·4772	80	8·9442	130	11·4017	180	13·4164
31	5·5677	81	9·0000	131	11·4455	181	13·4536
32	5·6568	82	9·0553	132	11·4891	182	13·4907
33	5·7445	83	9·1104	133	11·5325	183	13·5277
34	5·8309	84	9·1651	134	11·5758	184	13·5646
35	5·9160	85	9·2195	135	11·6189	185	13·6014
36	6·0000	86	9·2736	136	11·6619	186	13·6381
37	6·0827	87	9·3273	137	11·7046	187	13·6747
38	6·1644	88	9·3808	138	11·7473	188	13·7113
39	6·2449	89	9·4339	139	11·7898	189	13·7477
40	6·3245	90	9·4868	140	11·8321	190	13·7840
41	6·4031	91	9·5393	141	11·8743	191	13·8202
42	6·4807	92	9·5916	142	11·9163	192	13·8564
43	6·5574	93	9·6436	143	11·9582	193	13·8924
44	6·6332	94	9·6953	144	12·0000	194	13·9283
45	6·7082	95	9·7467	145	12·0415	195	13·9642
46	6·7823	96	9·7979	146	12·0830	196	14·0000
47	6·8556	97	9·8488	147	12·1243	197	14·0356
48	6·9282	98	9·8994	148	12·1655	198	14·0712
49	7·0000	99	9·9498	149	12·2065	199	14·1067
50	7·0710	100	10·0000	150	12·2474	200	14·1421

E E

Signs and Abbreviations.

= is the sign of Equality ; thus : A = 6 means that A is equal
to 6.

+ (plus) is the sign of addition ; thus : 6 + 4 means 6 added
to 4.

— minus or less ; thus : 5 — 2, which means 5 minus 2.

× signifies Multiplication ; thus : 5 × 2 means 5 multiplied by 2.
Multiplication is also indicated by writing the factors
when they are not numerals one after the other ; thus,
$c \, d \, e$, which means c multiplied by d and the product
therefore multiplied by e.

÷ stands for Division ; thus : 12 ÷ 3 means 12 divided by 3.
Division is also indicated by writing the divisor under
the dividend with line between them ; thus $\dfrac{a}{b}$, which
means that a is to be divided by b.

: : : : signifies Proportion ; thus : 2 : 3 : : 4 : x, which means, as
2 is to 3 so is 4 to x, or to the answer which when un-
known is usually indicated by x.

A small ² placed at the upper right-hand corner of a
quantity indicates that the square or second power of
the quantity is to be taken ; thus : 3^2 means the square
of 3, that is 3 multiplied by itself.

If the index figures is ³ instead of ² it indicates that
the third power or cube of the quantity is to be taken ;
thus : 6^3 means the cube of 6, × 6 × 6.

√ stands for root, and if used as shown without a small figure
attached represents the square root ; thus : $\sqrt{9}$ means
the square root of 9.

$\sqrt[3]{}$ represents the third or cube root.

——When a bar is appended to quantities it indicates that the
quantities are to be taken together ; thus : $\sqrt{20 \times 12}$
+ 6, which means that 20 is to be first multiplied by 12,
and the square root of the product is to be extracted and
added to 6. If the bar were not there it would mean
that the square root of 20 was to be extracted, and this
root multiplied by 12 and the product added to 6.
Instead of a bar being used, quantities which are to be
taken together are sometimes enclosed in parentheses ( )
or brackets [ ].

m. Minute of time = $\frac{1}{60}$th of an hour.
s. Second of time = $\frac{1}{3600}$th of an hour.
° Degree of Arc ; $1° = \frac{1}{360}$th of the circumference.
′ Minute of Arc = $\frac{1}{60}$th of a Degree.
″ Second of Arc = $\frac{1}{3600}$th of a Degree.

Trigonometrical Definitions.

A circle is divided into 360 degrees, and each quadrantal arc therefore contains 90. If a quadrantal arc is divided into two parts, as in the diagram, whatever number of the degrees is contained in the one part (A), the difference or complement of 90 is contained in the other part.

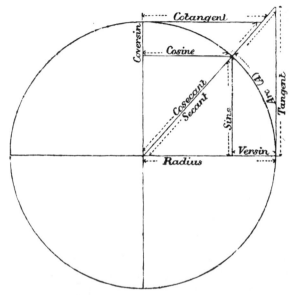

Taking the radius as 1, the value of the various functions of a given angle and of its complement may be found on page 468.

If the sine of the angle A in the diagram were produced till it touched the lower part of the circle, it would then form a chord of double the angle ; therefore, Sin. $\sqrt{\div 2} \times 2 = $ Chord.

Alloys of Nickel and Steel.—As related on page 24, Dr. Ch. Ed. Guillaume has observed remarkable results from alloying steel and nickel in various proportions. He found the co-efficients of dilatation and elasticity to vary as follows :—

Percentage of nickel.	Co-efficients of dilatation. Per cent.	Co-efficients of elasticity. Per cent.
0	10·3	22·0
24	17·5	19·3
35·7	0·88	14·7
44·4	8·5	16·7
100	12·5	21·6

468

Table of Sines, Tangents, &c.

	Sine.	Cover.	Cosecnt.	Tangt.	Cotang.	Secant.	Vrsn.	Cosin.	Deg.
0	·00	1·00000	Infinite.	·0	Infinite.	1·00000	·0	1·00000	90
1	·01745	·98254	57·2986	·01745	57·2899	1·00015	·0001	·99984	89
2	·03489	·96510	28·6537	·03492	28·6362	1·00060	·0006	·99939	88
3	·05233	·94766	19·1073	·05240	19·0811	1·00137	·0013	·99862	87
4	·06975	·93024	14·3355	·06992	14·3006	1·00244	·0024	·99756	86
5	·08715	·91284	11·4737	·08748	11·4300	1·00381	·0038	·99619	85
6	·10452	·89547	9·5667	·10510	9·5143	1·00550	·0054	·99452	84
7	·12186	·87813	8·2055	·12278	8·1443	1·00750	·0074	·99254	83
8	·13917	·86082	7·1852	·14054	7·1153	1·00982	·0097	·99026	82
9	·15643	·84356	6·3924	·15838	6·3137	1·01246	·0123	·98768	81
10	·17364	·82635	5·7587	·17632	5·6712	1·01542	·0151	·98480	80
11	·19080	·80919	5·2408	·19438	5·1445	1·01871	·0183	·98162	79
12	·20791	·79208	4·8097	·21255	4·7046	1·02234	·0218	·97814	78
13	·22495	·77504	4·4454	·23086	4·3314	1·02630	·0256	·97437	77
14	·24192	·75807	4·1335	·24932	4·0107	1·03061	·0297	·97029	76
15	·25881	·74118	3·8637	·26794	3·7320	1·03527	·0340	·96592	75
16	·27563	·72436	3·6279	·28674	3·4874	1·04029	·0387	·96126	74
17	·29237	·70762	3·4203	·30573	3·2708	1·04569	·0436	·95630	73
18	·30901	·69098	3·2360	·32491	3·0776	1·05146	·0489	·95105	72
19	·32556	·67443	3·0715	·34432	2·9042	1·05762	·0544	·94551	71
20	·34202	·65797	2·9238	·36397	2·7474	1·06417	·0603	·93969	70
21	·35836	·64163	2·7904	·38386	2·6050	1·07114	·0664	·93358	69
22	·37460	·62539	2·6694	·40402	2·4750	1·07853	·0728	·92718	68
23	·39073	·60926	2·5593	·42447	2·3558	1·08636	·0794	·92050	67
24	·40673	·59326	2·4585	·44522	2·2460	1·09463	·0864	·91354	66
25	·42261	·57738	2·3662	·46630	2·1445	1·10337	·0936	·90630	65
26	·43837	·56162	2·2811	·48773	2·0503	1·11260	·1012	·89879	64
27	·45399	·54600	2·2026	·50952	1·9626	1·12232	·1089	·89100	63
28	·46947	·53052	2·1300	·53170	1·8807	1·13257	·1170	·88294	62
29	·48480	·51519	2·0626	·55430	1·8040	1·14335	·1253	·87461	61
30	·50000	·50000	2·0000	·57735	1·7320	1·15470	·1339	·86602	60
31	·51503	·48496	1·9416	·60086	1·6642	1·16663	·1428	·85716	59
32	·52991	·47008	1·8870	·62486	1·6003	1·17917	·1519	·84804	58
33	·54463	·45536	1·8360	·64940	1·5398	1·19236	·1613	·83867	57
34	·55919	·44080	1·7882	·67450	1·4825	1·20621	·1709	·82903	56
35	·57357	·42642	1·7434	·70020	1·4281	1·22077	·1808	·81915	55
36	·58778	·41221	1·7013	·72654	1·3763	1·23606	·1909	·80901	54
37	·60181	·39818	1·6616	·75355	1·3270	1·25213	·2013	·79863	53
38	·61566	·38433	1·6242	·78128	1·2799	1·26901	·2119	·78801	52
39	·62932	·37067	1·5890	·80978	1·2348	1·28675	·2228	·77714	51
40	·64278	·35721	1·5557	·83909	1·1917	1·30540	·2339	·76604	50
41	·65605	·34394	1·5242	·86928	1·1503	1·32501	·2452	·75470	49
42	·66913	·33086	1·4944	·90040	1·1106	1·34563	·2568	·74314	48
43	·68199	·31800	1·4662	·93251	1·0723	1·36732	·2686	·73135	47
44	·69465	·30534	1·4395	·96568	1·0355	1·39016	·2806	·71933	46
45	·70710	·29289	1·4142	1·00000	1·0000	1·41421	·2928	·70710	45

	Cosin.	Versin.	Secant.	Cotang.	Tangt.	Cosecnt.	Covsn.	Sine.	

Properties of Metals and other Substances.

DESCRIPTION.	Weight of a cubic inch in lbs	Specific Gravity *	Tenacity in lbs. per square inch	Crushing force in lbs. per square inch	Melting point, Fahr.	Expansion between 32° & 212° ¶	Conducting power †	Specific heat.
Aluminium	·092	2·56	27·00	...	1400°	...	...	...
Ditto Bronze (90 per cent. copper)	·276	7·68	..	...	...	...	...	...
Antimony, cast ...	·242	6·7	1,066	...	810°	·0011	..	·0507
Bismuth	·35	9·82	3,250	...	497°	·0014	...	·0288
Brass, cast	·3	8·4	17,978	10,300	1800°	·002	...	...
Ditto, wire	...	8·5	49,000	...	...	...	...	...
Copper, cast	·32	8·9	19,072	11,700	1996°	·0017	...	·0449
Ditto, sheet ...	...	8·85	33,000	...	...	...	898	...
Ditto, wire	...	9·9	61,000	...	..	...	...	...
Dry Deal Rod ...	·025	·68	..	...	...	·0004	...	...
Do. Ebony do. ...	·043	1·18	...	...	...	...	...	...
Gold pure, hammered	·7	19·35	20,400	...	2616°	·0016	1000	·0298
Ditto, cast	...	19·25	...	...	...	...	...	...
§Ditto, standard ...	·638	17·724	...	...	...	...	...	...
Gun Metal	·3	8·4	36,000	...	...	...	...	...
Glass	·1	2·83	...	...	...	·00082	...	...
‖Iron, wrought ...	·28	7·7	60,000	38,000	...	·0012	347	·1100
Ditto, cast	·26	7·18	19,000	92,000	2786°	·0011	...	...
Lead, cast	·41	11·35	1,824	7,000	612°	·0028	180	·0293
Ditto, sheet	..	...	3,328	...	...	...	...	...
Mercury...	·49	13·59	...	...	39°	018	...	·0333
Nickel	·31	8·82	...	...	...	...	...	...
Palladium	·46	12	...	...	...	·001	...	...
Platinum	·78	21·4	.	...	3080°	·00085	...	...
Silver, pure	38	10·47	41,000	...	1873°	·0019	973	·0557
‡Ditto, standard ...	·371	10·312	...	...	...	...	...	...
Steel	·282	7·82	120,000	...	...	·0011	...	...
Teak	·03	·86	...	...	...	...	...	...
Tin	·263	7·29	5,000	15,000	442°	·0021	304	·0514
Water (distilled 50°)	·036	1·00	...	...	32°	·0477	...	1·00
Zinc...	·253	7·	8,000	...	773°	·0028	363	·0927

* The specific gravity of a body is the proportion which its weight bears to the weight of an equal bulk of water.

† Gold 1,000 is taken as the standard.

‖ Iron wire has a greater tenacity than wrought bar, and may be taken as 85,000 lbs. per square inch.

¶ The figures given represent the proportion of its length which the substance expands in the range of temperature mentioned, except in the case of mercury, in which the increase of volume is taken. In practice, allowance must in the case of mercury be made for the expansion of the containing vessel. In a glass pendulum jar, mercury may be taken to expand in length 5·8 times as much as the steel rod.

§ Gold is alloyed with silver or copper. Standard, or 22-carat gold, contains 22 parts of gold and 2 parts of alloy.

‡ Silver is alloyed with copper. 12 ozs. Troy standard silver contains 11 ozs. 2 dwts. of pure silver.

Thermometers.

Comparison of the scales of Fahrenheit, Réaumur, and the Centigrade, between the freezing and the boiling point of water :—

Fahr.	Raum.	Cent	Fahr.	Raum	Cent.	Fahr.	Raum	Cent.
212	80	100	110	34·7	43·3	68	16·0	20·0
203	72	96	107	33·3	41·7	65	14·7	18·3
194	76	90	104	32·0	40·0	62	13·3	16·7
185	68	85	101	30·7	38·3	59	12·0	15·0
176	64	80	98	29·3	36·7	56	10·7	13·3
167	60	75	95	28·0	35·0	53	9·3	11·7
158	56	70	92	26·7	33·3	50	8·0	10·0
149	52	65	89	25·3	31·7	47	6·7	8·3
140	48	60	86	24·0	30·0	44	5·3	6·7
131	44	55	83	22·7	28·3	41	4·0	5·0
122	40	50	80	21·3	26·7	38	2·7	3·3
119	38·7	48·3	77	20·0	25·0	35	1·3	1·7
116	37·3	56·7	74	18·7	23·3	32	0·0	0·0
113	36·0	45·0	71	17·3	21·7			

NOTE.—Zero Fahrenheit corresponds with *minus* 17·78 Centigrade and *minus* 14·22 Réaumur. Any other number of degrees, Centigrade or Réaumur, may be converted into degrees Fahrenheit by the following rules.

Let F = No. of degrees Fahrenheit.
,, C = ,, ,, Centigrade.
,, R = ,, ,, Réaumur.

$$\text{Then } F = \frac{C \times 9}{5} + 32. \text{ Or } F = \frac{R \times 9}{4} + 32.$$

INVAR.—The title given to steel alloyed with about 36 per cent. of nickel is marketed in three grades, with co-efficients of expansion, as follows :—

No. 1, 0·0000010 per degree centigrade.
No. 2, 0·0000016 ,, ,,
No. 3, 0·0000025 ,, ,,

CHROMATED NICKEL.—The title used for the composition of the inner portion of the Guillaume non-magnetic compensation balance. In Dr. Guillaume's English Patent (1904 No. 4698) at least 7 per cent. of chromium is prescribed and an addition of manganese is mentioned as a desirable constituent.

FERRO-NICKEL.—The title given to a material used for balance springs. Its composition appears to be closely allied to that of Invar.

The English patent of P. Perret for balances and balance springs (1897 No. 25,142) says the balance spring is made of an alloy the elasticity of which increases with a rise of temperature; such an alloy is made of nickel and steel in the proportions of 27 or 28 of the former and 73 or 72 of the latter. A balance which only expands slightly is composed of 35 to 36 parts of nickel and 65 to 64 parts of steel.

Clock Trains and Lengths of Pendulums.

Wheels	Pinions	Escape Wheel	Vibrations of Pendulum @ Min	Length of Pendulum in Inches	Wheels	Pinions	Escape Wheel	Vibrations of Pendulum @ Min	Length of Pendulum in Inches
120 90 75	10 10 9	Double 3-Leg. ged.	*30	156·56	96 76	8	30	114	10·82
120 90 90	10 9 9	Do.	*40	88·07	115 100	10	30	115	10·65
128 120	16	30	60	39·14	84 78	7	26	115·9	10·43
112 105	14	30	60	39·14	96 80	8	30	120	9·78
96 90	12	30	60	39.14	84 70	7	30	120	9·78
80 75	10	30	60	39.14	84 78	7	27	120·3	9·73
64 60	8	30	60	39·14	90 84	8	31	122	9·46
64 64	8	30	68	30·49	84 78	7	28	124·8	9·02
70 64	8	30	70	28·75	100 80	8	30	125	9·01
72 64	8	30	72	27·17	90 84	8	32	126	8·87
75 60	8	32	75	25·53	100 96	10	40	128	8·59
72 65	8	32	78	23·15	84 78	7	29	129·3	8·42
75 64	8	32	80	22·01	100 78	8	32	130	8·34
84 64	8	30	84	19·97	84 77	7	30	132	8·08
86 64	8	30	86	19·06	84 78	7	30	133·7	7·9
88 64	8	30	88	18·19	90 90	8	32	135	7·68
84 78	7	20	89·1	17·72	84 78	7	31	138·2	7·15
80 72	8	30	90	17·39	84 80	8	40	140	7·18
84 78	7	21	93·6	16·08	120 71	8	32	142	6·92
94 64	8	30	94	15·94	84 78	7	32	142·6	6·9
84 78	8	28	95·2	15·52	100 87	8	32	145	6·69
108 100	12 & 10	32	96	15·28	84 78	7	33	147·1	6·5
84 84	9 & 8	30	98	14·66	100 96	8	30	150	6·26
84 78	7	22	98	14·66	84 78	7	34	151·6	6·1
84 78	8	29	98.6	13·44	96 95	8	32	152	6·09
80 80	8	30	100	14·09	84 77	7	35	154	5·94
83 72	8	32	102	13·54	104 96	8	30	156	5·78
84 78	8	30	102	13·54	84 78	7	35	156	5·78
84 78	7	23	102·5	13.4	120 96	9 & 8	30	160	5·5
105 100	10	30	105	12·88	84 78	7	36	160	5·5
84 78	8	31	105·4	12·67	84 78	7	37	164·9	5·15
84 78	7	24	107	12·3	132 100	9 & 8	27	165	5·17
96 72	8	30	108	12·14	84 78	7	38	169·4	4·88
84 78	8	32	108·8	11·9	128 102	8	25	170	4·87
88 80	8	30	110	11·64	84 78	7	39	173·8	4·65
84 77	7	25	110	11·64	36 36 35	6	25	175	4·6
84 78	7	25	111·4	11·29	84 77	7	40	176	4·5
84 80	8	32	112	11·22	45 36 36	6	20	180	4·25
84 78	8	33	112·2	11·15	47 36 36	6	20	188	3·92

* These are good examples of turret clock trains; the great wheel (120 teeth) makes in both instances a rotation in three hours. From this wheel the hands are to be driven. This may be done by means of a pinion of 40 gearing with the great wheel, or a pair of bevel wheels bearing the same proportion to each other (three to one) may be used, the larger one being fixed to the great wheel arbor. The arrangement would in each case depend upon the number and position of the dials. The double three-legged gravity escape wheel moves through 60 degrees at each beat, and therefore to apply the rule given for calculating trains it must be treated as in an escape wheel of three teeth. A turret clock train for 1¼ sec. pendulum is given under the head of " Turret Clock."

Circumferences and Areas of Circles advancing by Tenths.

The circumference of a circle = diameter × 3·14159. (π is used by mathematicians to represent 3·14159.)
The area of a circle = the square of the diameter × ·7854.
The diameter of a circle × ·886226 = the side of an equal square
The diameter of a circle × ·7071 = the side of an inscribed square.
An arc of 57·29578° = the radius.

Diam.	Circum.	Area.	Diam.	Circum.	Area.
1	3·1416	·7854	5	15·7080	19·6350
·1	3·4557	·9503	·1	16·0221	20·4282
·2	3·7699	1·1309	·2	16·3363	21·2372
·3	4·0840	1·3273	·3	16·6504	22·0618
·4	4·3982	1·5393	·4	16·9646	22·9022
·5	4·7124	1·7671	·5	17·2788	23·7583
·6	5·0265	2·0106	·6	17·5929	24·6301
·7	5·3407	2·2698	·7	17·9071	25·5176
·8	5·6548	2·5446	·8	18·2212	26·4208
·9	5·9690	2·8352	·9	18·5354	27·3397
2	6·2832	3·1416	6	18·8496	28·2744
·1	6·5973	3·4636	·1	19·1637	29·2247
·2	6·9115	3·8013	·2	19·4779	30·1907
·3	7·2256	4·1547	·3	19·7920	31·1725
·4	7·5398	4·5239	·4	20·1062	32·1699
·5	7·8540	4·9087	·5	20·4204	33·1831
·6	8·1681	5·3093	·6	20·7345	34·2120
·7	8·4823	5·7255	·7	21·0487	35·2566
·8	8·7964	6·1575	·8	21·3628	36·3168
·9	9·1106	6·6052	·9	21·6770	37·3928
3	9·4248	7·0686	7	21·9912	38·4846
·1	9·7389	7·5476	·1	22·3053	39·5920
·2	10·0531	8·0424	·2	22·6195	40·7151
·3	10·3672	8·5530	·3	22·9336	41·8539
·4	10·6814	9·0792	·4	23·2478	43·0085
·5	10·9956	9·6211	·5	23·5620	44·1787
·6	11·3097	10·1787	·6	23·8761	45·3647
·7	11·6239	10·7521	·7	24·1903	46·5663
·8	11·9380	11·3411	·8	24·5044	47·7837
·9	12·2522	11·9459	·9	24·8186	49·0168
4	12·5664	12·5664	8	25·1328	50·2656
·1	12·8805	13·2025	·1	25·4469	51·5300
·2	13·1947	13·8544	·2	25·7611	52·8102
·3	13·5088	14·5220	·3	26·0752	54·1062
·4	13·8230	15·2053	·4	26·3894	55·4178
·5	14·1372	15·9043	·5	26·7036	56·7451
·6	14·4513	16·6190	·6	27·0177	58·0881
·7	14·7655	17·3494	·7	27·3319	59·4469
·8	15·0796	18·0956	·8	27·6460	60·8213
·9	15·3938	18·8574	·9	27·9602	62·2115

Diam.	Circum.	Area.	Diam.	Circum.	Area.
9	28·2744	63·6174	·5	45·5532	165·1303
·1	28·5885	65·0389	·6	45·8673	167·4158
·2	28·9027	66·4762	·7	46·1815	169·7170
·3	29·2168	67·9292	·8	46·4956	172·0340
·4	29·5310	69·3979	·9	46·8098	174·3666
·5	29·8452	70·8823	15	47·1240	176·7150
·6	30·1593	72·3824	·1	47·4381	179·0790
·7	30·4735	73·8982	·2	47·7523	181·4588
·8	30·7876	75·4298	·3	48·0664	183·8542
·9	31·1018	76·9770	·4	48·3806	186·2654
10	31·4160	78·5400	·5	48·6948	188·6923
·1	31·7301	80·1186	·6	49·0089	191·1349
·2	32·0443	81·7130	·7	49·3231	193·5932
·3	32·3580	83·3230	·8	49·6372	196·0672
·4	32·6726	84·9488	·9	49·9514	198·5569
·5	32·9868	86·5903	16	50·2656	201·0624
·6	33·3009	88·2475	·1	50·5797	203·5835
·7	33·6151	89·9204	·2	50·8939	206·1203
·8	33·9292	91·6090	·3	51·2080	208·6729
·9	34·2434	93·3133	·4	51·5224	211·2411
11	34·5576	95·0334	·5	51·8364	213·8251
·1	34·8717	96·7691	·6	52·1505	216·4248
·2	35·1859	98·5205	·7	52·4647	219·0402
·3	35·5010	100·2877	·8	52·7788	221·6712
·4	35·8142	102·0705	·9	53·0930	224·3180
·5	36·1284	103·8691	17	53·4072	226·9806
·6	36·4425	105·6834	·1	53·7213	229·6588
·7	36·7567	107·5134	·2	54·0355	232·3527
·8	37·0708	109·3590	·3	54·3496	235·0623
·9	37·3840	111·2204	·4	54·6038	237·7877
12	37·6992	113·0976	·5	54·9780	240·5287
·1	38·0133	114·9904	·6	55·2921	243·2855
·2	38·3275	116·8989	·7	55·6063	246·0579
·3	38·6416	118·8231	·8	55·9204	248·8461
·4	38·9558	120·7631	·9	56·2346	251·6500
·5	39·2700	122·7187	18	56·5488	254·4696
·6	39·5841	124·6901	·1	56·8629	257·3048
·7	39·8983	126·6771	·2	57·1771	260·1558
·8	40·2124	128·6799	·3	57·4912	263·0226
·9	40·5266	130·6984	·4	57·8054	265·9050
13	40·8408	132·7326	·5	58·1196	268·8031
·1	41·1549	134·7824	·6	58·4337	271·7169
·2	41·4691	136·8480	·7	58·7479	274·6465
·3	41·7832	138·9294	·8	59·0620	277·5917
·4	42·0974	141·0264	·9	59·3762	280·5527
·5	42·4116	143·1391	19	59·6904	283·5294
·6	42·7257	145·2675	·1	60·0045	286·5217
·7	43·0399	147·4117	·2	60·3187	289·5298
·8	43·3540	149·5715	·3	60·6328	292·5536
·9	43·6682	151·7471	·4	60·9470	295·5931
14	43·9824	153·9384	·5	61·2612	298·6483
·1	44·2965	156·1453	·6	61·5753	301·7192
·2	44·6107	158·3680	·7	61·8895	304·8060
·3	44·9248	160·6064	·8	62·2036	307·9082
·4	45·2390	162·8605	·9	62·5178	311·0252
			20	62·8320	314·1600

Difference between Greenwich Mean Time and Local Time
at the Principal Places throughout the World.

The word " fast " after any place indicates that the local time is fast of Greenwich time, and the word " slow " that it is slow of Greenwich time.

NOTE.—Four minutes in time = a degree of longitude. In estimating the time of any place by comparison with Greenwich, ADD to G.M.T. for EAST and SUBTRACT for WEST.

ENGLAND AND WALES.

			M.	S.				M.	S.
LONDON—					Burton-on-Trent	..	slow	6	28
Aldgate Church	..	slow	0	17	Bury (Lanc.) ..	..	slow	9	4
Bank of England	..	slow	0	20	Bury St. Edmunds	..	fast	2	48
Battersea Church	..	slow	0	42	Cambridge	..	fast	0	23
Berkeley Square	..	slow	0	35	Canterbury	..	fast	4	30
Blackfriars Bridge	..	slow	0	24	Cardiff ..	..	slow	13	0
British Museum	..	slow	0	30	Cardigan	..	slow	18	20
Fulham Church	..	slow	0	50	Carlisle ..	..	slow	11	44
Hampstead Church..		slow	0	43	Carmarthen	..	slow	17	16
Highgate Church	..	slow	0	35	Carnarvon	..	slow	17	0
Horological Institute		slow	0	24	Chatham	..	fast	2	10
Kensington Palace	..	slow	0	45	Chelmsford	..	fast	2	0
Muswell Hill	..	slow	0	29	Cheltenham	..	slow	8	20
St. Paul's ..	..	slow	0	23	Chester ..	..	slow	11	36
Streatham Church	..	slow	0	31	Chichester	..	slow	3	10
Westminster Abbey		slow	0	30	Chippenham	..	slow	8	20
Abingdon	..	slow	5	4	Christchurch	..	slow	7	2
Alnwick	..	slow	6	52	Cockermouth	..	slow	13	32
Andover	..	slow	5	52	Colchester	..	fast	3	30
Appleby	..	slow	9	52	Coventry	..	slow	6	0
Arundel..	..	slow	2	20	Darlington	..	slow	6	8
Ashford, Kent	..	fast	3	22	Deal	..	fast	5	30
Ashton-under-Lyne	..	slow	8	24	Derby ..	..	slow	5	58
Aylesbury	..	slow	3	20	Devizes	..	slow	7	56
Banbury	..	slow	5	12	Devonport	..	slow	16	48
Bangor ..	..	slow	16	32	Dewsbury	..	slow	6	28
Barnstaple	..	slow	16	20	Dorchester	..	slow	9	48
Bath ..	..	slow	9	28	Douglas, Isle of Man	..	slow	17	46
Bedford..	..	slow	1	52	Dover ..	..	fast	5	16
Berwick-on-Tweed	..	slow	8	0	Droitwich	..	slow	8	44
Beverley	..	slow	1	44	Dudley ..	..	slow	8	26
Bideford	..	slow	16	56	Durham	..	slow	6	20
Birkenhead	..	slow	12	4	Eastbourne	..	fast	1	20
Birmingham	..	slow	7	36	Epsom ..	..	slow	1	4
Blackburn	..	slow	9	52	Exeter ..	..	slow	14	12
Bodmin	..	slow	19	0	Falmouth	..	slow	20	38
Boston ..	..	slow	0	6	Faversham	..	fast	3	36
Bradford, Yorkshire	..	slow	7	0	Flint ..	..	slow	12	31
Brecknock	..	slow	14	0	Folkestone	..	fast	4	36
Brentford	..	slow	1	20	Gateshead	..	slow	6	24
Bridgnorth	..	slow	9	40	Gloucester	..	slow	9	0
Bridgwater	..	slow	12	0	Grantham	..	slow	2	40
Brighton	..	slow	0	30	Gravesend	..	fast	1	40
Bristol ..	..	slow	10	22	Grimsby	..	slow	0	16
Buckingham	..	slow	3	56	Guernsey	..	slow	10	22
Burnley..	..	slow	8	56	Guildford	..	slow	2	10

	M. S.			M. S.
Halifax	slow 7 8	Ramsgate	fast 5 40
Harrogate	slow 6 8	Reading	slow 3 55
Hartlepool	slow 4 40	Reigate	slow 0 48
Harwich	fast 5 8	Rochester	fast 2 0
Hastings	fast 2 24	Runcorn	slow 11 0
Hereford	slow 11 0	Salford	slow 9 4
Hertford	slow 0 20	Salisbury	slow 7 8
Holyhead	slow 18 36	Scarborough	slow 1 35
Horsham	slow 1 20	Sheerness	fast 2 59
Huddersfield	slow 7 10	Sheffield	slow 5 50
Hull	slow 1 20	Shields (North) ..	slow 5 46
Huntingdon	slow 0 44	Shields (South) ..	slow 5 37
Ilfracombe	slow 16 28	Shoreham	slow 1 8
Ipswich	fast 4 40	Shrewsbury	slow 10 56
Isle of Wight (Newport)	slow 5 5	Southampton	slow 5 36
Jersey (St. Helier's) ..	slow 8 28	Southport	slow 12 0
Kendal	slow 10 40	Stafford	slow 8 30
Kew Observatory ..	slow 1 15	Staleybridge	slow 8 15
Kidderminster ..	slow 9 0	Stamford	slow 1 55
King's Lynn	fast 1 40	Stockport	slow 8 40
Lancaster	slow 11 5	Stockton-on-Tees ..	slow 5 10
Leamington	slow 6 0	Stroud	slow 8 50
Leeds	slow 6 4	Sunderland	slow 5 28
Leicester	slow 4 33	Swansea	slow 15 40
Leominster	slow 11 0	Taunton	slow 12 25
Lewes	fast 0 5	Tavistock	slow 16 35
Lichfield	slow 7 10	Teignmouth	slow 13 46
Lincoln	slow 2 6	Tiverton	slow 14 0
Liskeard	slow 17 50	Torquay	slow 20 10
Liverpool	slow 12 0	Truro	slow 14 0
Llanelly	slow 16 40	Tunbridge Wells ..	fast 1 4
Louth	.. — —	Wakefield	slow 5 50
Macclesfield	slow 8 30	Walsall	slow 7 55
Maidstone	fast 2 10	Wareham	slow 8 25
Malvern (Great) ..	slow 9 16	Warrington	slow 10 20
Manchester	slow 8 52	Warwick	slow 6 15
Margate	fast 5 32	Wednesbury	slow 8 10
Merthyr Tydfil ..	slow 13 30	Wenlock (Much) ..	slow 10 10
Middlesborough ..	slow 4 56	Westbury	slow 9 30
Monmouth	slow 10 56	Weston-super-Mare ..	slow 11 54
Montgomery	slow 12 40	Weymouth	slow 9 45
Morpeth	slow 6 46	Whitby	slow 2 24
Newcastle	slow 6 24	Whitehaven	slow 14 24
Northampton	slow 3 32	Wigan	slow 10 30
Norwich	fast 5 12	Winchester	slow 5 20
Nottingham	slow 4 30	Windsor	slow 2 25
Oakham	slow 2 50	Wisbech	fast 0 45
Oldham	slow 8 30	Wolverhampton ..	slow 8 25
Oxford	slow 5 2	Woodstock	slow 5 52
Pembroke	slow 19 40	Worcester	slow 8 52
Penzance	slow 22 10	Worthing	slow 1 30
Peterborough	slow 0 56	Yarmouth	fast 7 0
Plymouth	slow 16 30	York	slow 4 16
Poole	slow 7 50		
Portsmouth	slow 4 24	**SCOTLAND.**	
Preston	slow 10 50	Aberdeen	slow 8 20

		M.	S.
Ayr	slow	18	36
Banff	slow	10	5
Dumbarton	slow	18	16
Dumfries	slow	14	24
Dundee	slow	11	52
Edinburgh	slow	12	44
Elgin	slow	13	20
Forfar	slow	11	20
Glasgow	slow	17	20
Greenock	slow	19	1
Inverness	slow	16	54
Kilmarnock	slow	18	0
Kirkcaldy	slow	18	0
Kirkwall	slow	11	48
Leith	slow	12	36
Montrose	slow	9	52
Paisley	slow	17	40
Perth	slow	13	40
Stirling	slow	15	46
Wick	slow	12	26

IRELAND.

Note.—Dublin time is kept throughout Ireland.

		M.	S.
Dublin	slow	25	22
Armagh	slow	26	36
Bandon	slow	34	48
Belfast	slow	23	46
Cork	slow	33	56
Downpatrick	slow	22	52
Drogheda	slow	25	20
Dundalk	slow	25	30
Enniskillen	slow	30	40
Galway	slow	36	12
Kilkenny	slow	29	0
Limerick	slow	34	30
Lisburn	slow	24	8
Londonderry	slow	29	20
Queenstown	slow	33	0
Sligo	slow	33	52
Waterford	slow	28	30
Wexford	slow	25	56
Wicklow	slow	24	8
Youghal	slow	31	24

OTHER COUNTRIES.

Special Time Standards are established in Europe, America and Australasia. (See article on "Time," page 400.)

		H.	M.	S.
Adelaide	fast	9	14	30
Alexandria	fast	1	58	20
Algiers	fast	0	12	10
Amsterdam	fast	0	19	33
Athens	fast	1	34	55
Auckland	fast	11	39	4
Barbados (Bridgtown)	slow	3	59	0
Berlin	fast	0	53	35
Bombay	fast	4	51	36
Boston, U.S.A.	slow	4	42	0
Brisbane	fast	10	11	40
Brussels	fast	0	17	29
Buenos Ayres	slow	3	53	30

		H.	M.	S.
Cairo	fast	2	5	6
Calais	fast	0	7	28
Calcutta	fast	5	53	46
Canton	fast	7	32	56
Cape Town	fast	1	13	54
Chicago	slow	5	50	38
Christiania	fast	0	52	54
Colombo	fast	5	19	24
Constantinople	fast	1	56	0
Copenhagen	fast	0	50	19
Geneva	fast	0	24	37
Gibraltar	slow	0	21	22
Hamburg	fast	0	39	53
Havre	fast	0	0	26
Jamaica (Kingston)	slow	5	7	12
Jeddo	fast	9	19	0
Jerusalem	fast	2	20	56
Lima	slow	5	8	10
Lisbon	slow	0	36	35
Madeira	slow	1	7	36
Madras	fast	5	20	57
Madrid	slow	0	14	45
Malta	fast	0	58	0
Melbourne	fast	9	39	54
Mexico	slow	6	36	0
Milan	fast	0	36	46
Montreal	slow	4	54	30
Moscow	fast	2	30	17
Munich	fast	0	46	26
Naples	fast	0	56	59
New York	slow	4	56	0
Palermo	fast	0	53	24
Paris	fast	0	9	21
Pekin	fast	7	45	52
Philadelphia	slow	5	0	40
Pietermaritzburg	fast	2	0	0
Quebec	slow	4	45	0
Rangoon	fast	6	25	10
Rio Janeiro	slow	2	52	36
Rome	fast	0	49	54
St. Helena	slow	0	25	10
St. John, N.B.	slow	4	24	24
St. John, Newfoundland	slow	3	30	52
St. Petersburg	fast	2	1	13
San Francisco	slow	8	10	0
Shanghai	fast	8	5	20
Singapore	fast	6	55	0
Stockholm	fast	1	12	14
Suez	fast	2	10	0
Sydney	fast	10	5	0
Teheran	fast	3	25	32
Tripoli	fast	0	52	44
Tunis	fast	0	40	44
Venice	fast	0	49	25
Vienna	fast	1	5	31
Warsaw	fast	1	24	7
Washington	slow	5	8	11
Wellington, N.Z.	fast	11	39	14

TABLE GIVING ONE CONVERSION WHERE THE SCAPE WHEEL OF VERTICAL HAS 13 TEETH, AND THE SCAPE WHEEL OF LEVER IS TO HAVE 15 TEETH. THE THICK FIGURES REFER TO THE VERTICAL.

[It will be seen that the number of scape pinion teeth is the same after as before conversion, but as a higher numbered pinion may be desired, it is well to know that any vertical train having 13 in scape wheel and a scape pinion of 6 may be converted into a suitable lever train by substituting a scape wheel of 15 and a scape pinion of 7. The effect will be to diminish the original train by one 91th part of its whole number, which in any train not exceeding 18000 will never amount to more than 198. Also, any train having 13 in scape wheel and scape pinion 7, would, by substituting a scape wheel of 15 and scape pinion of 8, have the original train increased by one 104th of its whole number, which in trains not exceeding 18000 could never exceed 173.]

Centre Wheel.	Third Wheel.	Fourth Wheel.	Third Pinion.	Fourth Pinion.	Scape Pinion.	Vibrations per hour.	No. of seconds for one revolution of fourth wheel.
52	52	52	6	6	6	16925*	47·92
...	...	45	...	...	..	16900	
54	50	50	6	6	6	16250	48·
...	...	43	...	...	...	16125	
54	52	48	6	6	6	16224	46·15
...	...	41	...	...	...	15990	
54	52	50	6	6	6	16900	46·15
...	...	43	...	..	...	16770	
54	52	51	6	6	6	17238	46·15
...	...	44	...	...	...	17160	
54	52	52	6	6	6	17576	46·15
...	...	45	...	...	...	17550	
54	53	50	6	6	6	1725	45·28
...	...	43	...	...	...	17092*	
55	51	51	6	6	6	17219*	46·20
...	...	44	...	...	:	17141*	
56	50	50	6	6	6	16851*	46·28
...	...	43	...	..	...	16722*	

Centre Wheel.	Third Wheel.	Fourth Wheel.	Third Pinion.	Fourth Pinion.	Scape Pinion.	Vibrations per hour.	No. of seconds for one revolution of fourth wheel.
56	52	50	6	6	6	17525*	44·50
...	...	43	...	...	...	17391*	
56	54	48	6	6	6	17472	42·85
...	...	41	...	...	...	17220	
58	48	52	6	6	6	17425*	46·55
...	...	45	...	...	...	17400	
58	48	50	6	6	6	16755*	46·55
...	...	43	...	...	...	16626*	
58	50	50	6	6	6	17453*	44·69
...	...	43	...	...	...	17319*	
60	48	48	6	6	6	16640	45·
...	...	41	...	...	...	16400	
60	50	48	6	6	6	17333*	43·20
...	.	41	...	...	..	17083*	
60	50	46	6	6	6	16611*	43·20
...	...	39	...	...	...	16250	
60	54	52	7	6	6	17382*	46·66
...	...	45	...	...	...	17357*	

Centre Wheel	Third Wheel	Fourth Wheel	Third Pinion	Fourth Pinion	Scape Pinion	Vibrations per hour.	No. of seconds for one revolution of fourth wheel.
56	50	51 / 44	6	6	6	17188* / 17111* }	46″·28
56	52	48 / 41	6	6	6	16824* / 16582* }	44·50
64	52	50 / 43	7	6	6	17168* / 17036* }	45·43
60	56	56 / 48	7	7	6	16640* / 16457* }	52·50
60	58	56 / 48	7	7	6	17234* / 17044* }	50·69
60	60	56 / 48	7	7	6	17828* / 17632* }	49·
62	56	56 / 48	7	7	6	17194* / 17005* }	50.80
63	60	52 / 45	7	7	6	17382* / 17357* }	46·66
64	60	50 / 43	7	7	6	16979* / 16848* }	45·93
63	60	60 / 52	7	7	7	17191* / — }	46·66
64	60	60 / 52	7	7	7	17464* / — }	45·93
65	62	58 / 50	7	7	7	17717* / 17623* }	43·77
68	60	56 / 48	7	7	7	17319* / 17128* }	43·23
70	60	56 / 48	7	7	7	17828* / 17632* }	42·
72	56	56 / 48	7	7	7	17115* / 16927* }	43·75

Centre Wheel	Third Wheel	Fourth Wheel	Third Pinion	Fourth Pinion	Scape Pinion	Vibrations per hour.	No. of seconds for one revolution of fourth wheel.
60	60	48 / 41	7	6	6	17828* / 17571* }	42″·
63	52	51 / 54	7	6	6	17238 / 17160 }	46·15
60	54	60 / 52	8	6	6	17550 / — }	53·33
60	58	56 / 48	8	6	6	17593* / 17400 }	49·65
60	60	54 / 46	8	6	6	17550 / 17250 }	48·
64	58	50 / 43	8	6	6	16755* / 16626* }	46·55
72	52	51 / 44	8	6	6	17238 / 17160 }	46·15
70	60	52 / 45	8	7	6	16900 / 16875 }	48·
70	60	50 / 43	8	7	6	16250 / 16125 }	48·
70	60	54 / 46	8	7	6	17550 / 17250 }	48·
72	64	56 / 48	8	7	7	17115* / 16927* }	43.75
64	64	72 / 62	8	8	7	17115* / 17005* }	56·25
70	66	64 / 55	8	8	7	17160 / 17015* }	49·87
72	64	64 / 55	8	8	7	17115* / 16971* }	50·
72	66	63 / 54	8	8	7	17374* / 17183* }	48·48

* These trains give a fraction more than the number of vibrations stated.

French Equivalents.

Accélération, acceleration.
Acier, steel.
Acier à pignons, pinion wire.
Adouci ondé, wavy surfacing.
Adouci mat, frosting.
Adouci moiré, wavy surfacing.
Adoucissage, greying, smoothing.
Adoucissage en colimaçon, snailing.
Adoucissage en taches, spotting.
Aiguilles, hands.
Aiguille de cadran solaire, gnomon.
Aiguille de raquette, pointer of index.
Aiguille de secondes, seconds hands.
Aile de pignon, pinion leaf.
Alidade, index.
Alésoir, round broach.
Ancre, pallet.
Ancre à chevilles, pin pallet escapement.
Ancre à lévee visibles, pallets with visible stones.
Ancre à rateau, rack lever escapement.
Ancre ligne droite, straight line lever.
Angle, angle.
Anneau, watch bow.
Arbre, arbor, turning arbor.
Arbre à vis, screw arbor.
Arbre excentrique, eccentric arbor.
Arbre tournant, live spindle.
Archet tournant, live spindle.
Archet, bow (stick and string).
Argent, silver.
Arc-boutement, butting.
Arrêtage, stop work.
Arrête, cutting edge.
Arrondi, addendum.
Axe, staff.
Axe de balancier, balance staff.

Balancier, balance (2) fly press.
Balancier compensé, compensation balance.
Baleine, whalebone.
Banc, bed.
Barette, bar, arm of wheel.
Barette à nom, name bar.
Barillet, barrel.
Barillet denté, going barrel.
Barillet suspendu, hanging barrel.
Barillet tournant, going barrel.
Bascule, pivoted detent.
Bati, body.
Bavure, burr.
Bec, nose, beak.

Belière, watch bow.
Benzine, benzine.
Bigorne, stake with beak.
Biseau, chamfer, bevel.
Bleuir, bluing.
Bloc à polir, polishing stake.
Bocfil, saw frame.
Bois carré, peg wood.
Bois de fusain, peg wood, charcoal wood.
Boîte, case.
Boîte à double fond, double bottom case.
Boîte double, pair case.
Boîte à glace plate, crystal case.
Boîte à musique, musical box.
Bonde, barrel arbor collet.
Boucher, to bush.
Bouchon, bouchon.
Boulon, bolt.
Bourrellet, flange.
Boursonflure, blister.
Braser, to braze.
Bride, collar.
Broche, runner.
Bronzage, bronzing.
Brosse, brush.
Brucelles, tweezers.
Brucelles pour joyaux, corn tongs.
Brunissoir, burnisher.
Burin, graver.
Burin fixe, slide rest.
Butoir, stop.

Cabinet d'horloge en marbre, marble clock case.
Cabron, buff.
Cadran, dial.
Cadran à centre rapporte, double sunk dial.
Cadran en trotteuse, seconds dial.
Cadran indicateur, set hands dial.
Cadran solaire, sun dial.
Cadre, frame.
Cage, frame.
Calendrier, calendar.
Calibre, calliper.
Calibre à coulisse, slide gauge.
Calibre aux carres, key pipe gauge.
Callotte, cap, dust cap.
Canon, pipe.
Capillarité, capillarity.
Carillon, chimes.
Carré, square.
Cassant, brittle.

Centre de gravité, centre of gravity.
Centre de rotation, centre of gyration.
Centre d'oscillation, centre of oscillation.
Chaîne, chain.
Chalumeau, blowpipe.
Champ, rim of wheel.
Champignon, taper washer.
Chanfrein, chamfering tool.
Chapeau, cap, dust pipe.
Chapeau pour fusée, fusee piece.
Chariot, chariot.
Charnerons, knuckles.
Charnière, joint.
Chaton, setting.
Chaussée, cannon pinion.
Chemin perdu, run (lever escapement).
Cheville, pin.
Cheville de bois carré, stick of pegwood.
Cheville d'impulsion, impulse pin.
Cheville de plateau, ruby pin, impulse pin.
Cheville, set hands arbor.
Chiffre d'huit, double ended callipers
Chronographe, chronograph.
Chronomètre, chronometer, an exact timekeeper.
Chronomètre de marine, marine chronometer.
Chronomètre de poche, pocket chronometer.
Chute, drop.
Ciment, cement.
Cire, wax.
Cisalles, shears.
Clavette, key.
Clef, key.
Clef de spiral, balance spring guard.
Clepsydre, clepsydra.
Cliquet, click.
Cloche, large bell.
Coche, notch.
Cœur, heart piece.
Colimaçonnage, snailing.
Collecteur, gathering pallet.
Collet, collet.
Compas aux engrenages, depthing tool.
Compas de proportion, sector.
Compensateur, compensation curb.
Compensation auxiliaire, auxiliary.
Compteur, dividing plate.
Contre-charnière, bearer (watch case)
Contre-poulet, lock nut.
Contrôle, hall mark.

Conversion, conversion.
Copeaux, shavings.
Coq, cock.
Coq de balancier, balance cock.
Coudé, overcoil.
Coulisse, slide.
Coup, beat.
Coup de feu, scale.
Couronne, button.
Couvercle, cover.
Crampon, cramp.
Cran de fermeture, snap.
Crémaillère, cremaillere.
Creuset, crucible.
Creuseure, hollow, sink.
Coqueret, endstone cap.
Corde à boyant, catgut.
Crin, horsehair.
Crique, crack.
Crochet, hook.
Crochet de chaîne, chain hook.
Crochet de fusée, fusee cap.
Crochet de ressort, mainspring hook.
Croiser, to cross out.
Croix de Malte, Maltese cross, stop work.
Cuivre, copper.
Cuivrot, ferrule (grooved wheel).
Cuivrot à vis, screw ferrule.
Cuvette, dome.
Cycloïde, cycloid.

Dard, guard pin, finger piece.
Dégrenage en fermant la boîte, shutting off (keyless work).
Demi chronomètre, half chronometer.
Demi platine, half plate.
Dent à talon, club tooth.
Détente, detent.
Détente, warning piece.
Détentillon, lifting piece.
Deux platines, full plate.
Diamantine, diamantine.
Dipléidoscope, dipleidoscope.
Distance des dentures, pitch.
Doigt, finger piece.
Doigt d'arrêtage, star finger.
Doigt de levée, cam.
Dorage, gilding.
Doublet-filet, double-bead.
Dragenoir de barillet, barrel groove.

Ebarber, to trim up.
Ébauche, a piece stamped or otherwise roughed out.
Ébiseler, to chamfer.
Ébiseloir, chamfering tool.

Ébiseloir creux, hollowing chamfering tool.

Échappement, escapement.

,, *à ancre*, anchor escapement, lever escapement.

,, *à coup perdu*, single beat escapement.

,, *à cylindre*, cylinder escapement.

,, *à demi repos*, half dead escapement.

,, *à détente*, chronometer escapement.

,, *à double rouleau*, double roller escapement.

,, *à friction*, frictional escapement.

,, *à gravité*, gravity escapement.

,, *à goupilles*, two pin escapement.

,, *à recul*, recoil escapement.

,, *à repos*, dead beat escapement.

,, *avec roue à goupilles*, pin wheel escapement.

,, *à verge*, verge escapement.

,, *Denison*, double three-legged gravity escapement.

,, *duplex*, duplex escapement.

,, *libre*, detached escapement.

,, *résilient*, resilient escapement.

Échoppe, cutter.

Écorce, shell

Écrou, nut.

Écrou à oreille, thumb screw or nut.

Écrouissage, cold hammering.

Émail, enamel.

Émeri, emery.

Émousser, to blunt.

Empierrage, jewelling.

Empierré, jewelled.

Encliquetage, click-work.

Enclume, stalk.

Entaille, notch.

Entretien, maintaining power.

Épicycloide, epicycloid.

F F

Équation, equation of time.

Équerre, square.

Équilibre, swing stool.

Erreur de temperature moyenne, middle temperature error.

Estrapade, mainspring winder.

Établi, bench.

Étain, tin.

Étainer, to tin.

Étau, vice.

Étau à main, hand vice.

Étau à queue, pin vice.

Étirer, to draw.

Étoile, star wheel.

Étrier pour pendule mercuriel, stirrup (pendulum).

Étuve, oven.

Éxcentrique pour équation, kidney piece.

Fer, iron.

Fer à souder, soldering iron.

Filet de vis, screw threads.

Filière, wire gauge.

Filière à coussinets, stocks and dies.

Filière à pivots, pivot gauge.

Filière à tirer, draw plate.

Filière aux vis, screw plate.

Filière pour cylindres, cylinder gauge

Fondre, to melt.

Force centrifuge, centrifugal tendency

Force constant, remontoire.

Foré, drilled.

Foret, drill.

Forgeage, hammer hardening.

Former, to shape.

Fouet, flirt.

Fourchette de pendule, crutch.

Fractions décimales, decimal fractions.

Fraise, cutter, milling cutter.

Fraise à crochet, fly cutter.

Fraise aux noyures, sinking tool.

Fraise aux novures de vis, screwhead sinking tool.

Fraise couteau, fly cutter.

Fraise Ingold, ingold fraise.

Fraise-lime, milling cutter, filling machine.

Fraiseuse, milling machine, shaping machine.

Friction, friction.

Frottement, friction.

Fusée, fusee.

Fusée à carré percé, hollow fusee.

Fusée d'entretien, going fusee.

Fusée et barillet transposés, reversed fusee, left-handed fusee.

Gauchir, to distort.
Gercure, flaw.
Glacière, ice box.
Gomme laque, shellac.
Goupille, pin.
Goupille renverseèment, banking pin.
Goupilles de raquette, curb pins.
Grains d'orge, barleys.
Grammaire, dividing plate.
Grand platine, pillar plate.
Grappe, rack hook.
Grattebrosse, scratch brush.
Grattoir, scraper.
Gravité, gravity.
Griffe, dog, clamp.
Griffes de serrage, dogs, clamps.
Guichet, demi hunter, cut hunter.
Guides cycloidaux, cycloidal cheeks.
Guillochis, engine-turning.

Heure, hour.
Heure sidérale, sidereal time.
Heure solaire, solar time.
Horloge, clock.
Horloge à carillon, chiming clock.
Horloge à maree, tidal clock.
Horloge à 3 rouages, three part clock.
Horloge à quarts, quarter clock.
Horloge à 2 timbres, ting-tang clock.
Horloge de clocher, turret clock.
Horloge de voiture, carriage clock.
Horloge équatoriale, equatorial clock.
Horloge électrique, electric clock.
Horloge pneumatique, pneumatic clock.
Huile, oil.
Huiler, oil sink.
Huit-chiffre, double ended callipers.
Hypocycloide, hypocycloid.

Impulsion, impulse.
Indicateur de haut et bas, up and down indicator.
Inertie, inertia.
Involute, involute.
Isochrone, isochronous.
Isolateur, isolater.

Jeu, end shake.
Jour, play, freedom.
Jumelle, gut-band hook.

Laiton, brass.
Lame, blade, saw blade.
Lame de bocfil, frame saw.

Laminoir, flattening mill.
Lampe, lamp.
Lampe à esprit, spirit lamp.
Lanterne, lantern.
Lapidaire, lap.
Lacque, lacquer.
Lentille, pendulum bob.
Levée de duplex, duplex hook.
Levée du balancier, balance arc.
Levée entrant, entering pallet.
Levée sortante, exit pallet.
Levier, lever, fly spring.
Levier à équerre, bell crank lever.
Levier à fusée, adjusting rod.
Lèvre, lip.
Ligne des centres, line of centres.
Ligne d'engrenage, pitch line.
Limaçon, snail.
Limaçon des quarts, quarter snail.
Limaille, filings.
Lime, file.
Lime à bord non taillé, safe-edge file.
Lime à carrelotte, potence file.
Lime à émeri, emery stick.
Lime à fendre, slitting file.
Lime à pilier, pillar file.
Lime à pivots, pivot file.
Lime à sabot, bull's-foot file.
Lime à 3 coins, three-cornered file.
Lime barrette, ridged-back file.
Lime en diamant, diamond file.
Lime queue de rat, rat-tail file.
Lime très fine, dead smooth file.
Limon de pierre noire, coomb.
Loupe, eye-glass.
Lunette, bezel, cone plate for lathes.
Lunette méridienne, transit instrument.

Machine à arrondir, rounding-up tool.
Machine à fendre, wheel cutting engine.
Machine à tailler, wheel cutting engine.
Magnétisme, magnetism.
Manchon, collar.
Mandrin, pin tongs (2) chuck.
Mandrin à vis, bell chuck with screws.
Mandrin à serrage concentrique, self-centring chuck.
Manivelle, cranked handle.
Manquement, tripping.
Marteau, hammer.
Martelage, forging.
Martinet, sledge hammer.

483

Masse, mass.
Mat, matted.
Matrice, die.
Mèche de foret, blade of drill.
Mercure, mercury.
Mesure à piston, pump cylinder.
Mesure aux pignons, pinion gauge.
Meule, mill, grindstone.
Meule chargée de diamant, diamond mill.
Meule lime, filing mill.
Micromètre, micrometer.
Minute, minute.
Minuterie, motion work.
Moëlle de sureau, pith.
Moiré circulaire, circular surfacing.
Molette, mill, milling tool, knurling tool.
Montre, watch.
Montre à grand sonnerie, clock watch
Montre à répetition, repeater.
Montre à secousses, pedometer watch.
Montre non magnetisable, non-magnetizable watch.
Montre pour les aveugles, blind man's watch.
Montre savonette, hunting watch.
Moulure, groove.
Mouvement, movement.
Mouvement Lépine, Lepine movement.

Nez, nose, beak.
Noyer, to cut out a sink.
Noyure de fusée, fusee sink.

Or, gold.
Oreille, wing.
Oreille du coq, wing of cock.
Outil, tool.
Outil à équilibrer, poising tool.
Outil à mesurer, gauge.
Outil à planter, uprighting tool.
Outil à tailler leg fusées, fusee engine.
Outil aux douzièmes, douzieme gauge.
Outil aux têtes de vis, screw-head tool.
Outil pour hauteur de pignons, pinion height tool.
Outil pour hauteur de cylindres, cylinder height tool.

Palette de recul, pad.
Pan, facet.
Panne, pane (of hammer).

Parachûte, parachute.
Pas de vis, pitch of screw.
Passage de fourchette, crescent, passing hollow.
Passage de renversement, overbanking.
Peau chargée de rouge, rough leather.
Peau de chamois, chamois leather.
Pédale, treadle.
Pédomètre, pedometer.
Pendant, pendant.
Pendant tournant, turning pendant.
Pendule, pendulum, pendulum clock.
Pendule à torsion, torsion pendulum.
Pendule aux guichets, chronoscope.
Pendule compensé, compensated pendulum.
Pendule conique, conical pendulum.
Pendule (ou montre) de contrôle, tell-tale clock.
Pendule de navire, ship's timepiece.
Pendule sans sonnerie, timepiece.
Percage, drilling.
Perce-droit, table tool.
Perceuse, drilling machine.
Perche, bar of turns or lathe.
Pèse spiral, balance spring gauge.
Pièce aux quarts, quarter rack.
Pied, steady pin.
Pied de minute, minute wheel stud.
Pied de cadran, dial foot.
Pierre à contre pivot, endstone.
Pierre à eau, bluestone.
Pierre à émeri, emery file.
Pierre à huile, oilstone.
Pierre à huile pilée, oilstone dust.
Pierre de levée, pallet stone.
Pierre de touch, touch stone.
Pierre pouce, pumice stone.
Pierres précieuses, precious stones.
Pignon, pinion.
Pignon à lanterne, lantern pinion.
Pignon d'échappement, escape pinion.
Pignon de renvoi, minute wheel pinion.
Pignon de volant, fly pinion.
Pignon ivrogne, sliding pinion.
Pignon percé, hollow pinion.
Pilier, pillar.
Pince à boucle, sliding tongs.
Pince à couper, nippers.
Pince à percer les ressorts, mainspring punch.
Pince à ressort, spring chuck.
Pince à tirer, wire drawing tongs.
Pince de tour aux vis, tongs for screwhead tool.

F F *

Pince fendre, split chuck.
Pince-platte, pliers.
Pinces, clams, pliers.
Piton, stud.
Pivot, pivot.
Pivot conique, conical pivot.
Pivot de secondes, seconds pivot.
Planer, to planish.
Plantage, uprighting.
Plateau à griffes, face plate for dogs.
Plateau à toc, carrier plate.
Plateau simple, table roller.
Platforme bercante, rocking bar, yoke.
Platforme divisée, diving plate.
Platine, plate.
Plattes de place, glass plates.
Plomb, lead.
Poids, weight.
Poinçon, punch.
Poinçon à chiffres, chapters.
Poinçon à coude, knee punch.
Poinçon creux, hollow punch.
Poinçoneuse, power press.
Polissage, polishing.
Pompe à centrer, pump centre.
Ponte, bridge cock.
Portée, shoulder.
Porte foret, drill stock.
Potance, potance.
Poulie, pulley.
Pompée, poppet head, back centre.
Pousse goupille, joint pusher.
Poussitte, push-piece.
Première roue, great wheel.

Queue d'aigle, dovetail.
Queue de marteau, hammer tail.

Rainure, recess, slit.
Raquette, index, regulator.
Rateau, rack.
Rateau d'heures, hour rack.
Rebattement, banking.
Rebord, edge border.
Recuit, annealing.
Réglage, timing.
Régulateur, regulator.
Remontoir, keyless winding work.
Remontoir au pendant, stem winding.
Renvoi, idle wheel, intermediate wheel.
Renvois, countershaft, intermediate spindle.
Repassage, examining, finishing.
Repos, locking.

Résistance au dégagement, unlocking resistance.
Resserre chaussé, cannon pinion tightener.
Ressort, spring.
Ressort d'encliquetage, click spring.
Ressort de fermeture, locking spring.
Ressort de suspension, pendulum spring.
Ressort de crochetment, all or nothing piece.
Ressort d'or, gold spring.
Ressort moteur, mainspring.
Ressort spiral, balance spring.
Ressorts de carrure, case springs.
Ressorts secrets, secret springs.
Retour d'équerre, bracket.
Réveil, alarum.
Revenir, to temper.
Revenoir, bluing pan.
Revenu, tempered.
Rhabillage, repairing.
Rivoir, riveting punch.
Rochet, ratchet.
Rochet des heures, repeating rack.
Rognure, scrap, shavings.
Roue, wheel.
Roue à goupilles, pin wheel.
Roue croisée, crossed-out wheel.
Roue de centre, centre wheel.
Roue de champ, fourth wheel.
Roue de compte, locking plate.
Roue d'échappement, escape wheel.
Roue d'heures, hour wheel.
Roue de renvoi, minute wheel, intermediate wheel.
Roue de rencontre, crown wheel.
Roue de toto, warning wheel.
Roue petite moyenne, third wheel.
Roues coniques d'égale grandeur, mitre wheels.
Rouage, train.
Rouage à gauche, left-handed movement.
Rouge, red-stuff.
Rouille, rust.
Rouleau, roller, barrel.
Rouleau de carillon, chiming barrel.
Rouleau de duplex, duplex roller.
Rude, coarse.

Sans clef, keyless.
Sautoir, jumper.
Scie, saw.
Scie en diamant, skive.
Seconde, second.
Seconde au centre, centre seconds.

Seconde indépendante, independent seconds.
Secondes rapportées, sunk seconds.
Secondes rattrappantes, split seconds.
Secteur, sector.
Seremontant de soi-même, self-winding.
Sertissage, jewel setting.
Sillon de fusée, fusee hollow.
Sommaire d'essai, trial number.
Sonnerie, striking work.
Soudage, soldering.
Soudure, solder.
Spirale à courbe Phillips, Phillips' spring.
Spirale coudé, overcoil spring.
Spirale sans raquette, free spring.
Spirale sous le balancier, sprung under.
Spirale sur le balancier, sprung over.
Spire, coil.
Support de mouvement, seat board.
Support pour pignons (outil), gallows tool.
Surprise, surprise piece.
Suspension de chronomètre, gymbals.

Tambour, drum.
Tampon de cylindre, cylinder plug.
Taquet, carrier.
Taraud, screw tap.
Taraudeuse, screw tapping machine.
Tas, anvil.
Tasseau, chuck.
Tasseau centrant de soi-même, self-centring chuck.
Tasseau fraise à trou, rose cutter.
Temps, time.
Temps moyen, mean time.
Temps perdu, backlash.
Tenaille pour ressort de suspension, chops.
Tenon de minuterie, minute wheel stud.
Terre Pourrie, Tripoli.
Tête de raquette, curb pin arm of index.

Tetine, nipple, the cone formed at the bottom of a hole in drilling.
Tige, staff, rod.
Tige de remontoir, winding stem.
Tige d'ancre, pallet staff.
Tige de carre, set hands arbor.
Timbre, gong, bell.
Tirage, draw.
Tirage de répétition, repeating slide.
Toc, carrier.
Tour, lathe, turns.
Tour à fileter, screw-cutting lathe.
Tour à gommer, wax lathe.
Tour à renvoi à corde, throw.
Tour Jacot, Jacot tool.
Tour universel, mandril.
Tourbillon, tourbillion.
Tourne-virole, balance spring collet tool.
Tourne vis, screw driver.
Tranchant, cutting edge.
Trempé, hardened.
Tremper, to harden.
Trois quarts de platine, three-quarter plate.
Trou, hole.
Trou en pierre, jewel hole.

Vélocité, velocity.
Verge, verge.
Verges de connexion, leading off rods.
Verges de marteaux, hammer rods.
Vernier, vernier.
Vernier à coulisse, vernier slide gauge.
Verre, glass.
Verre de montre ordinaire, lunette.
Virgule, virgule.
Virole, ferrule (metal ring).
Virole de spiral, balance spring collet.
Vis, screw.
Vis réglante, timing screw rating nut.
Volant, fly.

Wig wag, wig wag.

German Equivalents.

Abfall, drop.
Abgleichstange, adjusting rod.
Alles oder Nichts Sicherung, all or nothing piece.
Amboss, stake.
Anker, pallets.
Ankergang mit Rechen und Trieb, rack, lever escapement.
Ankerhemmung, lever escapement, anchor escapement.
Ankerhemmung in gerader Linie, straight line lever.
Ankerpalette (Pendeluhr), pad.
Ankerwelle, pallet staff.
Anlassblech, bluing pan.
Anlassen, tempering.
Anlaufpalette, warning piece.
Anlaufrad, warning wheel.
Anlaufwippe, flirt.
Antimagnetische Taschenuhr, non-magnetizable watch.
Antrieb, impulse.
Anzug (Ankerhemmung), draw.
Arrondierscheibe, Zapfenrollierstuhl, lantern.
Aufhängungsfeder, suspension spring.
Aufsatz, chuck.
Aufsetzen in einem Eingriffe, butting.
Auf und Ab-Werk, up and down work.
Aufziehrechen (Repetitionsuhr), cremaillere.
Aufziehzapfen, winding square.
Auskochtiegel, boiling out pan.
Auslösunghebel, lifting piece.
Auslösungswiderstand, unlocking resistance.
Ausschnitt in der Rolle, passing hollow.
Ausschwung, overbanking.
Ausschwungstift, banking pin (verge or cylinder).

Beisszange, nippers.
Benzin, benzine.
Blauanlassen, bluing.
Blei, lead.
Bohren, drilling.
Bohrer, drill.
Bréguet Spiralfeder, overcoil.
Bügelaufzug, stem winding.
Bürste, brush.

Chronograph, chronograph.
Chronometerhemmung, chronometer escapement.
Chronometer, chronometer.
Cremaillere, cremaillere.
Cycloide, cycloid.
Cycloidische Backen, cycloidal cheek.

Decimalbrüche, decimal fractions.
Deckstein, end stone.
Diamantfeile, diamond file.
Diamantine, diamantine.
Diamantscheibe zum Steinschneiden, skive.
Diamantschleifscheibe, diamond mill.
Dipleidoscop, dipleidoscope.
Doppelchronograph, split seconds.
Doppelrand (Gehäuse), double-head.
Drahtmass, wire gauge.
Drehbank, lathe.
Drehbank mit Lackscheibe, wax lathe.
Drehbarer Bügelknopf, turning pendant.
Drehbogen, bow (stick and string).
Drehstift, turning arbor.
Drehstuhl, turns.
Dreiviertelplatine, three-quarter plate.
Druckknopf, push piece.
Duplexfinger, duplex hook.
Duplexhemmung, duplex escapement.
Duplexrolle, duplex roller.

Ebbe und Flutuhr, tidal clock.
Edelsteine, precious stones.
Einfallschnalle, rack hook.
Eingriffslinie, pitch line.
Eingriffzirkel, depthing tool.
Einsatzbohrer, drill stock.
Einseitig unterstütztes Federhaus, hanging barrel.
Einsprengen, snap.
Eisschrank, ice box.
Eisen, iron.
Elektrische Uhren, electric clocks.
Email, enamel.
Endluft, endshake.
Erweichung durch Hitze, annealing.
Excenter, snail.
Excentrischer Zierschliff, snailing.
Excentrischer Drehstift, eccentric arbor.

Excentrische Scheibe (Equationsuhren), kidney piece.

Federbacken (Pendel), chops.
Federhaken, barrel hook, mainspring hook.
Federhaus, barrel.
Federhausdeckel, barrel cover.
Federhausausdrehung, barrel hollow.
Federlochzange (Zugfeder), mainspring punch.
Federstift, barrel arbor.
Federwinder, mainspring winder.
Federzirkel, spring callipers.
Feile mit feinstem Hieb, dead smooth file.
Feile ohne Seitenhieb, safe-edged file.
Feilkloben, hand vice.
Feingehaltsstempel, hall mark.
Feuerhärtung, hardening.
Fischbein, whalebone.
Flachzange, pliers.
Fliedermark, pith.
Fräse, cutter, milling cutter.
Freie Hemmung, detached escapement.
Futter, bouchon.

Gang, escapement.
Gangbeschleunigung natürliche, acceleration.
Gangfeder des Chronometers, detent.
Gangkloben, potence.
Gangnummer, trial number.
Gangrad, escape wheel.
Gangrad einer Spindeluhr, crown wheel.
Gangradstrieb, escape pinion.
Gegengesperr, maintaining power.
Gestell, frame.
Gehäuse, watch case.
Gehäusefedern, case springs.
Gehäuseknopf, pendant.
Gehäuse mit doppeltem Boden, double bottom watch case.
Gehäuserand, middle (watch case).
Geradbohrmaschine, table tool.
Geradestellen, uprighting.
Geradesteller, uprighting tool.
Geschwindigkeit, velocity.
Gezahntes Federhaus, going barrel.
Gips, plaster of Paris.
Glasreif, bezel.
Glättahle, round broach.
Glocke, bell.
Glockenspiel, chimes.

Glockenzughebel, bell-crank lever.
Gold, gold.
Goldfeder, gold spring.
Grahamhemmung, dead-beat escapement.
Grossbodenrad, centre wheel.
Guilloche, engine-turning.

Halbplatinenwerk, half plate.
Halbsavonette, *Taschenuhr*, demi-hunter, cut hunter.
Hammer, hammer.
Hammerhebel, hammer tail.
Hammerleitung (bei Turmuhren), hammer rods.
Handschwungrad, handwheel.
Handschwungrad, *englisches für Grossuhrmacher*, throw.
Hebescheibe, cam.
Hebestein (Anker), pallet stone.
Hebestein, ruby pin, impulse pin.
Hebestiftrad, pin wheel.
Hebungsflächen (Zylinder), lips.
Hebungsflächen, impulse faces.
Hemmung, escapement.
Hemmung mit einem verlorenen Schlage, single beat escapement.
Hemmung mit unvollkommener Ruhe, half dead escapement.
Hemmung von Savage, two-pin escapement.
Herz, heart piece.
Höhenmass für grosse Arbeit, pump cylinder.
Hohlfräse, rose cutter.
Hohltrieb, hollow pinion.

Indicationszifferblatt, set hands dial.
Ingoldfräse, Ingold fraise.
Involute, involute.
Isolationshebel, isolator.

Kalenderuhr, calendar clock.
Kapillarität, capillarity.
Kaliber, calliper.
Kapsel, cap.
Kette, chain.
Kettenhaken, chain hook.
Kitt, cement.
Klammerdrehbank, mandril.
Kleinbodenrad, third wheel.
Kloben, cock.
Klobenwerk, bar movement.
Krone (Aufzug), button.
Kolbenzahn, club tooth.

Kompensationsbogen (Rückerzeiger), compensation curb.
Kompensationsfehler, middle temperature error.
Kompensationspendel, compensated pendulum.
Kompensationsunruh, compensation balance.
Konstante Kraft, remontoire.
Konisches Pendel, conical pendulum.
Konischer Zapfen, conical pivot.
Kontrolluhr, tell-tale clock.
Körnung, matted.
Kornzange, tweezers.
Kronrad, contrate wheel.
Kupfer, copper.

Lack, lacquer.
Laternentrieb, lantern pinion.
Laufwerk, train.
Lederfeile, buff.
Linker Drehstift, screw arbor.
Linksangelegtes Uhrwerk, lefthanded movement.
Loch, hole.
Lot, solder.
Löten, soldering.
Lötrohr, blowpipe.
Luftdruckfehler, barometric error.
Lupe, eye-glass.

Mass, gauge.
Magnetismus, magnetism.
Malteserkreuz, Maltese cross.
Mantel, dome.
Marmorgehäuse, marble clock case.
Mattirung, frosting.
Messing, brass.
Messingfutter für den Schneckenzapfen, fusee brass.
Metallhärtung mittelst des Hammers, hammer hardening.
Mikrometer, micrometer.
Millimeter, millimeter.
Minute, minute.
Minutenrohr, cannon pinion.
Mitnehmer, carrier.
Mit Steinen versehen, jewelled.
Mittelpunktslinie, line of centres.
Momentum, momentum.

Namenplatte, name bar.
Normaluhr, regulator.

Oel, oil.
Oelgeber, oiler.

Oelsenkung, oil sink.
Oelstein, oilstone.
Ofen, für Temperatur Reglage, oven.

Parachute, parachute.
Pendel, pendulum.
Pendelgabel, crutch.
Pendellinse, pendulum bob.
Pendelschwingungsfehler, circular error.
Pfeiler, pillar.
Pfeilerplatine, pillar plate.
Pferdehaar, horsehair.
Platte, Platine, plate.
Pneumatische Uhren, pneumatic clocks.
Polieren, polishing.
Polierstahl, burnisher.
Polierstock, polishing stake.
Postament, seat board.
Präcisions Taschenuhr, half chronometer.
Prellen, banking.
Prellschraube, banking screw (chronometer).
Prellstifte, banking pins (lever).
Prellung, ausweichende (Ankerhemmung), resilient escapement.
Probierstein, touch stone.
Probiergestell, horse.
Proportionszirkel, sector.
Putzen, collet.
Putzholz, peg-wood.

Quecksilber, mercury.

Rad, wheel.
Rahmen (Quecksilberpendel), stirrup pendulum.
Rechen, rack.
Regulieren, timing.
Regulieren in den Lagen, timing in positions.
Regulierschraube (Pendel), rating nut.
Regulierschrauben, vier, der Unruh, quarter screws.
Reibung, friction.
Reiseuhr, carriage clock.
Remontoiruhr, self-winding watch (but commonly used for keyless watch)
Repetierrechen, repeating rack.
Repetieruhr, repeater.
Repetitionsschieber, repeating slide.
Rohr, pipe.
Rolle, ferrule.

Römische Zahlen, chapters.
Rost, rust.
Rotationsschwerpunkt, centre of gyration.
Rücker, index, regulator.
Rückerstifte, curb pins.
Rückfallende Hemmung, recoil escapement.
Ruhe, locking.
Rundfeile, rat-tail file.

Säge, saw.
Savonnetteuhr, hunter watch.
Scharnierstiftdrücker, joint pusher.
Scharnierteile, knuckles.
Scharnierträger, bearer (watch case).
Schenkelrad, crossed-out wheel.
Schiebzange, clams, sliding tongs.
Schiffsuhr, ship's timepiece.
Schlag, beat
Schlaguhr mit drei Federhäusern, Vierteluhr, three-part clock.
Schlagwerk, striking work.
Schleifglas, glass plates.
Schleifscheibe, lap, mill.
Schleifvorrichtung am Drehstuhl, swing tool.
Schliessfeder, locking spring.
Schlitten, chariot.
Schlosscheibe, locking plate.
Schlüssel, winder, key.
Schlüsselschraube, dog-screw.
Schmirgel, emery.
Schmirgelfeile, emery file.
Schmirgelrad, emery wheel.
Schnecke, fusee.
Schnecke mit versenktem Aufziehzapfen, hollow fusee.
Schneckenausdrehung, fusee sink.
Schneckenrad (Hauptrad), great wheel.
Schneckenschneidmaschine, fusee engine.
Schneckenzapfenunterdrehung, fusee hollow.
Schneidbohrer, screw tap.
Schneideisen, screw plate.
Schöpfer, gathering pallet.
Schräge, chamfer.
Schraube, screw.
Schraubenkopffeile, slitting file.
Schraubenpoliermaschine, screwhead tool.
Schraubenrolle, screw ferrule.
Schraubenzieher, screw-driver.
Schraubstock, vice.
Schrittzähler, pedometer.

Schublehre, slide gauge.
Schwalbenschwanz, dovetail.
Schwerkraft, gravity.
Schwerkrafthemmung, gravity escapement.
Schwingungsmittelpunkt, centre of oscillation.
Sekunde, second.
Sekunde aus der Mitte, centre seconds.
Sekundenrad, fourth wheel.
Sekundenzapfen, seconds pivot.
Sekundenzeiger, seconds hands.
Selbstschlag-Taschenuhr, clock watch.
Seechronometer, marine chronometer.
Selbstaufziehend, self-winding.
Senkerfeile, bull's-foot file.
Senkung im Zifferblatte für die Sekunde, sunk seconds.
Sicherheitsstift (messer), guard pin.
Sonnenuhr, sun dial.
Sonnenzeit, solar time.
Sonnenzeit mittlere, mean time.
Sperrkegel, click.
Sperrad, ratchet.
Spielwerk, musical box.
Spindel, verge.
Spindelhemmung, verge escapement.
Spirale mit theoretischen Endcurven, Phillips' spring.
Spiralfeder, balance spring.
Spiralfeder ohne Rücker, free spring.
Spiralfeder oberhalb der Unruh, sprung over.
Spiralfeder unterhalb der Unruh, sprung under.
Spiralschlüssel, balance spring guard.
Spiralzange, tweezers.
Spirituslampe, spirit lamp.
Spitzen (Drehstuhl), runners.
Springende Sekunde, independent seconds.
Springende Zahlen Uhr, chronoscope.
Springfeder, fly spring.
Stahlrot, red-stuff.
Standuhr, clock.
Staubhülse, dust cap.
Staubhütchen, dust pipe.
Steinarbeit, jewelled.
Steinloch, jewel hole.
Stellstift, steady pin.
Stellung, stop work.

Stellungshaken (Schnecke), fusee cap.
Stellungszahn, star finger.
Stern, star wheel.
Sternkegel (Schlaguhren), jumper.
Sternzeit, sidereal time.
Stichel, graver.
Stielkloben, pin vice.
Stift, pin.
Sonnenuhrzeiger, gnomon.
Stiften Ankerhemmung, pin pallet escapement.
Stiftengang, pin wheel escapement.
Stiftenwalze, chiming barrel.
Stiftenzange, pin tongs.
Stunde, hour.
Stundenrad, hour wheel.
Stundenrechen, hour rack.
Support, slide rest.

Tamponpunzen, knee punch.
Taschenchronometer, pocket chronometer.
Taschenuhr, watch.
Teilscheibe, dividing plate.
Teilung, pitch.
Tonfeder, gong.
Torsionspendel, torsion pendulum.
Tourbillon, tourbillion.
Trägheit, inertia.
Transit Instrument, transit instrument.
Trieb, pinion.
Triebhalter (Werkzeug), gallows tool.
Triebhöhenmass, pinion-height tool.
Triebmass, pinion gauge.
Triebstahl, pinion wire.
Triebzahn, pinion leaf.
Trippel, tripoli.
Trommel, drum.
Turmuhr, turret clock.

Uhr, timekeeper.
Uhr ohne Schlagwerk, timepiece.
Uhrbügel, watch bow.
Uhrglasform gewöhnliche, lunette.
Uhr mit flachem Glase, crystal case.
Uhr mit Glockenspiel, chiming clock.
Uhrwerk, watch or clock movement.
Uhrwerk eines Aequatoriums, equatorial clock.
Umarbeitung, conversion.
Universalaufhängung (Schiffscompass), gymbals.
Universalfutter, self centring chuck.
Unruh, balance.
Unrukhloben, balance cock.
Unruhwage, poising tool.

Unruhwelle, balance staff.

Vergoldung, gilding.
Vernier, nonius vernier.
Vernierschublehre, vernier slide gauge.
Versenken, to cut a sink.
Verzinnen, to tin.
Viereckig, square.
Viertelrechen, quarter rack.
Viertelrohr, cannon pinion.
Viertelstaffel, quarter snail.
Vierteluhr, quarter clock.
Vierteluhr mit zwei Glocken, ting-tang clock.
Virgülhemmung, virgule escapement.
Vollplatinen Werk, full plate.
Vorfall, surprise piece.

Wälzmaschine, rounding-up tool.
Wälzungshöhe, addendum.
Waschleder, chamois leather.
Waschleder mit Rot, rouge leather.
Wassersteinschmutz, coomb.
Wasseruhr, clepsydra.
Wechselrad, minute wheel.
Wechselradstift, minute wheel stud.
Wechseltrieb, minute wheel pinion or " nut."
Wechselwerk, motion work.
Wecker, alarum.
Welle, arbor, staff.
Welle, rotierende (Drehbank), live spindle.
Werk der Uhr, movement.
Werk einpassen, boxing in.
Werkzeug, tool.
Westminstergang, double three-legged gravity escapement.
Windfang, fly.
Windfangstrieb, fly pinion.
Winkel, angle.
Winkeleingriffräder (Bügelaufzug), mitre wheels.
Wippe, pivoted detent.
Wippe (Bügelaufzug), rocking bar.

Zahnfräse, fly cutter.
Zahnweite, width of tooth.
Zapfen, pivot.
Zapfen, konischer, conical pivot.
Zapfenrollierstuhl, Jacot tool.
Zapfenfeile, pivot file.
Zapfenmass, pivot gauge.
Zeiger, hands.

Zeigerverbindung (*Turmuhren*), leading-off rods.
Zeigerviereck, set hands square.
Zeigerwelle, set hands arbor.
Zeigerwerk, motion work.
Zeit, time.
Zeitgleich, isochronous.
Zeitgleichung, equation of time.
Zentrifugalkraft, centrifugal tendency.
Zentrierspitze, pump centre.
Zieheisen, draw plate.
Zierschliff, *rundlicher*, spotting.
Zifferblatt, dial.
Zifferblatt eingesetztes für Stunde und Sekunde, double sunk dial.

Zifferblattfüsse, dial feet.
Zink, zinc.
Zinn, tin.
Zirkel, callipers.
Zugfeder, mainspring.
Zurichten, to shape.
Zweigehäusig, pair case.
Zwischenrad, idle wheel.
Zwölftelmaos, douzieme gauge.
Zylinder, cylinder.
Zylinderhemmung, cylinder-escapement.
Zylinderhöhenmass, cylinder height tool.
Zylindermass, cylinder gauge.
Zylinderspunde, cylinder plugs.

INDEX.
